To E

11. 1

*Acquired Traits*

*Raissa L. Berg*

# ACQUIRED TRAITS

## *Memoirs of a Geneticist from the Soviet Union*

Translated by David Lowe

VIKING

VIKING
Published by the Penguin Group
Viking Penguin Inc., 40 West 23rd Street,
New York, New York 10010, U.S.A.
Penguin Books Ltd, 27 Wrights Lane,
London W8 5TZ, England
Penguin Books Australia Ltd, Ringwood,
Victoria, Australia
Penguin Books Canada Ltd, 2801 John Street,
Markham, Ontario, Canada L3R 1B4
Penguin Books (N.Z.) Ltd, 182–190 Wairau Road,
Auckland 10, New Zealand

Penguin Books Ltd, Registered Offices:
Harmondsworth, Middlesex, England

First published in 1988 by Viking Penguin Inc.
Published simultaneously in Canada

1 3 5 7 9 10 8 6 4 2



This work was originally published in Russian as *Sukhovei Vospominania Genetika* by Chalidze Publications, New York, New York. Copyright 1983 by Chalidze Publications.

*Acquired Traits* has been revised by the author for this English-language edition.

Grateful acknowledgment is made for permission to reprint the following copyrighted works:

Epigraph on page vi by permission of Lev Khalif.

Excerpt from "Fish in Water" by Joseph Brodsky, translated by George Reavy from *The New Russian Poets 1953–1966*. By permission of the publisher, Marion Boyars Publishers Inc.

LIBRARY OF CONGRESS CATALOGING IN PUBLICATION DATA
Berg, Raissa, 1913–
Memoirs of a geneticist from the Soviet Union.
Translation of: Sukhoveĭ.
1. Berg, Raissa, 1913– . 2. Geneticists—
Soviet Union—Biography. I. Title.
QH31.B465A313 1988 575.1'092'4 [B] 87-40458
ISBN 0-670-80254-9

Printed in the United States of America by
Arcata Graphics, Fairfield, Pennsylvania
Set in Garamond

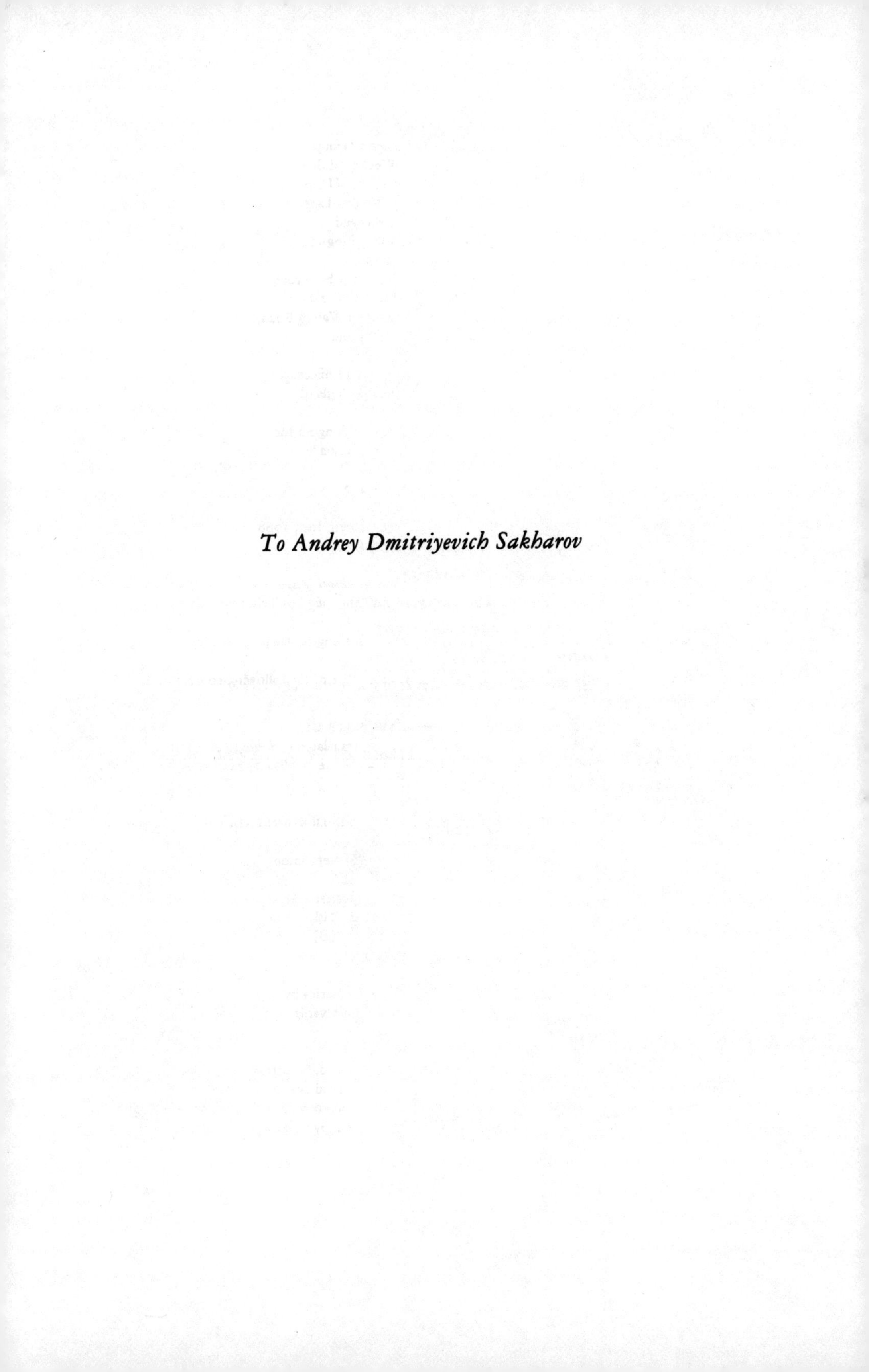

*To Andrey Dmitriyevich Sakharov*

*"Out of what is made your shell?"*
*I had asked the tortoise,*
*And I heard the tortoise tell:*
*"It's of the condensed fright*
*I have collected,*
*Best protection in our life's unending fight."*

—LEV KHALIF
Translated by R. L. Berg

# Contents

# *Introduction*

It was not without forethought that Solzhenitsyn chose for his narration a good day in the labor-camp life of Ivan Denisovich and created Ivan Denisovich not as a spoiled intellectual but a common Soviet fellow—a soldier, kolkhoznik, and hard laborer. Solzhenitsyn was not interested in the impact of gruesome events, but in picturing life in its normal flow. Using Ivan Denisovich's language, Solzhenitsyn presents a somewhat decorated everyday hell.

Ervin Zinner, a character in my narration, in comparing his camp experiences with those of Ivan Denisovich, told me about his camp commandant. This ruler of destiny forced camp inmates to dig their own graves in the summer when the soil had thawed and drove them past those graves all year long. Prisoner mortality reached a record high level.

The naked truth about life in the Soviet Union is found written on the pages of Solzhenitsyn, Yevgenia Ginzburg, and Nadezhda Mandelshtam. My fate and capacity for writing are insignificant when compared with theirs. Nevertheless, I have taken up a pen, and supported by Solzhenitsyn's example in his story about Ivan Denisovich, I have presented a most common life described in the most common language. Mine has been a most common fate. It was an exceptionally happy life: I have met and loved so many wonderful people. However, it is impossible to relate everything, and some of the most tragic events have been omitted. Moreover, this book was conceived as a story about others, not about myself, and only the play of chance has turned it into an autobiography.

I was born in Russia while the Tsar still ruled. I emigrated from Soviet Russia during its fifty-eighth year of existence. This book describes my life there. What happened to me abroad will, perhaps, be the subject of another book.

The verses about fear, which I have taken as an epigraph for my Memoirs, belong to a famous Russian poet and writer, now an emigré, Lev Khalif. I took them from the novel of Vasily Grossman, *Life and Fate*, published in the West. A personage of the novel, a prisoner, recites the verses to his cellmate. In the Russian-language edition of this book, I ascribe the authorship to Grossman by mistake.

The title of the Russian-language edition of this book is *Sukhovey*, the withering wind, chosen to symbolize the destructive nature of the Soviet regime. The word can't be translated into English. The expression "withering wind" does not reflect the *force majeure*, the insurmountability of the natural calamity perfectly indicated by my Russian title word.

The title *Acquired Traits* has been chosen for this edition with the hope that the English-speaking reader will grasp its bitter, ironic sense.

One of the targets in the battle for scientific truth I describe was the doctrine of the inheritance of traits acquired during an individual lifetime. This dogma was lauded to the skies by an all-powerful charlatan, Trofim Lysenko, Stalin's and Khrushchev's right hand. Resistance to the militant ignorance of Lysenko triggered and justified terror.

The most shameful part of the official Soviet ideology and the vehicle of its bloody policy is its claim to remake human nature, to force people to acquire traits fit for communism, for a hive of self-sacrificing equal creatures obedient to planning organizations. One of the reasons for Lysenko's dizzy career was his creation of a theoretical basis for this anti-human way of governmental management. This theoretical basis was the inheritance of acquired traits. My book shows that this policy failed. Human nature resisted remaking. Traits needed by the government are not acquired. If they could be acquired, they would not be inherited.

The diabolic interaction between science and politics under Stalin and under those who took Stalin's place, so similar to the ideological scenario in Nazi Germany, is crowned in genetics by the victory of scientific truth. Genetics has been "rehabilitated from above." The governmental dogma of the inheritance of acquired traits has been tacitly but not shamefacedly abandoned.

The rehabilitation of genetics in the Soviet Union is a victory attained at the cost of enormous suffering. Under a totalitarian regime even the victory of truth is ambivalent. Geneticists are now allowed to be attendants of the State. Together with representatives of all other branches of human culture, they are creating the State's spiritual and material power, and first of all its mighty war potential.

I am intensely thankful to Dr. Valery Chalidze for his offer to publish my Memoirs in Russian, and to Professor Jenny van Brink and Ms. Abigail Thomas for their efforts to win the attention of an English-language publisher for this book.

I have the pleasure of expressing my deep gratitude to the translator of the book, the literary scholar Professor David Lowe, who did the enormous work absolutely selflessly long before an agreement with a publisher started to loom on the horizon, and also to thank my dear friends, the proofreaders of the manuscript, Mr. Steven Holmes, Mrs. Ruby Hinze, Mr. Alvin Kraus, Mr. Thomas L. Neill, and especially Mrs. Jessica Burleigh.

I want to acknowledge my deepest gratitude to professors of the University of Wisconsin, Madison, Dr. Millard Susman and Dr. James F. Crow, for their generous support in bringing the manuscript to the point of publication. I wish to express my appreciation to their secretaries, who worked diligently in making the typewritten copy of the manuscript, especially to Mrs. Chris Onaga.

I am thankful to the editors, Ms. Stacy Schiff, Mr. André Bernard, and especially to Ms. Lisa Kaufman, for their efforts to bring my description of events in the realm of Russian culture nearer to the American reader.

I am most grateful to all of them for their encouragement of my writing and for their interest in the contents of the book.

# *At the Threshold of the Socialist Paradise*

Terror stood by my cradle. In the year when I appeared on the earth and in the country where that significant event took place, 287 children out of 1,000 died before their first birthday. The year was 1913, the country Russia. There was no crop failure, no famine cutting down human lives, peace still reigned, and statistics weren't falsified. Every third child died before leaving the cradle.

I was raised by my grandmother, Klara Lvovna Berg, the widow of a notary public, and by my father, Lev Semyonovich Berg, a professor at the Moscow Agricultural Institute. We lived in Moscow on Dolgorukov Street in a luxurious apartment house; we had electricity, and gas at a time when they were rare. There was a liveried doorman at the elevator.

Grandmother and Father had worked out strict rules for bringing up progeny with a minimum risk of failure. The progeny, besides myself, included my brother Simon—Simochka. I was Raissa. Not Rayechka. I was deprived of a pet name in honor of my father's youngest sister, for whom I was named. Not long before my birth she had died of tuberculosis in Lausanne, Switzerland, where she had been taken for treatment. From her earliest childhood on she had inspired in her parents, brother, and two sisters not only ardent love but respect as well. They called her by her full name, not its diminutive form. And I was to take her place in the hearts of her inconsolable relatives, even though my aggressive independence had nothing in common with her character.

Sim was seventeen months older than me. When our nanny spoke of him, she always called him a "heavenly angel." But this heavenly angel did not wish to live among mortals. When I was already eating by myself,

Sim was still spoon-fed as Nanny repeated over and over, "Chew, chew, chew." Otherwise he wouldn't chew. The gentleness of his disposition was extraordinary. You couldn't help loving him.

No one knew about vitamins back then and no one feared vitamin deficiencies. However, people were afraid of infections; they lived in fear of losing their children. Tap water, fresh milk, and unwashed fruit were strictly taboo. To avoid contagion, absolutely all contact with other children was forbidden to us. We weren't allowed to go into the kitchen—there were gas, irons, and a samovar with burning coal there, and we might inhale coal gas. When we set off for our summer house, we were always taken to the railway station in an open-topped wedding carriage, to get fresh air. We spent the major part of the day outside, taking walks with our nanny.

Our nanny, Maria Filippovna Maslova, an Old Believer from a wealthy peasant family, was the only human being with whom we associated. For all her religious fanaticism, Nanny didn't carry rules of hygiene to an extreme. When a lump of mud from under the hoof of a cabby's horse landed on a child's face, she would remove a handkerchief from her pocket, spit on it, and wipe off the mud. In contrast with Grandmother's habits, Nanny was enormously relaxed.

My father, a follower of Lev Tolstoy, a pacifist and a vegetarian, wanted to raise children in ignorance of evil. We were supposed to know that the life of a person, an animal, or a plant was inviolable. To destroy a plant for the sake of a moment's pleasure was just as reprehensible as torturing an animal. We had no Christmas trees and didn't know about the existence of that custom. We didn't walk on the grass and we didn't pick flowers. There were neither plants nor animals in the house. Death and the deprivation of freedom were considered comparable to each other. We weren't taken to the zoo—the innocent captives were not to serve as a source of amusement for us.

We never ever went to a toy store. Wooden soldiers and rifles might have led us to thoughts of war or murder. Books and toys were supposed to inculcate a love of nature in us.

We weren't supposed to know that there was a war going on.

Father and Grandmother weren't always successful in enforcing this physically and morally sterile regime. I was four years old when, moved by compassion for all tiny and helpless things, I kissed the doorman's two-year-old granddaughter—and she coughed in my face. I came down with whooping cough.

Except for that, neither Sim nor I was ever sick. We never saw a doctor. Our temperatures were taken constantly. When I caught the whooping cough, though, my brother was isolated from me so he wouldn't come down with it. Nanny walked with him, Grandmother with me. "Don't come near this girl, she has whooping cough," Grandmother told the girls in the park when they came too near to me. I wasn't then, and I am not now, struck by the fact that Grandmother worried about the other children, but it did seem strange that she didn't call me by name and spoke of me, in my presence, in the third person. In Russia that's considered impolite. Broken pine branches were brought into the dining room, where I had been moved from the nursery. The pine fragrance was supposed to help my recovery. Apparently the prohibition against breaking tree branches was not absolute. There existed a certain hierarchy of values.

We knew about the war, too. We knew from Nanny. My lace shirt, made in Switzerland, was full of holes. Nanny said that it had been shot up by the Germans.

In the hierarchy of things Father thought we should be shielded from, a Christmas tree, toy stores, the zoo, even the war were minor matters. What we particularly weren't supposed to know was that in addition to fathers, other children had mothers, too. On this point, as well, Nanny was our first enlightener. She put it like this: "It wouldn't be punishment enough to shoot a mother who abandons her children."

Nanny was the victim of disinformation. The story of my brother's and my orphan status went back to seemingly ancient times. Father and Mother had met and, one assumes, come to love each other in Bendery, in Bessarabia, within the Pale of Jewish Settlement. I can reconstruct the history of their relations from fragmentary information: a couple of phrases involuntarily emitted by my father; his sister's, Aunt Musinka, very sparing words; and tirades, seething with hatred for my father, pronounced by Mother's two other daughters (whose acquaintance I made just before leaving the Soviet Union forever).

My father, an eighteen-year-old youth who had just graduated from high school in Kishinev, faced a dilemma: either to stay with the beautiful Polina or to devote all his energies to science. My father's father, a notary public in Bendery, was on the side of education. There was a quota for Jews who wished to attend secondary schools, and Father had turned out to be among the few lucky ones who were accepted. Jews were not allowed the chance to enroll in a university, and so, like his relatives who had received a medical education, he would be forced to study abroad; or, if he wanted to enter Moscow University where he could pursue the academic route that attracted him, he would have to convert to Christianity. He chose the Lutheran faith.*

Father entered Moscow University, and my mother married and moved to St. Petersburg with her husband. In Tsarist Russia money served Jews as a substitute for Christianity when it came to obtaining permission to leave the Pale of Settlement. Many Jews lived in the capital, where there were eight synagogues. By the time that my father, a renowned scientist and explorer, got the chance to settle in St. Petersburg, my mother had been widowed and was raising two adolescent daughters, the same ones who more than a half-century later told me the sad story of their mother's second marriage. She was thirty-six, Father thirty-four when they finally wed. Sim arrived in 1911, and I in 1913.

A born tyrant, whether he be an executioner or a vegetarian, can be either a tyrant or a slave. While he would later become my stepmother's obedient slave, in the marriage to my mother, my father was a tyrant. The fervent advocate of Tolstoy's teachings forbade dealing with doctors. A conflict erupted when it came time to bring me into the world. In that age women gave birth at home. Father suggested that Mother follow the example of the animal kingdom and do without anyone's help. Mother suggested that the fervent advocate of Tolstoyan principles leave her. He did so, forever, but he did not do so alone. Our orphan state was Father's revenge on Mother: He stole her children.

During a walk one day an empty carriage drove up alongside Nanny, Sim, and me. The driver loaded us all, including the high-wheeled baby carriage, into his carriage and drove us to Father's apartment. I was six weeks old at the time, and Sim was a year and a half. My father's mother—Grandmother—was to take the place of our own mother. The nanny kidnapped along with us, after a foiled attempt to take us away with her, went back to Mother.

Mother sued, but the Russian Orthodox Church decided such cases at that time. A woman's priority in raising children had not yet been elucidated by revolutionary legislation. The case was decided in favor

* Before the Revolution, when Jews converted to Christianity, they adopted Protestantism. The Lutheran Church, as I understand it, imposed the least number of rituals on them. Nowadays, in the Soviet Union, when protest against an imposed ideology brings a young man or woman to a church, they choose Orthodoxy or Catholicism. None of my acquaintances became Lutherans, but among my young friends there are many Catholics and members of the Orthodox Church.

of my father, a Christian. Grandmother went to Finland to become a Lutheran, too.

And the heavenly life according to Tolstoyan rules began.

I listened to the adults' conversations with wide-open ears. Grandmother and Father were people of the most liberal views. The names of Trotsky and Lenin were mentioned with hope and goodwill. Within the family, however, democratic principles were not really observed. The cook did not leave the kitchen; Sim and I never saw her. Nanny slept in our room but ate in the kitchen with the cook. Father called her "Nanny" and used the polite form of address with her. "You, Nanny, are a fanatic," he would say, reproaching her for her excessively zealous observance of fasts. And she in turn called Father and Grandmother "master" and "mistress." In Russia before and after the Revolution, there was and there remains to this day the custom of calling people by their name and patronymic. Such a mode of address is universal and respectful. Father did not begin calling Nanny "Maria Filippovna" after the Revolution, but caution suggested to him that a servant's calling someone "master" was fraught with danger. He told her: "Don't call my mother and me 'master' and 'mistress.' New times are upon us."

"What do you mean, master?" Nanny responded. "I'm not a tsar murderer." Nanny lived with my father's family for twenty-five years after the Revolution. And for twenty-five years she called my father "master."

On the first of May 1918, little red flags were made for Sim and me. We walked along Tverskoy Boulevard with Grandmother and I put my flag on a bench and ran around. A young Red Army soldier brought the flag to me and said, "Miss, you forgot your flag."

The idyll of our physical and moral hygiene ended that same year. We were starving. Father decided to send us with Grandmother to the Ukrainian city of Melitopol, where his sister Maria Semyonovna Raikh lived. Aunt Musinka was married to the proprietor of a pharmacy, a wealthy man and a great humanitarian, Grigory Moiseyevich Raikh—Uncle Grisha.

The Ukraine was occupied by the Germans. Compared to hungry Moscow, there was an unimaginable abundance of food. I remember that a cube of butter decorated with little rosettes was served to us with breakfast the day after our arrival. How large could the side of that cube have been? Certainly no more than five inches. But I was little and hungry, and until then had never seen a cube of butter. In my recollections the side of the cube is no less than a foot long.

Still and all, it would have been much safer to stay in Moscow. Epidemics were raging in Melitopol. People were very afraid of spotted fever. I heard a neighbor screaming as he lay dying of cholera. Aunt Musinka said that he left two orphans.

There was no possibility of safeguarding Sim and me against the hideous impressions of the war with the Germans, and the Civil War.

The Germans carried out judgment and retaliation in the public squares. Citizens suspected of theft were flogged there. When the Germans disappeared, the Civil War began. The city changed hands seven times. Whites, Reds, Makhnovites, and Greens occupied the city. It grew more and more difficult to observe rules of good hygiene, or any other rules for that matter.

Neither the Whites nor the Reds touched Uncle Grisha. For the Whites—defenders of the old ways—he was a bourgeois who could be counted on to be hostile to the Reds. For the Reds—with their slogan of "peace to the huts and war to the palaces"—he was what he was in fact: a person ready to help any poor fellow in need, a great benefactor of any just cause.

But then Maruska appeared in Melitopol. She was the leader of a band of "Makhnovites," as they called themselves, though they had no connection with the anarchical movement linked to the name of Makhno. She was not an anarchist, either Red, White, or Green. She was simply Maruska—the leader of a gang of bandits who were using Makhno's name as a cover. These gangs compromised the most sensible of all social movements in Russia—anarchism, whose teachings were closest to that of populism in the United States. The essence of anarchism was not anarchy, not the tyranny of evil, but instead the rule of the laws of nature, of good, and of the mutual assistance on which nature's harmony is founded. It was precisely on these positions of good that Prince Kropotkin, a relative of the tsar, a great aristocrat, and a great revolutionary, founded his teaching and so tragically and ineptly named it anarchism.

Maruska showed up to rob Uncle Grisha. Later they told us that she ripped a gold and diamond broach—a family treasure—from Aunt Musinka's chest. Maruska didn't take junk. She held up the Dutch linen—tablecloths and sheets—to the light to see whether each was worn and whether it was worth taking.

What I remember is this: Sim and I were sleeping. Grandmother woke us up. With her were two brawny Cossacks wearing felt cloaks and tall sheepskin hats. Their rifles were in their hands, and the bayonets were fixed in position. There was no electric light since the town's electric station had broken down, but it was bright in the room. The city was burning. We weren't the least frightened. Grandmother was absolutely calm. She took each of us into her arms, carefully, one after the other, and carried us over to the couch. The Cossacks ripped open our mattresses with their bayonets. Their suspicion that golden treasures were concealed in the children's mattresses was not confirmed, and they left. I remember the silhouettes of people, the squares of their heads and shoulders, enormously exaggerated by these famous sleeveless black cloaks and angular hats, and the rifles and bayonets against a background of large windows illuminated by fire. But what I remember most is the stiff sofa upholstery against my skin. Grandmother hadn't pulled down my nightshirt, and the couch was dirty, too dirty for a child who all her life had been bathed every evening before being put to bed. My grandmother's neglect that my naked skin had come in contact with the dirty upholstery upset me more than the Cossacks. Now I knew something horrible was happening.

My uncle, aunt, and grandmother weren't afraid of robberies. They didn't value things. They lost them and so what? They were hard workers and would make it back. Besides, they didn't need much. But there came a time when they too began to tremble. They were afraid for us.

There were battles in the streets. We could hear artillery fire. Rooms with windows opening onto the street became uninhabitable. Boris Pasternak has described what was happening:

*But these days*
*Even the air smells of death*
*Opening a window is the same*
*As opening one's veins.*

A rumor spread that either the Chinese or the Latvians were on their way to Melitopol. Opinions differed as to the ethnic designation of the troops, but according to all versions their goal was one and the same—to knife Jewish children. Sim and I had been baptized into the Lutheran faith, but you couldn't very well show rampaging Chinese a certificate of baptism. At first Uncle, Aunt, and Grandmother decided to resort to mimicry and put gold crosses on us. But then they decided that the likelihood of losing a child was still too great and that it would be better to leave Melitopol. The whole family went to Bakhchisarai, a small city in the steppe area of the Crimea. There are many Tatars there and few Jews; if one has set out to destroy Jews, then Melitopol is the most appropriate place to begin; Bakhchisarai the least appropriate.

In my recollections Bakhchisarai is a veranda enveloped in twining grapevines, poppies in bloom clear up to the horizon, low hills, sheep herds, sheep dung, countless shells littering the remains of an ancient sea, the steppe, the abandoned palace of the Tatar Khan with its fountain of tears and in the midst of the palace park an ancient map—a little pool in the shape of the Black and the Azov seas. Sim fell into the pool and cried.

We spent the summer in Bakhchisarai. In the fall we returned to Melitopol. All the Jewish children who hadn't been struck down by the epidemic had survived. The rumor had been false.

In Melitopol I saw Wrangel, the commander-in-chief of the White Army, in a car, entering the city he had just won back from the Reds. I saw his rotting army of degenerates and drunkards. Part of Uncle Grisha's apartment was taken away from him and there, in the best rooms, resided a band of reveling White Guards, victorious for that short moment.

My uncle was hiding a Red Army man, with his wife and sick child, in the room behind the kitchen. Grandmother took food there.

From the large windows in our best room, once the White Guard had vacated, I saw a Red Army parade on the main square of Melitopol. The soldiers marched in an unending chain; not in ranks, but single file. The parade lasted a long time. The troop numbers seemed huge. I watched and kept waiting for the march to come to an end. Then I noticed a small, red-headed soldier. He was the same one I'd seen passing fifteen minutes earlier, in the same procession. I counted the soldiers in the parade and realized there were only a few hundred. The endless procession turned out to be a camouflage, a military trick.

I was six years old when I read a huge poster with enormous lettering—it was the agrarian program of the Left Social Revolutionaries, which, for a short stretch of time before Stalin set his sights on the complete enserfment of the peasantry, became the Bolsheviks' program: "Land—to the Poorest Peasants." I thought it was very good to help the poorest ones. But then the Bolsheviks came to power on the bayonets of the victorious Red Army, and the Soviet regime was established.

---

Uncle Grisha was thrown out of his marvelous apartment. He himself transferred the pharmacy to the state and stayed on in it as a pharmacist. He died in 1935. Aunt Musinka was killed by the Germans in 1941. She was one of the 15,000 Melitopol Jews annihilated by Hitler's hordes during the Ukrainian occupation.

Father came to get Grandmother, my brother, and me in 1921, when the Civil War in the Ukraine had ended and before the terrible famine broke out.

Father's disappointment knew no limits when he again found his children, who, under his roof, had been raised in ignorance of evil and amid cleanliness and order. We ran up and down the street with the neighbors' children, throwing rocks at each other, climbing over fences, speaking a hideous Russo-Ukrainian lingo; we called each other Rayka and Simka, and we had a big watchdog from whom we were inseparable. We were extremely dirty and vulgar. At home, however, the same ideal cleanliness reigned.

Filled with woe, Father took Grandmother and us away, without any hope of drawing us nearer to his ideal.

In Moscow we were reunited with Nanny and soon all five of us set off in a cattle car for Petrograd, where Father was at work both as a geographer and a zoologist. St. Petersburg was called Petrograd at that time. Later, in 1924, it was renamed again and became Leningrad. It took us eight days to cover the 400 miles separating Moscow from Leningrad. We went on foot from Moscow Station to the corner of Anglisky Prospekt and the Moyka River embankment, where we were to live. All the palaces in the city had been painted red. In the square in front of the train station stood a monument to Alexander III, a marvelous work of art by the sculptor Trubetskoy, which later would be taken down. The tsar sat motionless on an enormous cart horse. The horse's tail was cropped. The autocrat was dressed in a military uniform, Russian boots, and a fur cap. The peak of that cap can be seen even today behind the wall in front of the Russian Museum, formerly called the Alexander III Museum. The monument was erected by the last Russian monarch, Nicholas II, as a symbol of the stolid Russia of his father's reign. In the brilliant sculptor's execution the colossus looked like a caricature.

Monuments to tsars and military commanders have been retained in Leningrad. Two Peter the Firsts, Nicholas I, Catherine II, Kutuzov, and the symbolic monument to Suvorov still stand where they always did. They stand because their execution doesn't contradict the canons of socialist realism—the art of idolaters. But the innovative monument to Alexander III and to Russian conservatism was taken down in 1938. If Trubetskoy had observed tradition his work would be standing on the railway station square to this day.

Not long before the October Revolution the first Geographic Institute in Russia had been established by a decree of the provisional government. My father was one of the institute's organizers and professors. The institute was situated at the corner of Anglisky Prospekt and the Moyka River embankment in the Grand Duke Alexey Alexandrovich's palace. The grand duke fled abroad during the early days of the Revolution, and his palace, built in Mauritanian style, with wings, stables, and a park, became the residence of the Geographic Institute.

My father probably took part in the room distribution; and did so in accordance with his ascetic principles. I say that because he got the yardman's apartment, which was quite wretched. We lived there until 1923, when the Geographic Institute was transformed into the Geography Department of Leningrad University and its students were moved to the university dormitory. Then my father received part of the apartment in an outbuilding, previously occupied by the major in charge of the Grand Duke's attendants.

At the time we came to Petrograd the grand duke's stables looked as if the horses would come back. The fumed oak flooring of the stable was washed away from its proper place by the flood in 1924, and we used it as a raft to float across the flooded park.

The old English park, with its paths of centenarian linden trees, blossomed in spring with blue snowdrops and lilac crocuses. But marble statues and a granite fountain that never spouted were already beginning to deteriorate. We were surrounded by the memorials of the collapse of an aristocratic estate. Its erosion had begun even before the Revolution, when the lives of the inhabitants were endangered by rebels, many of whom worked at the huge ship-building yard across the river from the estate. The linden walks were partitioned off by an ugly wall crowned with barbed wire. Part of the estate had been sold to a chocolate factory owner, George Borman, and the air was spoiled by the sweet smell of chocolate.

An enormously high iron gate and a row of linden trees still surrounds the square in front of the palace facade. The gate was decorated by big gilded monograms of the grand duke, the two crossed A's. Now the gate and its monograms are covered with black paint. Names scratched on a black surface of a monogram stay written in gold until the next repair. Many people do it, so their names cover the huge A's like a golden net. The linden trees are thriving even now. The old ash trees along the Moyka embankment with its famous iron gate have been replaced by young poplars. Replaced also are the thick hexagonal wooden tiles used to pave the most aristocratic streets of St. Petersburg. They lasted only until 1924, when the flood washed them away. First came cobble and then asphalt. The clatter of hoofs became noisy, and then died away together with the horses. The streets became more and more noisy, but that was with another noise.

Next to the gaudy palace of the grand duke is the marvelous palace of the abandoned Dutch embassy. I lived for more than a quarter of a century in a house fronting them both. Through the windows of my room I could see the uncurtained big windows of Father's study, his green lamp, his snow-white head. The house I lived in had also been deserted during the Revolution. The late husband of my unforgettable baby-sitter, Maria Nikolayevna Andreyeva, was a skilled worker. He was one of the first members of the Communist party, and he and Maria Nikolayevna had close links with revolutionaries. She had some experience in the class struggle. "Look at that," she said once, showing me a call button hidden under the kitchen windowsill. "Those who lived here before the Revolution had these call buttons to call for help when they were being attacked."

Having arrived from Melitopol, fleeing the terrible famine that struck the Ukraine at the end of 1921, we leapt from the frying pan into the fire. There was absolutely nothing to eat. Nanny traveled to the outlying village and exchanged her skirts, which she had gotten under the tsarist regime, for potatoes and milk. She was the main provider of food in the house. It was the time of war communism. The government issued paper money in huge quantities and you couldn't buy anything with it. There was a decree to give out wages* in goods rather than in money. Once Father brought home a box of nails. We were literally starving to death and we suffered terribly from the cold. A decree was issued to provide scholars and scientists with food, but Father was not included in the list of beneficiaries. I don't know if this was because he was late in making an application or because he ceded his place to someone else.

* Father invariably called it "grant," *zhalovanie*. He did not accept the revolutionary slang. In that jargon "wages" were transformed into *zharplata*, an abbreviation of two words, *zharabotnaya plata*, "work payment." Father considered these novelties as crimes against culture and never used the hybrid words.

Salvation came from an organization called the "ARA." All I knew then was that America had sent provisions to the starving Russians. Now I know that the ARA was the American Relief Administration: The United States had come to the aid of Soviet Russia.

Father received flour, sugar, and cans of concentrated milk. Grandmother opened a can, gave me a spoon, and said, "Eat from the can." I wouldn't eat. Eating with one's spoon from a common can was just the same as wiping one's lips with someone else's napkin or one's hands with someone else's towel. "Eat," said Grandmother, observing my confusion, "just like that. If you pour it into a glass, there will be some left on the sides." Grandmother was sacrificing principles of hygiene for the sake of economy. Economy was reckoned not in ounces, but in molecules. You could lick the spoon clean. You couldn't do that with a glass.

As the Soviets' power became firmer and the Revolution faded away into the past, it became more and more obvious that conversations about freedom were just conversations.

The government that had won made people fear it. The herd mentality was implanted through fire and sword. There was no longer a tsar, an aristocracy, or capitalists—it wasn't with them that the battle was being waged. The enemies were declared to be individualism, apoliticism, and the refusal to give up the right to one's opinions.

Father hung two lithographs in his cold, damp study, in that wretched, long room. One was of Luther Burbank, the American horticulturist, a man of the common people, who held a huge flower in his hand. The other was of Archimedes leaning over geometric figures sketched in the sand. The murderer's sword was already raised above the old man. According to legend, Archimedes' last words were "Don't touch my sketches." Those words were printed under the picture in Latin. Burbank was a foreigner. Archimedes and his words, "Don't touch my sketches," were the apotheosis of creative thought completely removed from daily life.

The pictures hung and hung in my father's study and then disappeared. Apparently it had become dangerous to advertise, even in such a modest fashion, one's nonparticipation in the bloody vicissitudes of the building of communism. But an implacable spirit that affirmed its independence remained, and it was revealed by a tiny lithograph Father left hanging. It was a work of art, not someone's portrait, a romantic depiction of a beautiful young woman boldly looking ahead. Her hair loosened, she was dressed in rags, and her hands were tied behind her back. The picture bore the French words *La Sorcière* (The Witch). Darwin, Kessler, Knipovich, and my father's father, Simon Grigoryevich Berg, kept company with that witch, who had been condemned to death.

We made it safely to the NEP period, the New Economic Policy instituted by Lenin in 1923.* A time of great plenty ensued. The magnificent orchards, owned by Germans, Finns, Latvians, and Estonians inhabiting the environs of Petrograd, produced the most exquisite vegetables on earth. As soon as the market was permitted, and money started to play its routine role, the luxury produce of these mini-Edens could be had. The supermarkets in Italy and America would never impress me with their selection of vegetables. At the market Nanny bought cauliflower and Brussels sprouts, asparagus and shallots, short carrots and pods of green peas. German and French confectionaries and bakeries opened. The words napoleon, éclair, and praline began to be heard and the forgotten taste of pastries, Vienna rolls, petit fours, and cream horns could have become an everyday thing in our family.

They could have, but didn't. At that time, great changes took place in our family. My father remarried. We parted with Grandmother, who went off to Melitopol. I never bore a greater loss in my entire life. Father made an attempt to rid himself of us as well. (He told me that later, quite unexpectedly.) He proposed to Mother that she take us. She turned up. She didn't see Father, either because she came while he was gone or because he was sitting out her visit in his study. I saw her twice. Each time she was accompanied by one of her daughters. In the end she refused to take us, which was largely my fault. She was unable to establish contact with me; I was afraid that she would separate me from Father. I hadn't the slightest idea that the initiative had come from him. And because of what he had told us, I thought Mother had rejected us long ago.

Although I often thought that Kaspar Hauser, who lived from early childhood on in seclusion, without intercourse with the outside world,

* The New Economic Policy was instituted to replace the policy of War Communism and its strict centralism. New laws gave to peasants the right to treat their holdings as their own and even to use hired labor to cultivate them. Private trade was legalized. Financial methods of the past were restored. Small industrial enterprises were released from nationalization. The government kept firmly under its control over 80 percent of the working force. This included all large-scale industry, all transportation, all foreign trade, and all banking and credit institutions. Introduction of NEP was greatly beneficial to the economic rehabilitation of the country. At the end of the 1920s Stalin returned to strictest centralism and the NEP was abolished.

was my prototype, things weren't quite that bad. There were two of us and we were friends. Nanny was still with us and we studied at a German school. Since the time of Peter the Great many Germans had populated St. Petersburg. There were three schools in the city where the teaching was conducted in German and where English, French, and of course, Russian were taught. Father sent us to one of these schools; under the yoke of his new wife, Maria Mikhailovna Ivanova (whom we called Marmikha, using the Revolutionary slang), he didn't have the option of paying private tutors. We'd had a little preparation in German at home and we entered the German department of the Reformation School. Now it was called the Soviet Workman's Nine-Year School No. 34. The customs remained old-fashioned. Instead of addressing our teachers by their name and patronymic, we called them Herr Sadovsky, Fräulein Held, Fräulein Ludvig, Herr Seewald. When greeting the teachers the girls were supposed to perform a half-curtsy. We wrote in the Gothic script.

I was a prankster. "Warum hast du eine Palme auf dem Kopf?" (Why do you have a palm tree on your head?), Herr Sadovsky, the physical education teacher, asked me. The hair caught up at the top of my head by a ribbon really did look like a palm tree. "Das ist keine Palme," I answered, "das ist ein Springbrunnen." (This is not a palm tree, it's a fountain.)

I also got away with defending Tolstoy. The Russian literature teacher, Herr Strizheshkovsky, was quoting Lenin: "Tolstoy is an intellectual sniveler." I said that I'd get up and leave if he didn't stop. I remember that teacher and our mathematics teacher, Olga Yulyevna Ludvig, with enormous gratitude. It's from that time that I remember by heart the verses of Igor Severyanin, Esenin, Kluyev, and Gastev, poets who were either forgotten, forbidden, or persecuted, who had either destroyed themselves or been destroyed.

The instruction was first-class. We studied four languages, as before and in the tiled workrooms we also received a workman's education. We were taught draftsmanship, topography, and cartography, so that we'd graduate with a profession. Any governmental decree, and they followed one after the other at an insane gallop, was carried into effect in our school, and invariably with positive results. Our school thoroughly proved the universal belief that there is no way to make a German a do-nothing.

At home, though, things were very bad. I was forbidden by Stepmother to draw—that was considered idleness. When drafting was introduced at school, my school girl friends bought all the things I needed to draft with and on. I didn't have occasion to draft anything, but a bottle of India ink turned up in my possession. Now there was something to draw with but absolutely nothing to draw on. And then Father came to my aid. From his desk he would take a tiny little sheet of good paper left from the tsarist times and give it to me. On that scrap of paper I would make an extremely involved design with India ink. It was intricate not only because my artistic taste demanded such an opposition, such a harmony of black and white forms flying and interweaving in flight; its design was dictated by the limitations of my life. Since I would not receive the next one for a long time, a tiny scrap of paper had to serve me for many hours instead of just one or two.

Father made fun of my abstract creations, and my stepmother would ask sternly whether my stockings had been darned. You may be assured that they had been—and, moreover, skillfully. But it was only for my stepmother's sake that Father pretended not to think my creations worth anything. I know this today because of one picture, an oval. I drew it on a rectangular sheet of paper and then cut it out. When Father died and my stepmother, transferring his archive to the academy, was removing all evidence of the fact that my father had a daughter at all, she returned to me, along with other proof of my existence, an envelope. It contained the scraps from the sheet of paper from which I had made the oval. Father had gathered them up and kept them in his desk.

At the same time this prohibition was imposed on my drawing, my father forbade my brother to act in the school theater, even though the school's drama club was run by an actor from the city's best theater. The People's Artist of the Republic, B. A. Freundlich, came from that club. Sim possessed an outstanding talent. He easily outdid Freundlich. He played old men in Pushkin's *Gipsies* and Molière's *The Doctor in Spite of Himself*. But my father even went to school to ask them to have Sim excluded from the drama club. That was his only visit to the school and the only demonstration of his concern for our education.

I was in a music appreciation club—they had one of those, too. A professional musician, a teacher from the Conservatory, played the piano for us on the top story, where the acoustics were marvelous. But I kept quiet about that at home. Sim would have kept quiet too about drama club, of course (in general he was keeping more and more quiet), but the actors' costumes had to be provided by their parents.

My fellow students had entirely taken the place of my family for me. Great friendship, solidarity, and loyalty united all the students in our grade. Between us and our wonderful teachers, whom in spite of everything we loved, there was a constant struggle, genuine class warfare. There was no need to agree on strategy, work out a secret code, or make use of any conspiratorial methods. All we had to do was not give our comrades away; not tell our parents what was going on at school. The parents might betray us to the teachers. And things were going on that were rather unusual and possible only because of the students' great friendship. We not only withstood our teachers' tyranny, but that of the state as well. Submitting to orders from above, our teachers were trying to impose a profession on us. We resisted.

I didn't want to be either a draftsman or a topographer. The necessity of wasting time on blueprints and maps restricted my freedom and hindered me from developing in the direction that I wanted. I wanted to write compositions. I didn't prepare a single map or blueprint during all the years that I was a student. But blueprints and maps signed with my name hung at regional exhibits of the best student work. The students who were to become first-class draftsmen and topographers, Alya Neumann and Yura Valter, did all my work for me. I wrote Russian, German, French, and English compositions for others, and not necessarily for those who helped me; I wrote for whoever needed help. No one was ever caught. There wasn't a single instance of betrayal. Forgery wasn't the only form of mutual assistance. We helped each other in all possible ways.

The citizens of the Soviet Union are involved in governmental organizations from childhood to the grave. Every educational facility has a "Pioneer" organization and a "Komsomol" organization. The word "Komsomol" is an abbreviation of "Union of Communist Youth." Children become Pioneers at ten, Young Communists at fifteen. Every growing citizen of the Soviet Union twice goes through this purportedly voluntary, but in reality mandatory, entrance into the clique of the orthodox.

The Pioneer and Komsomol organizations at our school didn't enjoy our respect and I wasn't a member of either. But in my heart I was an ardent Communist. "I wish they'd take over the whole world," I thought, "undergo great deprivation and build communism, embodying in life great humanitarian ideals for the good of all mankind and future generations." I believed what they wrote in the newspapers. "When I enter the university, I'll join the Komsomol," I told myself.

# *An Idealist's Daughter at Leningrad University*

I finished school in 1929, the year of the Great Break with the Past, as Stalin later called it in the *Short History of the Bolshevik Party*. I was sixteen years old. You had to be seventeen to enter the university. I began working as a statistician at the Hydrological Institute, where I calculated the average wind force based on data from weather stations, without using any kind of adding machines. I left home with a girl friend, rented a disgusting room on the Moyka embankment. Father had no objections, and gave me a small monthly allowance.

The entrance requirements for the university gave me no hope that I'd be accepted. The examinations were conducted differently for the representatives of the various classes. For workers, poor peasants, and their descendants, the exams were easy, but for children of white-collar workers and the intelligentsia they were difficult. Only one out of forty of those unfortunates had a chance of acceptance. Concerned parents hired tutors.

Without help or encouragement from Father, I took lessons and paid the tutors out of my own niggardly salary. There was hardly anything left for food. But no matter how I prepared for mathematics or the social sciences, I had no chance of being admitted. Unavoidable failure awaited me at the examination in Russian language and literature—and all entrants were required to take it. In my German school I'd been taught to write literate German, but not Russian.

In 1930 I applied to Leningrad University's Biology Department and was accepted. I didn't have to take any exams. On a countrywide scale they had been done away with. Universities had begun to accept students strictly according to the class principle. The entrant had to present a diploma from a school, factory workers' department, or peasant youth school and proof of his or her class affiliation.

How did I end up among this new sort of intelligentsia? This happy turn of events came about because the authorities feared sabotage by university and institute teachers whose children were deprived of the opportunity to taste of the fruit of enlightenment.

The Leninist principle of placating the "enemy" in order to use him before destroying him, worked. "Enemies" were needed to enlighten offspring of workers and farmhands to create the new intelligentsia. Of course, the enemy didn't exist in reality, only in the inflamed minds of those who had come to power. Fear and power greatly facilitate inflammation of the brain.

I had wound up among the elect. It was permitted for enemies' children to become students. Nowadays, newly accepted students are taken out into the country to dig potatoes on collective farm fields. Back then we were taken to the port for a "Leninist Saturday" that went on for no fewer than ten days. We loaded ballast, blocks of wood, onto foreign vessels. For any delays the state had to pay in hard currency. For our services we were given a piece of sausage and a roll; no other sort of payment was expected. It was assumed that we were full of enthusiasm. You don't pay enthusiasts. As far as I was concerned, the organizers were not mistaken.

Then there were military studies. The men and women were separated. I found myself in extraordinarily refined company. The majority of the men were from worker or peasant families, while the majority of the women were from the intelligentsia. I don't know whether that was purely coincidental or whether the lesser mobility of women played a role there. You could count the young women from out of town on the fingers of one hand.

Here was my chance finally to enter the Komsomol. At the end of our studies of the Red Army's administrative organization, a most repulsive woman came in and asked the Komsomol members to remain. No one left—everyone belonged to the Komsomol—everyone except me. Experiencing pangs of conscience because of my deceit, I remained. Now, I thought, I'll ask this woman to help me enter the Komsomol.

"Your task as members of the Komsomol, zealous builders of communism, is to conduct the class struggle and expose class enemies," said the woman. "You must talk to your comrades, with everyone around you, and let the party organization know about all their ideological waverings."

The girls were silent. Everyone left, and so did I. That was the beginning and end of my party career.

Our "studies" began. The students ruled, and they managed the affairs of the department in such a way that simply spending time within the walls of the university gave them the right to a diploma. The whole system of higher education in those unfortunate years was aimed at a battle against the intelligentsia, at the eradication of learning and of people capable of creating.

We were supposed to master our subjects by uniting into production brigades. Examinations were abolished. The day was planned from 9:00 A.M. until 11:00 P.M. It was obligatory to follow the schedule. Tardiness could be punished by expulsion from the university. The decision about expulsion was made by the Komsomol cell.

Testing knowledge was accomplished with the staging of academic "battles." It went like this: A group of students was divided into two subgroups and they conducted a battle with each other. A representative of one subgroup asked a question. Let's say that an exam in invertebrate zoology was being conducted. The question might be about the structure of the jellyfish. Any member of the opposing subgroup could answer. Then it was that subgroup's turn to ask a question. The academic battle ended up being a competition between the subgroups' two best representatives. Both sides received credit, however, all the members of both subgroups without exception. In our group there was a very strong student, Nina Ryabinina. She and I were always put in different subgroups. The side that she was on was assured of victory.

Neither the administration nor the students themselves strove for equality. The class principle was limited by the party and Komsomol committees to a division of students into the "clean" and the "unclean." To be considered clean, one needed to participate in the destruction of the "bourgeois intelligentsia." Some previous great service in unmasking hidden class enemies was highly appreciated. Among the chosen ones were especially chosen ones—shockworkers (*udarniki*)—students who rose above their milieu. They were small fry compared to the whales—the promotees (*vidvizhentsi*). A shockworker earned privileges through zeal, if not in his or her studies, then in the exposing of alien ideologies. The promotees were nominated by higher organs outside the university, so these promotees' futures were guaranteed. They obviously performed some sort of services for the authorities and were rewarded by the right first to be students, then graduate students, and then to occupy the post of director in various institutions. These promotees occasionally turned out to be capable people, but as often as not they were complete idiots.

The first promotee whom I met at the university, Davydov, was quite stupid. He failed to make a career for himself. The remnants of capitalism in his consciousness did him in. He was approaching thirty, it was time to get married, and girls from the intelligentsia were practically throwing themselves at him. He married Zoya Toporova. Her mother and father were doctors—teachers at the Medical Institute. Davydov moved from the dormitory to their apartment.

The remnant of capitalism in his consciousness was jealousy. He struck his young wife on the temple with an axe. She was saved and she left the Biology Department at the university and transferred to the Law Department.* Davydov was arrested. He was back at the university two years later.

Another promotee, Koverga, was much more successful. He initiated class warfare within his very first days at the university. His target was the professor of botany, the well-known explorer and future president of the Academy of Science of the USSR, Vladimir Leontyevich Komarov.

At his introductory lecture Komarov spoke about the multiple benefits that plants provide man; listening to him was a real joy. In discussing full-bodied food he mentioned rye bread with onion—what the Russian peasant ate. Koverga made his move. This was obviously counterrevolutionary propaganda, a camouflaged diversion aimed at demonstrating the high level of peasant life in prerevolutionary Russia and thereby discrediting the Revolution.

* I saw her thirty-four years later, on March 13, 1964, in the courtroom. In theory the trial was only a public hearing, but it was decked out like a real trial and the accused was taken from it straight to prison and sent from there into exile. Zoya Toporova fought against crime, but it wasn't the accused who was a criminal but the state itself and its inhuman laws. It was against the state that Zoya Toporova did battle as a lawyer. She defended the brilliant poet, Brodsky. Brodsky was not tried for the violation of any of the hundreds of articles in the Criminal Code, none of which he had violated in the course of the twenty-three years that had passed since the time of his birth. Rather, he was tried on the basis of a decree about social parasites that had just issued from the pen of that giant of all learning and absolutely all the arts, the poorly educated Tsar Nikita, Premier of the Land of Soviets: Nikita Sergeyevich Khrushchev. Brodsky is a born poet and a first-class translator. He had a contract with a publishing house that was preparing his translations for publication. First the KGB gave the publishing house an order to cancel the contract and then the author was tried as a social parasite. Brodsky learned about the fact that the contract had been canceled from the judge in the courtroom. Zoya Toporova lost her fight with crime. I'll have more to say about Brodsky later.

Komarov received a reprimand. But he was a party member, too, and giving lectures was an act of great selflessness on his part, since during his travels through the Far East he had been very ill, and the aftereffects of his illness were still apparent—tormented by eczema, he read his lectures wearing black gloves. He demanded an apology from his slanderer.

After the apology was given, he refused to read lectures to us. Koverga continued to flourish. After finishing graduate school he was appointed director of the Nikita Botanical Gardens on the Black Sea shore. I saw him in 1968. He was retired and complained that he'd lived his life in vain. He remained a convinced Stalinist.

Everything was required. Community work was required. I participated in the "liquidation of illiteracy," as my activity was described in official government language, teaching reading and writing to women workers at a food plant and male workers at a Kazitsky tool plant. No one had to force me to do this. I loved it. Going to meetings was required, meeting after meeting. One day they asked us to back the condemnation of a group of wreckers—members of the Industrial party were being tried—and we were asked to vote for the death penalty.*

I was seventeen. I sat at those meetings more dead than alive, not raising my hand for or against, nor to abstain. I realized what monstrous miscarriages of justice were going on right before my eyes, in which I was forced to participate, under penalty of expulsion from the university. No one protested; the vote for the death penalty was always unanimous.

Many years later, when, unbeknownst to my stepmother, my friendship with my father was renewed, I told him that I didn't understand why no one ever protested. Father told me that he knew of one such

* The era of political trials began long before 1930. The officially sanctioned falsification of accusations and the recruitment of false witnesses date from 1922, when at Lenin's order members of the Social Revolutionary party were tried as traitors to their country. They were incriminated by links with foreign intelligence agents and with enemies abroad on whose commands they allegedly acted. The year 1928 opened a new era in Soviet legal procedures, an era of show trials of the intelligentsia, an era of a special class of terror. Arrests were made, and then prisoners were bunched together into a group that would be tried as a whole for collective sabotage committed, of course, at the command of the hostile West. Indispensable links in a show trial were the public confessions of guilt by the accused. Much later it became known that the confessions were literally dragged from the accused. People who refused to give them died under torture. The first case of this type was the so-called Shakhty Case, a trial of engineers. The victims of the second trial were also engineers, whom the secret police bunched together into a phantasmagorical Industrial party.

case. It happened at a meeting at the Academy of Sciences. They were also voting for the death penalty for some innocent persons. When the chairman asked who was abstaining, a single hand was raised. It belonged to the brilliant scientist Vladimir Ivanovich Vernadsky, the founder of biogeochemistry, a new branch of science dealing with the role of living beings in the distribution of elements in all spheres of our planet. He was immediately asked why he was abstaining. "I'm opposed to the death penalty in principle," Vernadsky said. "And what about you?" I asked Father. "I didn't go to those meetings," he said.

For us, attendance was mandatory. The most valued extracurricular work was seeing to it that everyone attended everything that was required. Approximately one-fifth of the students spent their time looking after just that. For each fifteen or twenty people there was a party organizer, Komsomol organizer, union organizer, and a monitor. They were all watchdogs. The community trial of Mordukhay-Boltovskoy, which followed the trial of the soi-disant Industrial Party by one day, took place in the university's assembly hall. The hall seated 900 people. The student Mordukhay-Boltovskoy, the son of a well-known professor at Rostov University, had written his father in Rostov from Leningrad. He dropped one of the letters, it wound up at the Komsomol office, and its author wound up in the defendant's box.

In order to make it resemble the Industrial party trial, the organizers knocked together a group of accused, mostly from intelligentsia families. Professor I. I. Prezent and E. Sh. Ayrapetyanets, at that time a graduate student, acted as public accusers. The improvised prosecutors hurled thunder and lightning, accusing the defendants of moral degeneracy, drunkenness, sabotage, anti-Soviet propaganda, and slander of the Soviet regime. There were no defenders at the trial. The whole revolting farce was a flank attack on the university professors, who were allegedly encouraging debauchery and an anti-Soviet disposition on the part of the students. No one rose to defend the accused, and only one of the accused defended himself. That was the geneticist Dmitry Mikhaylovich Kershner.

At the point where the prosecutor, Prezent, raging against debauchery, exclaimed wrathfully, "the table was littered with bottles," one of the accused raised his hand. The chairman of the trial had to recognize the defendant. Kershner rose very slowly, turned to the chairman, and said: "Citizen chairman of the community trial against Mordukhay-Boltovskoy and his group, allow me to give factual testimony by way of clarifying some of the details in the speech given by the public accuser,

Professor Isay Israylevich Prezent." The chairman gave him permission to clarify the details. Kershner turned to Prezent: "Citizen public accuser at the trial of Mordukhay-Boltovskoy and his group, Professor Isay Izraylevich Prezent, allow me to firm up the information cited in your speech for the prosecution. You said that the table was littered with bottles. There were only two bottles." Then Kershner sat down.

Devoured by wrath over the defendants' anti-Soviet disposition, Prezent claimed that at the group's debauched gatherings, enemies of the Soviet regime sang anti-Soviet songs. Once again Kershner raised his hand and, in a syrupy voice lacking intonation, asked permission to firm up details. Receiving it, he addressed Prezent with the same drawn-out formality, finishing by saying: "We didn't sing anti-Soviet songs. We sang the Latin hymn 'Gaudeamus igitur.' Translated, it means 'Make merry, friends, while youth is with you.' " The defendants were thrown out of the university. Kershner immediately wound up in the army, was reinstated at the university four years later, and graduated four years late. His clarification of details cost him dearly.

I. I. Prezent was one of the main destroyers of the cream of the Russian intelligentsia. The same year that I entered the university, he aimed his sights at the biogeochemist V. I. Vernadsky and at my father.

A great blossoming of Russian culture coincided with the beginning of the twentieth century. The struggle for freedom is an integral element in a nation's spiritual rise. Revolutions were accomplished in music and painting, in literature and the theater, in the exact sciences and the humanities.

Vernadsky, a geologist, gave luster to the age in which he lived. Berg's work as a geographer made it worthy of the designation "the Russian Renaissance." Had a planned economy been the goal of the authorities, my father's book, *Landscape Zones*,* would have become a standard reference work for the State Planning Board instead of a pretext for his persecution. Father examined man in his geographical surroundings, asserting that they needed to be known and protected.

At the beginning of the twentieth century, geography, as a science, had gone to pieces. Climatology, geomorphology, and hydrology flour-

* Leo S. Berg, *Fisico-Geograficheskie (Landshaftniye) Zony SSSR (Geographical Zones of the USSR)*. 2nd edition. (Leningrad: Publishing House of the Leningrad State University, 1936).

ished. Geography had ceased to exist. My father revived it. Geography became the science of the interaction of living and nonliving elements in the landscape, including man.

At that time, in that year of the "Great Break," the year of the "Revolution from the Top," scientific polemics degenerated into political denunciations followed by punishment without virtue of trial or investigation. In the few cases when trials were held, they were fabricated farces. Prezent was a head torturer in these polemics. At the beginning of 1931, an article by him appeared in the university newspaper, *Leningrad University*. In it, he tore the mask from a class enemy—the geographer Berg. Berg's study of landscapes, he argued, was nothing but a hidden struggle with Marxism, a denial of the class struggle, idealism, obscurantism, and the propagation of peace with the aim of enslaving the working class and the poor peasants to capitalists, to landlords, and to the petit bourgeoisie. A caricature depicted Father as a muzhik wearing Russian boots and a Russian blouse with a cord belt. His enormous figure rose above an impoverished little village of ramshackle huts.

Father was praising two books to the heavens—his own *Nomogenesis* and Remarque's *All Quiet on the Western Front*. Remarque symbolized pacifism. *Nomogenesis*, the theory of evolution based on law, was Father's scientific credo with which he opposed Darwinism. Father refused to recognize the Darwinian principle of natural selection and of the struggle for existence as the reason underlying the progressive evolution of the organic world. In criticizing my father, Prezent was especially zealous in unmasking his Lamarckian errors. To be a follower of Lamarck was the equivalent of sedition.

Lamarck was the first in the history of science to create the theory of evolution, the first to give a scientific explanation for similarities and differences among living creatures. The cause of evolution, according to Lamarck, is the principle of gradation, which is the innate ability for self-perfection. Lamarck did not speak about competition for space and resources reigning in nature. According to him, organisms of each species help each other to live and to reproduce themselves. In the eyes of the Soviet methodologists, the great French materialist and encyclopedist Lamarck had no merits. To the representatives of the manipulated Soviet science, the principle of gradation was God's power of creation. This was creationism hidden by scientific phraseology. Lamarckism was condemned for its idealism.

Darwin, on the other hand, had been made a Marxist saint and to criticize him was to encroach on Marxism's holy of holies. You weren't even to mention the mutual help that rules in nature. Idealism: the weapon of the enemy, the subversion of Marxism, the camouflaged desire to discredit the idea of class struggle—that's what discussions of mutual help were.

I think that Father was wounded most of all by the accusation of idealism. Besides the word's philosophical meaning, Father defined idealism as selfless service to an ideal. Prezent used the word as a term of abuse. Materialism, i.e. dialectical materialism, was good, and idealism was bad. Father resigned from the departmental chairmanship and left the university. Had he done so a few months earlier, I would never have had a university education. Fortunately, Father's resignation did not result in my automatic expulsion.

Prezent, a present for the freshmen, as my brother and I called him, knew that I was among the university's students. He delivered an introductory lecture on the philosophy of dialectical materialism, ardently agitating in favor of the class basis of human consciousness and the necessity for a struggle with idealism from class positions. At the very first lecture one of the class enemies named was Berg. In a strange and incomprehensible way the professor addressed his lecture not to the almost 1,000 students in the university's Great Hall of Physics, but to a single young woman, the beautiful Natalya Vladimirovna Eltsina. Botticelli could have used her as a model for his gentle Madonnas. It so happened that after the lecture, she and I went up to Prezent at the same time to ask about recommended readings. Prezent asked who I was. I told him my name. He turned sharply to the Madonna and exclaimed: "Do you mean that you're not Berg?" He had thought he had a clear idea of what an idealist's daughter should look like.

My first threat of expulsion from the university, however, came about without Prezent's help. The class struggle was at its height. Everyone was battling everyone: students against teachers and against each other, teachers against students. You didn't even need to expose ideological waverings. The persecutors operated on the principle of the rigid determinism of human consciousness based on social conditions. Literally translated, this meant that if you or your parents had lived well before the Revolution, you were an enemy of the Revolution and should be destroyed. If you proclaimed that you were ready to serve the working people, you were a wolf in sheep's clothing from whom the costume needed to be torn. It was as if there had never been a self-sacrificing, socially conscious Russian intelligentsia, great Russian literature, great social movements of the past century and the present one, movements in which the intelligentsia led the worker and peasant masses. It was as if there had never been a problem of fathers and sons, a generation gap.

In a conversation with a comrade, Bukin was his name, I said that the intelligentsia had played a large role in the formation of the proletariat's revolutionary consciousness. I was warned by the group's party organizer, Zoya Feodorovna Feodorova, that I would be "worked over" at the next Komsomol cell meeting and that I was threatened with expulsion. And I would have been expelled if I hadn't been saved by Nina Ryabinina, who invariably defeated me in our academic battles.

Nina was knowledgeable about more than the structure of the jellyfish and the alternation of sexual and asexual generations in hydroid polyps. Parodying the revolutionary slogan "the end justifies the means," she said, "I lie rarely, and when I do, it's in the interests of the proletariat." But she said that to me, not to Bukin. They didn't make trouble for her. She knew how to keep her mouth shut. But she too had an ideal, one that was perverted in practice every second—the building of communism. Her father, a professor of geology, and especially her uncle, a paleontologist, were great admirers of my father, and Nina had no reason to conceal what was going on at the university from her parents. Those representatives of the bourgeois intelligentsia provided her with a copy of Lenin's book *What Is to Be Done?* and there, on page 72, in black and white—no, black and gray, because it was an edition from the 1920s, and paper in the Soviet Union was made out of those same blocks of wood that we, moved by enthusiasm in 1930, had loaded onto foreign ships—in that poorly done edition, it said that revolutionary consciousness had been introduced into the proletariat by the bourgeois intelligentsia. What's more, the leading role of the intellectuals in the Revolution is the basic idea of Lenin's strategy.

I took that book to the meeting. I hardly knew anyone there. I knew one student—at the beginning of the year she had sent me to the Kazitsky plant and the food factory to help eradicate illiteracy. I wasn't asked about anything at that community trial. They themselves spoke. It transpired that I had said that the intelligentsia was the hegemon of the Revolution. I didn't even know the word "hegemon." Moreover, they claimed, I hadn't done any community work, even though that was all that I had been doing. They voted. Who was for expulsion? It was a unanimous vote for expulsion.

Without observing the formalities, which Kershner had so carefully fulfilled at the Mordukhay-Boltovskoy trial, I then took the floor. "People are given the death sentence for murder for gain," I said, "and even then the criminal has the right to the last word at the trial. It's not true that I haven't done community work. Ask her," I said, pointing to the student who had sent me to Kazitsky. "She knows. The workers from both factories took the initiative to send the university a notice of their gratitude for my lessons. Why doesn't she speak up?" As for the intelligentsia's role in the Revolution, I said that the intelligentsia had played a role in the formation of the proletariat's revolutionary consciousness, and Lenin said that it was introduced into the proletariat by the bourgeois intelligentsia. And I read from Lenin's book. "And how could it have been any other way?" I said. "After all, the proletariat was cut off not only from means of material production but from spiritual culture as well, and the idea of Revolution is the highest achievement of spiritual culture." No one raised any objections. "I'll leave the university myself so as not to serve as a practice target for class warfare," I said.

I wasn't expelled from the university. And I did not leave.

Unlike St. Petersburg University, which produced graduates with a broad education, the students of the Leningrad University were given a specialty beginning in their freshman year. As soon as students were accepted to the Biological Department they were distributed among the department's various branches. These were zoology, botany, and physiology. Each branch offered different courses. The zoology branch had no offerings in geology and paleontology, microbiology, nor in botany. Political subjects were taught every semester in every branch: political economics, historical materialism, dialectical materialism, and the dialectics of nature. Students from all branches heard Prezent's lectures, and the introduction to the philosophy of dialectical materialism was required for all students in all departments. The period of study was shortened from five years to four years, and to three years for promotees. The new type of intelligentsia, i.e., specialists within a very narrow range of knowledge, were created at top speed.

On entering our sophomore year we had to select an even narrower specialty. The freedom of selection was limited. Only three departments within the zoology branch accepted students: hydrobiology, genetics, and vertebrate zoology. At that time the department in which I dreamed of receiving an education didn't exist at Leningrad or anywhere in the world. At the university I had hoped to receive an education in the field of bionics, as that field of knowledge—the study of the evolution of animal engineering—is now called. I wanted my bionics to be of the evolutionary sort. There was no bionics. The closest thing to the theory of evolution was genetics. I chose it.

Nikolay Nikolayevich Medvedev conducted a course in genetics. The fruit fly served as our laboratory animal. Over the entire course of human existence the study of only two animate objects—man and the fruit fly—have ever been prohibited; man during the Inquisition, and the fruit fly in Stalin's time. Actually, the study of man is forbidden in the Soviet Union to this day. History, sociology, and pedagogy enjoy a miserable existence, while ethology, the science of behavior, and medical genetics are only now recovering from the blows inflicted on them back then. At that time it wasn't just man, but animals as well that couldn't be studied from any of these seditious ethological, psychological, or genetic points of view. And there was no point in so much as thinking about the genetic bases of animal behavior. But in 1931, when Nikolay Nikolayevich was conducting his course on genetics in the Genetics and Experimental Zoology Department of Leningrad University, the fruit fly was not yet under prohibition.

In 1933, when at the invitation of N. I. Vavilov, the director of the Academy of Sciences Institute of Genetics, the distinguished American geneticist, H. J. Muller, arrived in Leningrad, Nikolay Nikolayevich recommended me to him as a laboratory assistant. I reported to Muller more dead than alive, overwrought at the prospect of seeing the great discoverer of the laws of nature. He asked me whether I needed a salary, and I lied and said that I didn't. I wasn't receiving a scholarship—the children of rich parents weren't eligible for them. I earned money by preparing tables and illustrations for lectures.

Muller took pity on me and instead of becoming his assistant, I received a topic and a working place at the Academy of Sciences. Three other students in our group received topics: Rapoport, now one of the most distinguished Soviet geneticists; Kovalev, a very capable person and a good comrade who died in the Second World War; and my group's party organizer, the former plumber and then promotee Feodorova.

In 1934, when I was a senior working on my diploma at the Institute of Genetics, Prezent gave a course called, after Engels's book of the same name, the Dialectics of Nature. Prezent subjected each branch of biology to review from class positions, exposed what had been the product of bourgeois or clerical ideologies, and anathemized it together with the names of great representatives of Russian science who were accused of all the mortal sins. When he got to genetics, Prezent argued that life conditions determine not only the characteristics of an organism, but the character of the transmission of those traits from generation to generation. "I assert the bold hypothesis that the linkage and crossing-over of genes are determined by external factors," said Prezent.

We were not broadly educated biologists. We were narrow specialists within our own field of genetics. But we knew genetics. The time of academic battles had long since passed, the five-year course of study had been reinstated, knowledge was expected of us at our exams and we possessed it. First-class teachers, disciples and co-workers of Yu. A. Filipchenko, the late founder of the Genetics Department, conducted our courses. We had first-class textbooks on general and specialized genetics, written by Filipchenko. We knew perfectly well that the linkage and crossing-over of genes depended on their position in the chromosomes. The genes in a single chromosome are linked to each other and transferred all together, while the genes located in different chromosomes combine freely. The genes within a single chromosome are linked the more tightly the closer they are to each other. Their recombination is the result of the crossing-over. This means that the partners of one and the same pair of chromosomes exchange parts with each other. The closer together the genes are located, the rarer it is for them to be involved in the process of recombination. Prezent's statement betrayed elementary ignorance. If we said anything like that at an examination on the cytology of heredity—a course taught by the very learned Ivan Ivanovich Sokolov—we would have been flunked without any questions.

After the lecture, I went up to Prezent and asked him when he had first proposed his hypothesis about the influence of external factors on linkage and crossing-over. "What do you have in mind?" asked Prezent.

"I have in mind Plough's experiments with the influence of temperature on the magnitude of crossing-over in the fruit fly," I said, "and Muller's and Altenburg's experiments with chromosome rearrangements under the influence of X-rays. External influence creates new linkage groups."

"I don't read specialized literature, I only criticize science's fundamental orientations," said the professor.

"Plough's experiments have been described by Morgan in his book on the physical bases of heredity—there's a book in Russian. Filipchenko translated it and published it in 1926," I said.

Prezent wasn't the least bit embarrassed. "Do you mean to say that linkage and crossing-over, independent of external factors, are the only ammunition in the arsenal of absurdities?" he asked rhetorically. "And what can you say about such nonsense as the linear arrangement of genes in chromosomes?"

"But that's a fact," I said, "and it's recently been confirmed yet again—and brilliantly. Painter discovered giant chromosomes in the salivary glands of the fruit fly larvae. We can now point out the location of a gene within a chromosome right in a laboratory preparation, looking through a microscope. You should come to the Institute of Genetics and I'll show you." Our conversation was diplomatically cut off by Prezent's assistant.

The thunderstorm broke at Prezent's next lecture. As always, he was reading in the Great Hall of Histology. He was about to begin, but suddenly broke off his reading and addressed himself to the auditorium. "And now let's have Berg tell us about the Marxist-Leninist theory of cognition."

I knew what Prezent was asking—the interrelation of subject and object, the cognizability of the world, the existence of objective truth, practice as a criterion of truth. And I also knew that not only being, but social being determined consciousness, and therefore that consciousness was determined by class. I knew all of that. Nina Ryabinina used to say that the university was an institute for well-born maidens where we were instructed in contemporary manners. I knew Lenin's *Materialism and Empirico-Criticism* by heart, and I certainly had enough knowledge to speak about the theory of cognition. The ability to react at lightning speed was what I lacked. I couldn't get up and recite in front of an audience of no fewer than 200 people without having had so much as a minute to get my thoughts together. "I can't do it without preparation," I said.

"Do you refuse to answer?" asked the professor.

"Yes," I said.

"Who will answer?" asked Prezent.

The promotee L. E. Khodkov rose to answer. He yelled that wreckers and saboteurs had asserted that you couldn't build the Volkhovstroy, the first hydroelectric station built after the victory of communism, without formulas and without a lot of other junk, but the workers had gone ahead and built it without formulas and all that other junk. One of the female students piped up: "What's that got to do with the theory of cognition?"

Prezent cut off his well-wisher and said: "I ask the department's social organizations to slap the hands of the class enemy who has gotten out of control, the student Berg, who disrupts the active method of teaching that I have used and which I will continue to use in the future."

He was lying—that was the first instance he had used the "active method" by addressing himself to the auditorium. I broke off the torrent of abuse unleashed on my innocent head and said loudly and firmly: "Professor Prezent, you are devoting too much attention to me in your lecture." A deathly silence followed. Then the lecture was resumed.

Prezent subsequently demanded my expulsion from the university. An article appeared in the university newspaper repeating Prezent's abusive remarks about me word for word, but it failed to produce the desired effect. I wasn't expelled from the university. But the party organization at the Academy of Sciences Institute of Genetics, where I was finishing up my graduation thesis and where I planned to show Prezent the enormous chromosomes of fruit fly larvae, demanded my dismissal from the institute. Institute director Vavilov himself came to see me late in the evening and told me that he hadn't managed to retain me, but he had not tried very hard. The institute was being moved to Moscow in the near future. I transferred my work to the university laboratory.

One of the reasons that I was not thrown out of the university at Prezent's insistence was the fact that the remarkable man was extraordinarily busy. The poorly educated agronomist Trofim Denisovich Lysenko had ascended to the post of scientific leader of the All-Union Institute of Genetics and Selection in Odessa and was elected a full member of the Ukrainian Academy of Sciences, and Prezent rushed off to place himself under Lysenko's banner so as to be able to offer him his great talents as an executioner. While Prezent was in Odessa, the examination for his course was given by his assistant, Kirill Mikhaylovich Zavadsky. Looking like an investigator, he asked me questions and suggested that I answer sincerely, that I say what I really thought.

He thought that I would immediately disclose my contempt for the theory, my negative attitude toward the dialectics of nature, and my attachment to the empirical rather than the dialectical method of cognition. He would easily make me give myself away. He had a very wrong idea of me. I wasn't contemptuous of the theory, and the traps that he'd set didn't work. After the exam, I went out into the hall and with curiosity opened the examination book where Zavadsky had recorded the grade. The grade was a "B." A few days later the university notified me of an official severe reprimand with a warning for ignoring the course on the dialectics of nature. A little misunderstanding had occurred. I applied to the administration to have the reprimand removed on the grounds of the grade that I had received on the examination. The reprimand remained in force.

The last time that they tried to drive me out of the university was in my fifth year. My graduation thesis had been finished and published in the *Reports of the Academy of Sciences*. I was the chairman of the genetics section of the Student Research Society and was working with a group of students on a topic that Muller had proposed. We were studying the correlation between the radiation dosage and the frequency of intra-chromosomal rearrangements in the fruit fly. We worked outside of town in the Peterhof Biological Institute in the former palace of Count Leichtenbergsky, who had fled abroad during the Revolution. The Peterhof Institute served the university as a summer base for practical studies. But this was in the winter. The city authorities had been threatening to take the palace away from the university since it went empty during the winter, so laboratories were started up, including ours. The university kept the palace. There was no electricity inside. We worked by the light of kerosene lamps. Two flasks filled with water tinted with blue vitriol, two circles of light coming together on a china plate, can, I assure you, easily replace an illuminator made by the best company in the world. You could even work with binocular microscopes under such conditions.

On his way from Moscow to Paris, where he was going in order to lead the International Conference on Radiation Genetics, Muller came to Leningrad and visited our laboratory in Peterhof. The dining hall was prepared for our important guest: The meat patties were gigantic that day. We were the first in the world to show that intrachromosomal rearrangements arose under the influence of X-rays as a result of two breaks. Muller's report in Paris included our data.

But it was at precisely that time that the university decided to give a Russian language examination to all students who had originally entered without an examination. The university didn't want to graduate specialists who weren't particularly literate. Out of our group of sixteen people, two of them wrote flawless dictation—Rapoport and Kovalev. I was the best among those who had failed. We were assigned a Russian course in which we were taught spelling and punctuation. Course attendance, as in all cases, was strictly required. Peterhof gave me the right to cut lectures, but that permission didn't extend to Russian. I wasn't planning to use that right anyway, though; the time had finally come for me to fill in the gaps left by my German education. I attended classes enthusiastically. But I had to miss several lessons. My group comrades, Feodorova among them, went into action to force me out. I was expelled from the university for failure to attend Russian language classes.

For the first time I believed that I'd been expelled and that there was no point in bothering to try to get reinstated. Perhaps the sword of Damocles hanging over my head on a thin thread had brought about a nervous breakdown. I went to the Lomonosov china factory, where I hoped to become a china decorator. In 1934, the administration of the factory—impressed by my drawings—agreed to take me and didn't even ask about my education. When the factory director saw my pictures he telephoned the art department to send some personnel to show me around. The sculptor Danko and the artist Skvortsov came. "Come here, I want to show you something," he told them. "This work is much better than what our Vorobyevsky does." I was taken to the studio and to the museum and saw Vorobyevsky decorating a dish with the traditional Russian-style waves. I looked at them and thought that being much better than Vorobyevsky wasn't hard.

Before I began decorating cups, however, I found out that I had not in fact been expelled from the university. The news of my expulsion, which had been transmitted by the comrades in my group, was a false rumor designed to confuse me. Had anyone cared to think about it, there was now reason to expel me from the university. In getting the job lined up at the Lomonosov factory I had missed classes for several days.

I was defending my graduation thesis and a comrade from the group, Chemekov (I had helped him all five years in what he found a difficult task—acquiring knowledge), sat in the front row of the auditorium where the defense was taking place and kept his eye on the clock so as to cut me off when my time expired. But I reached the finish line at the appointed time, right on the dot. I was given an "A." My graduation thesis was three articles published in the American journal, *Genetics*, and three articles published in three different journals in the USSR.

Alexander Petrovich Vladimirsky, the head of the department, went through a great battle to keep me on as a graduate student. The social organizations vetoed my candidacy. A lampoon appeared in the Leningrad University newspaper. I was reproached once again for a lack of community work. And again it wasn't true. I taught genetics and general biology to students in the lower classes who were behind in their studies because of illness or family considerations. When my students read the paper and learned the reason for my rejection, they wanted to write a protest, but I dissuaded them. I sensed that everything would work out in the end.

A commission came from Moscow to check the class makeup of those who had been proposed for graduate school. I was called in. The science rector for the university was E. Sh. Ayrapetyanets, one of the public prosecutors at the community trial of Mordukhay-Boltovskoy and his group. He introduced the future graduate students to the members of the commission.

At that time, the national pride of the builders of communism was being fed bountifully by Chkalov's airplane flights, investigations of the stratosphere, and the drifting station of the Arctic researchers. Drifting on the iceberg was P. P. Shirshov, a hydrologist, a member of the Mordukhay-Boltovskoy group. National pride, the consciousness of the state's technological and military strength, and patriotism based on power were conceived then and are still conceived today as a substitute for personal freedom and humane living conditions. Nonstop flights and Arctic explorations diverted worldwide public opinion and camouflaged starvation and bloody retribution taken upon millions of innocent victims. One had to be up to date in one's knowledge about these achievements. Ignorance of them exposed you. I was asked about the nonstop flights. You couldn't not have known about them—they talked about them everywhere you turned. But I had had enough. I said that I didn't have any solid knowledge in that field since I was busy with my own affairs—flies. I was told that, despite my ignorance, the commission wouldn't object to my staying on as a graduate student, since they bore in mind my father's meritorious service. "If my father has performed such meritorious service, then take him as a graduate student," I said and exited. I was denied entrance into graduate school.

Vladimirsky and two other teachers announced that they would leave the university if I wasn't retained as a graduate student. But no one and nothing would have made a difference if the university's Geography Department hadn't fallen apart completely just then. The university decided to ask my father to return to the department, and in trying to lure him back they decided to retain me. Father returned, and I stayed on.

# *Bronze and Golden Knights*

My acceptance into graduate school inaugurated four years in which I could blissfully study science full-time and without harassment. I tried to narrow the boundaries of my general biological ignorance by taking courses in comparative invertebrate anatomy, vertebrate anatomy, and general protistology.

Roza Andreyevna Mazing, a teacher in the department, conducted her research in close contact with me. This was a precious gift of fate. It does not happen every day that an experienced university lecturer joins the research of a graduate student exploring his or her own topic. Roza Andreyevna did it, and her investigations were crowned by discoveries about the interrelations between the mutation process and natural selection worthy of a Nobel Prize. The cruel time deprived her of all well-deserved recognition.

The main events of those years were Muller's departure and my heading down the path of population genetics. Both these things happened in the same year, bloody 1937. The terror of that year eclipsed and exceeded anything imaginable; it exceeded all the monstrous bloodlettings of years past.

The Institute of Genetics moved to Moscow, but I kept up ties with Muller, traveled to see him, and showed him the results of my experiments. Muller's contribution to world science is enormous. He was one of the founders of the chromosome theory of heredity. He was the first person in the world to use ionizing radiation for the artificial creation of hereditary changes. He was the first to invent quantitative methods of mutation studies, and he put the mutation process under strict quantitative control. And only when that had been done did he subject flies to the effects of temperature and radiation. The use of X-rays in his experiments increased the frequency of mutations by hundreds of times.

Muller was the first to point out the danger of radiation, not only for the person who is exposed to it, but for his or her offspring as well. It was at his suggestion that measures were taken to reduce the use of radiation for diagnostic and medical purposes to a minimum. His experiments in the artificial production of mutations were the source of a new branch of science—radiation genetics. He was awarded the Nobel Prize in 1946 for founding it.

But no matter how great Muller's contribution to science was, the investigation of nature was not his life's goal. The young scientist was consumed by the desire to better the human race, to rid humanity of physical and spiritual misery. He was a fierce enemy of any limitations on human freedom, and fought against those racist tendencies which had divided German society and found adherents in other countries as well, including America. In a society of class contradictions in which the very assessments of good and evil, perfection and defect are distorted in favor of the wealthy, the improvement of the human species was, according to Muller, impossible. Fearless among the fearless, he believed that only the restructuring of society on a socialistic basis would allow improvements to be made in man's private life. He thirsted for a revolution that would overthrow capitalism. Only the proletariat could be the hegemon of the restructuring of the new life.

In 1936, Muller went to Spain to help the freedom fighters. He was in the Canadian subdivision, where he supervised blood transfusions. In Madrid, he saved the library in a burning university building. During a short stay in the United States, Muller was attacked by members of the Ku Klux Klan. They tried to run him over with a car. They knocked him off his feet. His coat was ripped. He told me about that himself.

Muller's sights were set with hope on the Soviet Union. He wanted to be on that point of the globe where the fate of communism was being decided. He went to Germany, which seemed to him to be on the verge of great progressive changes, but in 1933, when Hitler came to power, he left. That same year he was elected a foreign corresponding member of the USSR Academy of Sciences, an honorary title he accepted with gratitude, and was invited to head the Department of General Genetics at the USSR Academy of Sciences Institute of Genetics. Muller and the institute's director, Nikolay Ivanovich Vavilov, were linked by bonds of close friendship. The American geneticist Bentley Glass, who had studied with

Muller in Berlin, told him that all totalitarian regimes were alike and that he'd find unleashed Hitlerism in Russia. Muller would not listen to him.

Muller arrived in Leningrad at a time when two mighty destructive forces—Lysenko and Prezent—had linked up, and genetics was poised on the edge of the abyss into which it was to plunge fifteen years later.

At the beginning of the 1930s, Muller wrote a book called *Out of the Night*. In it he called on humanity to do everything possible to hasten the arrival of the bright future—communism. Man himself was to be changed. Goodness and intelligence were to become the criteria for evaluating human perfection. A means that could be used in the near future was artificial insemination with the semen of those who conformed to the criteria of perfection.

After arriving in the Soviet Union, Muller managed to have a translation of his book shown to the authorities. The authorities reacted extremely negatively to his plan to better humanity. It's not difficult to understand why.

In the first place, they already thought they possessed the means for bettering the population of the whole country. To create a new Soviet type of man, Soviet authorities used terror. Not only did enemies of the regime have to be destroyed physically, but potential enemies also became victims of mass terror. Artificial selection of the best, of those who corresponded to the socialist standard as the powers that be deemed it, was combined with environmental influences. The best had to live being endangered. That was the way things worked out in practice.

In the second place, and this was only the official line, the Soviet workers who were building communism and the collective-farm peasantry had no need of improvement. Filled with enthusiasm and love for their government, they had just fulfilled the first Five-Year Plan in only four years. They achieved general collectivization and were heroically engaged in the country's industrialization. From within their midst they had destroyed the enemies of the Soviet regime: millions of prosperous peasants (kulaks and bloodsuckers), and many thousands of bourgeois intellectuals, wreckers, and saboteurs. They untiringly conducted a class struggle that was growing more and more heated. They were destroying the remnants of capitalism in people's consciousness. All of this was the creation of the new man. Based on the principle of the inheritance of traits acquired during life, collective labor was supposed to create a collective conscience, collective consciousness, group will, and universal enthusiasm. And here came Muller with his genes and jars of frozen semen taken from the best representatives of the human race.

Defeat followed defeat for Muller. He plunged into polemics with Lysenko and Prezent.

In 1936, at a debate on questions of genetics, to which a session of the Lenin Academy of Agriculture was devoted, Muller spoke along with Vavilov and A. S. Serebrovsky. These great scientists' opponents were Lysenko, his hanger-on Prezent, and their clique. A caricature version of Lamarckism was their banner. It was forbidden by the Ideology section of the Communist Party Central Committee to touch upon questions of human genetics. The debate was limited to questions of agriculture.

Serebrovsky and Vavilov submitted to the prohibition. Muller was the only one who broke it. He said that the principle of the inheritance of acquired traits was well-suited for racists. The representatives of wealthy races and the exploiting classes, enjoying the fleshly delights of existence, living in well-appointed quarters, were perfecting themselves from generation to generation. According to this same principle, the poor peoples of the earth and all the exploited elements in developed countries were doomed to degradation.

In addition to his attack on Lamarckism, Muller cited new proofs of the chromosomal theory of heredity and showed depictions of the gigantic chromosomes in the cells of the salivary glands from fruit fly larvae. On the screen appeared the same pictures that I had planned to show to Prezent at the USSR Academy of Sciences Institute of Genetics.

Lysenko's rebuttal speech was a profanation of science. He hadn't the slightest need to conceal his ignorance.

I was at that debate. The audience applauded wildly, and I was among those who applauded when A. S. Serebrovsky told about how he had bought genes. He had needed rabbits of the "Rex" variety for the Institute of Rabbit Breeding, where he headed the Department of Genetics and Selection. That breed of rabbit had to be brought from Germany. They were expensive and not hardy. They had to be paid for in hard currency. Serebrovsky bought two common, inexpensive, hardy rabbits, because he knew that they were hybrids and that in their genetic makeup there was the gene that gave rabbit fur the valuable qualities of the "Rex" breed. In Moscow, these hybrids were bred. Some of the baby rabbits had the desired fur. "Traits are not inherited. Traits develop all over again in each generation as a result of the interaction of genetic makeup and the environment," said Serebrovsky. "Genes, not traits, are transmitted from generation to generation. The probability that a gene will be transmitted to the next generation does not depend on whether the corresponding trait appeared or not. We not only knew that the hybrids would give birth to the baby rabbits that we needed, we could even predict how many of them there would be."

The audience was on the geneticists' side. The press was against them. Muller's presence in the Soviet Union jeopardized not only himself, but Vavilov as well. In 1937, Vavilov told Muller that it would be better for him to leave.

In September of that year, I had just returned from an expedition and left for Peterhof with my flies, where I sat at the binocular microscope day and night. I'll tell you later how it was that I moved from radiation genetics to population analysis. My father called the office at the Peterhof Institute and left a message for me to come to Leningrad right away. Vavilov was inviting me to his place at 6:00 P.M. My father told me that Muller was leaving and that Vavilov had invited me to see him off.

I arrived at Vavilov's apartment and found Muller there alone. Vavilov was at a meeting somewhere. When Vavilov came into his study, I was lying on the sofa on my stomach and Muller was on his knees next to the sofa. We were plotting curves and charts of interbreeding on the same piece of paper, both of us engrossed in a scientific polemic. This scene would have been any father's undoing, but for me and obviously for Muller, embarrassment was out of the question.

There was a small banquet. Then Vavilov took us to the movies, to see *Peter the First*. Later we walked around Leningrad. Muller spent his last night in Russia at the Hotel Angleterre—now called Leningradskaya—where the poet Sergey Esenin committed suicide. We parted at the entrance to it. It was past midnight. "We meet here tomorrow at five in the morning," Vavilov said as we all shook hands.

At five o'clock, I was at the appointed place. Vavilov brought apples and treated us. He looked as though he'd had enough sleep; he only needed three hours to feel completely rested. That was a family trait. As a farewell gesture, Vavilov wanted to show Muller the institute that he had founded in 1924. In a car from the Institute of Plant Breeding, where Vavilov was director, the three of us headed off for Detskoye Selo, the institute's branch outside of town. He led us around garden plots where samples of cultivated plants were sown. They were gathered by Vavilov from, quite literally, all the corners of the earth. It was a living plant collection. There was Abyssinian flax with its magnificent large blue flowers. Heaps of pink tubers, resembling little piglets, lay in the potato field. We were shown the harvest from individual plants.

In the cytology laboratory Grigory Andreyevich Levitsky showed us slides. In the laboratory where the baking qualities of cereal cultures were studied there were little loaves of bread that resembled Eucharist wafers.

It was in one of those laboratories that we were served breakfast: tea, white bread, smoked fish, and chocolate bars. The driver who had brought us had breakfast with us. I was very struck by that. My father, for all his Tolstoyanism, would not have seated a driver at the same table with himself, nor would any other director have done so. In this regard, as in many others, Vavilov was an exception.

We returned to Leningrad. Vavilov went to his next meeting and I was the only person who saw Muller off as he left Russia. In Vavilov's study, Muller left an official letter explaining his departure. He was about to slam the apartment door shut, but stopped, went back into the study for a moment, and then we left. He explained the reason for the delay: "In the letter, I wrote that I'd come back in two years. I corrected that. I'll return in a year."

He never returned.

I continued to study natural populations. The first observations on the mutation process in wild flies were made by me. A small Ukrainian town, Uman, was the cradle of my discoveries. It was there that I found the yellow mutation among wild flies, and that led me to ponder the high frequency of mutations. Fruit flies are marvelous. Looking at them through a binocular microscope is sheer pleasure. Their red, faceted eyes look like burning, pomegranate-colored bonfires, their translucent wings shimmer like a rainbow, and the bristles that cover their bodies seem to be made of nylon. (That comparison isn't mine. One of Leningrad University's graduate students, a Chinese named Shao, made it in 1956.) The flies' bodies are the color of honey or bright aged bronze. The yellow mutation makes them gold-colored. The stubble and the little hairs that cover the wings as well seem to be made of gold, and the wings of the yellow flies shimmer like gold instead of the rainbow. The yellow mutations became my guiding star, my star of Bethlehem. I wanted to find out why the mutation that made the fly's body golden had begun to occur so often, and why it was precisely that that mutation was repeated so frequently. Was the Uman population, with its high frequency of occurrence of mutations, an exception?

After having seen Muller off, I went to the Nikita Botanical Gardens on the Black Sea shore. I caught flies at its wine distillery and at the Magarach wine complex that was located by the shore among cypress trees. Yellow flies turned up there, too, and the mutation appeared there with great frequency as well. I set up a base with my binocular microscope and vials full of flies at the institute's cytological laboratory; the Nikita Gardens were a branch of the Institute of Plant Breeding and were under Vavilov's authority.

Obstacles arose one after the other. Yeast is an ingredient of fly food. I hadn't been able to find dry yeast in Leningrad before setting off on the expedition, so I took the fresh, compressed kind, and asked that it be kept on ice in the institute's dining hall's icebox. In the course of thirty-five years I visited the Nikita Gardens' flies twelve times and each time I stayed for a month or longer. Never and nowhere have I encountered a worse eating establishment. But this time, on the day after my arrival, they served marvelous rolls for breakfast. When I went to the icebox to get the yeast, it turned out that the substance had disappeared. "Where's the yeast?" I asked. "Didn't you eat the rolls?" was the response. I couldn't work. Of the endless list of items that were difficult to find in the Crimea, yeast was the scarcest of all.

Living in the Soviet Union would be impossible if it weren't for the existence of a great lever in Soviet life. That lever is connections, acquaintances. A woman who worked at the institute was a daughter of the manager of a grocery warehouse in Yalta. The woman wrote a note to her father. I set off for Yalta on the bus. No sooner had we departed than there was a stop. There had been a landslide and the road was buried by an avalanche. The dust hadn't even settled yet. Two lengthening lines of light and heavy trucks and buses formed on both sides of the jam. A herd of cows climbed up the avalanche and one of the cows, standing with its back to us, lifted its tail and added a little something to the pile of rocks and earth. I made my way over the avalanche after the cows.

At the end of the line, trucks were turning back toward Yalta. I reached the grocery warehouse on one of them. While climbing into the truck, I tore my skirt. I had a needle and thread with me, but I needed a patch. One of the passengers gave me a tie. I presented myself to the warehouse director with a tie on my thigh. There was yeast, a whole crate of it, but it was in the basement, and no one would get it because it was alleged that a horde of rats lived in the basement. The rats were a lie, the usual method of getting a bribe under the cover of payment to the person who would risk his life and go down into the basement. I was only twenty-four, and I of course believed that there was mortal danger lying on my path to the yeast. I went down into the basement anyway and found what I was looking for without encountering a single rat and without paying a bribe. I walked the twenty miles back to the Nikita Gardens.

A married couple, workers at the Gardens, undertook to prepare the fly food and to wash and sterilize the vials. When it came time for the grape harvest, however, they suddenly refused to work unless I gave them a raise. Their demands were beyond my means. A young Tatar woman, a cytologist and an assistant in the laboratory where I spent days and nights counting flies, saved me from disaster. She agreed to take the place of the couple and do what they did without payment. In return she asked me to recite to her the poems that I had been reciting aloud when I thought that I was alone in the laboratory. I recited Mandelshtam, Esenin, Blok, Akhmatova, Pasternak, Mayakovsky, Gumilev, Kluyev, and Lord only knows whom else. Poetry saved me.

At the end of the first year's research, I had enough material for a comparison of two populations of fruit flies—those of Uman and of the Nikita Gardens, and a laboratory strain called "Florida." The high incidence of mutation distinguished the wild flies, while the laboratory strain was stable. Fate gave me three superb assistants. They were C. F. Galkovskaya, V. T. Alexandriyskaya, and E. B. Brissenden. I'll have more to tell about Brissenden later.

Would the mutated gene be kept, or would it be thrown out of the population as one generation replaced another? What does its fate depend on? What facilitates the increased numbers of mutants and what forces prevent the mutants from crowding out the former norm? These traditional questions of population genetics interested me. I studied the numbers of yellow males on the site where they were hatched and among migrating flies as well. Yellow males occurred more rarely among the nomads.

My assistant, Alexandriyskaya, was doing research on how yellow males and normal males competed with each other for females. The golden knights had poor chances when they were matched against the bronze ones.*

* It isn't any kind of armor that gives me reason to call the male fruit flies knights, but rather their manner of courting the females. Their chivalry has nothing in common with the coarse manners of houseflies. A courting *Drosophila* male acts similarly to a rooster. He pursues the escaping female closely. After the female stops, he faces her as if trying to win her over to believing him, his spread wings constantly vibrating. Sometimes only one wing is moved further apart, making the resemblance to the rooster even more apparent. The rhythm of wing vibration, it has recently been discovered, is a song different in different species and even varieties of the same species. The *Drosophila* knights have their own serenades.

In the company of males of the wild type in their natural habitat, the shimmering gold mutants didn't have any chance of leaving offspring. This was because they were less sexually active. Therefore the mutation process, rather than natural selection, had to be the reason that yellow males turned up among wild flies with such striking frequency.

But these questions were not our main ones or the ones that distinguished our work from what had been done before us. For the first time in the history of genetics, we investigated the characteristics of normal genes in the population and we studied these characteristics in their interconnection.

The frequency with which mutations occur is determined by the qualities of the normal gene, on its stability. We studied the other side of the coin—mutability. But not just mutability alone. The fate of a mutant gene depends on a multitude of factors. One of them is the interrelation between the mutant gene and its nonmutant, normal partner.

Each gene we possess is represented twice in our hereditary makeup, in our genotype. It is true for all living beings except the most primitive ones, those that are lucky or unfortunate enough to have no sex. One of our genes is inherited from the mother, the other from the father. Both genes can be similar to each other, but they can be different as well, the one normal, the other one mutant. If they are different, the manifestation of the trait determined by the normal gene and modified by its mutant counterpart depends on the interaction of both of them during development of the organism. Suppression of manifestation of one partner by the other one is called "dominance." The suppressing gene is dominant. The gene incapable of manifesting itself in the presence of the other gene of a pair, provided this other gene is normal, is called "recessive." If the suppression is complete, the recessive gene is hidden and is out of reach of natural selection, however harmful or beneficial for its carrier its manifestation would be. The recessive gene manifests itself only in the absence of the dominant one. Two recessive genes inherited from both parents have no hindrance to manifesting themselves and are exposed to natural selection. In man, albinism is a good example of a recessive trait.

As a rule, a normal gene suppresses the manifestation of a mutant one. But that suppression isn't total. The more extensive it is, the more concealed from the effect of selection and, moreover, from our eyes, are the mutations. We established that normal genes of flies living in their natural habitats and the genes of captive flies of the old laboratory strains are distinguished from each other by their ability to suppress the harmful influence of mutant genes. The genes of wild flies had a lesser capability of suppressing the effect of mutations than the genes of flies kept as the laboratory strains. To a certain extent, this difference didn't deepen the distinction between hordes of wild flies and the small number of their laboratory kin; instead, and paradoxically so, it made their similarity possible. The hidden harmful mutations saturated them in identical quantities. They often arose in the wild flies, but mutant genes were eliminated more by natural selection frequently, while in the laboratory everything was just the opposite. The mutability was small, but so was the elimination. The quantity of hidden mutations ought to have been approximately the same. And that's exactly what turned out to be the case.

I spent the summer and fall of 1938 in the Nikita Botanical Gardens. Unlike the year before, when I had stayed in a hotel, I lived in the Tatar village of Nikita to which Nikita Gardens owed its name. The day that I arrived, when I was standing in line at the institute dining hall, a Tatar woman asked me, "Do you work as a serving woman here?" When I said I did not, she offered to rent me a room.

I settled in at her place. First of all I had to master her rules of hygiene. Shoes had to be left outside, and you couldn't brush your hair or wash in your room. I could come home as late at night as I pleased, however, even at three in the morning, and she would open up for me without the slightest rebuke. Once I took a sponge and soap with me and said that I was going to wash in the shower, which was near the laboratory I occupied with my flies. I came home very late—it was almost morning. Zaynep—that was her name, but she asked me to call her Zina—threw herself tearfully on my neck and began kissing me. "I cried all night," she said. "I thought that Raya had drowned. I pounded myself in the head and said that I was a fool, crazy, that Raya had washed in the shower, that she couldn't have drowned."

Once I was dragging around boxes of test tubes in the heat and I got a nosebleed. I left the blood-soaked handkerchief in the pocket of my suit. When I got around to taking it out to wash it, it turned out to be perfectly clean. "How could that be?" I asked Zina.

"I went through all your pockets and cleaned out all the dirt. I washed it, ironed it, and put it back," she said. "Do you think it was difficult for me to do it?"

Another time she asked, "Raya, do you have a lover?"

"Why?" I said.

"Have him buy me five meters of cloth." I had no lover, and those five meters of cloth that I wasn't ever able to supply torment me even now.

Once, in the middle of the night, I awoke with a terrible headache. I was lying there and crying. Zina came in. "Don't cry," she said. "If you don't have any money, my husband and I will give you some."

"I have a headache," I said. "Thanks for the offer, but I don't need it."

She gave me a pill of some sort, made a hot poultice for my head. "You feel kind of bad, huh? You mean cold water wouldn't help? Let us use the hot one. Do you think it is difficult for me to do it?"

All that was spoken in a terrible lingo, terrible from an academic point of view, beautiful as real poetry. Untranslatable. Never in my life shall I forget these words repeated so many times, as a refrain in a song, this question put at the end of her every offering: Do you think it is difficult for me to do it?

The Crimea then was the Autonomous Tatar Republic. I encountered Tatar nationalism only once. Zina's nephew told me that all the land from the Tatar Sea to the Crimea had been conquered by the Tatars and should belong to them. During the Second World War, traitors were found among the Tatars. The entire population—women, children, and old people, including the families of heroes of the Fatherland War, everyone, without a single exception—was exiled to Central Asia.* The Crimea became a part of the Ukraine.

In 1956, Stalin's crimes were partly denounced. Most of the exiled national minorities accused of collaboration with the invading Germans were allowed to return to their national territories, but the Crimean Tatars were not. In the sixties a stubborn campaign on the part of the Tatars was launched with a demand to permit the Tatars to return. The Tatars' defenders were arrested, and they were not allowed to return.

Let us imagine for a minute that they had received the permission to regain their country. To take advantage of the permit, they would have to be offered apartments and work. But their houses were occupied by Ukrainians so they would have nowhere to stay. Any daredevil who

* The deportation of Tatars was undertaken in accordance with the resolution of the State Defense Committee passed on May 11, 1944. All crimes committed by Nazis in the Crimea were slanderously ascribed to Tatars. The resolution and the false accusations were repeated on July 24, 1987, in *Pravda* in response to the Tatars' demands to restore to them their rights to have their autonomous republic in the Crimea and to inhabit it.

would try to stay could not register his passport there and so would be seized by the militia and, at best, sent back. You may not believe what I am saying. Let me explain. One registers one's passport at the office of the apartment house where one's "housing space" is located. You ought to know that Soviet citizens don't live just anywhere, but in space provided by the state. There's a stamp for that in one's passport. The stamp is placed there by the precinct militia. According to administrative rules that are not recorded in any law code and are applied arbitrarily, the sentence for a person staying nine days somewhere in the Soviet Union if he or she isn't registered is a year in prison.

I'll make a slight digression. Not only people, but cities too are divided into the clean and the unclean. The division is determined by the size of the populated area and the presence or absence of high-ranking authorities. The privileged cities are better supplied than other ones, and one can move to them only with the permission of the authorities. The privileged grant such permission only to the privileged or to those whose work is essential; for instance, youths who have completed their term of service in the army. There is a shortage of young workers everywhere. For others the city is "closed." There are enormous, terrible, hungry cities and they nonetheless are in the ranks of the "closed." Such a place is Novosibirsk, an enormous city with a population of 1 million.

In 1968, on the thirtieth anniversary of the events being described, I took an English language course. I was the head of the Laboratory of Population Genetics at the Cytology and Genetics Institute of the Siberian branch of the Academy of Sciences. That institute was located in Siberia, in Science City, about forty miles from Novosibirsk. The great Russian authority on English and the innovator in methods of teaching it, the great wit and joker Tankred Grigoriyevich Golenpolsky, decided to assemble all those who wanted to learn English or to improve their knowledge of it. Thirty-six such people turned up. He hired eleven teachers, bought out the second class on a passenger ship, invited the female students from the Pedagogical Institute—future English teachers—to serve as waitresses in the restaurant so that we could be waited on in English, and also invited the amateur jazz orchestra from Science City so that we would have something to dance to. We set off on the ship *Mikhail Kalinin* for a twelve-day excursion from Novosibirsk to Surgut and back. Novosibirsk is located on the Ob, the westernmost of the three great Siberian rivers, in the very heart of Siberia. And Surgut, the center of the new petroleum region, is also located on the banks of the Ob, to the northeast of Novosibirsk. As the crow flies it would be about 600 miles, but along the arc formed by the Ob the journey is about 800 miles, or twice that if one sails the river—as we were to do—in both directions.

We sailed past low banks, accompanied by gulls and swarms of gnats and mosquitoes. We studied English. In the intervals between classes I read a book in English about group selection. I was lying on the hammock in my cabin. The ship's engine pounded away rhythmically. Voices carried down from the deck. A man and a woman were speaking softly, but every word could be made out. "We should have waited," said the man. The intonation alone was enough to break one's heart: There wasn't a hint of protest or rage. They were discussing a phenomenon that you didn't argue with. "We should have waited. If we'd waited for the head person, he might have given us a registration permit. His deputy refused us. We should have waited some more. The deputy might not have refused us. You won't write to me, I know. And my letters probably wouldn't reach you, anyway."

"The woman who was the head of our factory section got letters," she said. "Maybe I'll get them, too." Those were the only words she spoke. She didn't promise to answer him.

"It's all because of you that we were denied registration. It's you who didn't want to wait for the main person," his voice spoke out again. "And the deputy would have agreed if we'd waited some more." The ship's engine pounded away and pounded away. It was obvious that the man, the possessor of the soft voice, was seeing his girl friend off out of the city. She worked at a factory and lived in a worker's dormitory.

The fortunate possessor of living space in a closed habitat has the right to register his parents, a wife, or a husband. A huge number of juridical and administrative rules and ruses restrict that right. The possessor of living space must have an excess of living space, and furthermore, the excess must not be less than the established dimensions. Each person has a right to have 9 square meters. Let's say that you have a room that contains 27 square meters. Your parents may move in with you. But what if you have 26 meters instead of 27? There is legitimate reason for denying registration to both your parents. But you can get anything you want for a bribe. Without a bribe you may be denied what is legally yours. The way that people in the Union of Soviet Socialist Republics give and accept bribes was brilliantly described by Gogol in *Dead Souls*.

In short, then, a Tatar who came to the Crimea to settle in his homeland would, because of the Central Committee's decree, wind up in jail after nine days. To be quite accurate, after three days he would be seized by the militia and fined for "residence without a permit," three days later he would be seized and fined again, and in nine days he would be imprisoned.

The title of the dissertation that I defended for the advanced degree which corresponds exactly to the American Ph.D. is a long one: "A Comparison of the Genetic Properties of Natural Populations and Laboratory Strains of Fruit Flies: A Hypothesis of Genetic Correlations."

In the spring of 1939, A. P. Vladimirsky, the head of the Department of Genetics and Experimental Zoology, a wonderful teacher and scientist, the son of a priest and a member of the Communist party, a great master of silence when lawlessness by decree took place under his own eyes, suddenly died. I was defending my dissertation that same year, just after his death. His successor, promotee Mikhail Yefimovich Lobashev, became the de facto decider of fates for the department. Lobashev would throw obstacles my way all his life. In 1935, when Vladimirsky was proposing candidates for graduate status, he had named five people: Novikov, Barancheyev, Kovalev, Rapoport, and me. Lobashev had contested my candidacy and Rapoport's. In abilities, Rapoport was head and shoulders above the rest of us. I've already described how Vladimirsky succeeded in keeping me on. He wasn't successful with Rapoport. But Rapoport hadn't been aiming for graduate studies in Leningrad. He was accepted as a graduate student at the Institute of Experimental Biology, and under the guidance of its director, Nikolay Konstantinovich Koltsov, one of the founders of genetics in the Soviet Union, he began his famous experiments on the artificial mutagenesis through chemical preparations. In 1939, when Vladimirsky died, Lobashev gave me to understand that my time in the department had come to an end.

# *The Morality of the Future and the Bombs*

After Muller's departure in 1937, I had been very ill and had stayed in bed reading scholarly books. When I read I. I. Schmalhausen's *Paths and Laws of Evolution* I had realized where my place was. I wanted to go to Moscow and work for the Academy of Sciences at the A. N. Severtsov Institute of Evolutionary Morphology, which was headed by I. I. Schmalhausen.

Schmalhausen's path to genetics had crossed through embryology. The law of heteronomous growth* was discovered by Julian Huxley and Schmalhausen simultaneously, but independently of each other, and in his book, the great English scientist gives his Russian co-discoverer his due. The law of heteronomous growth concerns the independent rate of growth of certain parts of an embryo in relation to all the other parts and was the basis for Schmalhausen's theory of stabilizing selection. When Schmalhausen came out with that theory in 1938, he was poorly understood. Now such things are no trouble at all to understand.

I'll explain. Don't be frightened by terminology; I'll explain it in such a way that you'll understand. Imagine three cities. In each one there are mice. In each one there are mousetraps, but the mousetraps aren't all alike. In the first city the mousetraps have a small aperture. Tiny mice get caught, but large ones survive—they can't get in. An order comes from the State Planning Board to enlarge the openings. They're enlarged a tiny bit. The smallest of the large mice perish. The largest ones survive.

* The term *heteronomous growth* belongs to Schmalhausen. Huxley called the change in proportions during development "allometric growth."

The mice in the city grow larger and larger. This is a progressive form of selection since it progressively increases the average size of a trait. The weight distribution curve underwent a change, but its shape didn't change.

Now let's take the second city. The doors on the mousetraps admit the average-sized mouse and anything smaller. The large mice survive. But the baited hooks that the mice are supposed to pull to make the doors slam shut are placed so high that only the average mice can reach them—the little ones can't. The average-sized mice die, while the large and small ones enjoy life. This is a disruptive form of selection. It creates two norms—tiny mice and large ones. The weight distribution curve is twin-peaked. There's a gap where the former norm was.

And now the third city. It has two types of mousetraps. Some of them have tiny doors and low hooks. They only catch small mice—the average-sized and larger ones can't reach the bait. The other mousetraps have a large opening—all the mice can get in, but only the large ones perish. The hooks hold the bait up too high and the average mice can't reach it. The selection process is in favor of the average-sized mice. Their average size doesn't change. The weight distribution curve remains single-peaked, just as it had been, but its shape changes. It is steep and rises to a high peak. The extremes at both ends have been lopped off by selection. This is precisely what selection in its stabilizing form is—the selection of the standard.

Under the influence of stabilizing selection, changes in embryo development take place. Progress lies in increasing the probability of survival. Stabilizing selection is one of its primary movers.

You're fed up with the imaginary examples. All right. Let's turn to nature. Take the number of eggs that a nightingale or a chiffchaff lays. Very few. What's good about that? The chances for leaving progeny are small. What if there are a lot of eggs? There are a lot of baby birds, but there are still only two parents. There's always the risk that all the fledglings will perish from lack of food. Selection favors the golden mean and it's not for nothing that it's called "golden."

To be stable is to possess a maximum likelihood of survival, to be safe among life's adversities, and to maintain a standard in spite of accidental changes in the environment. The independence of the growth rate of certain parts of an organism from the growth rate of others led Schmalhausen to the idea of the adaptive value of the standard. Much later he formulated his theory of stabilizing selection. The paths by which the standard is achieved in evolution is elimination by selection of all those who deviate from it. Independence is the advantage of survivors. To be fit to the environment you have to be able to resist its chance fluctuations.

Even before I had defended my dissertation I went to Moscow to the Institute of Evolutionary Morphology. I took a bound copy of my dissertation, reprints of the articles published in *Genetics* and *Reports of the Academy of Sciences,* donned my best dress, and presented myself to the institute's director. My dress had cost me a great deal of work and trouble. Among valuables of prerevolutionary vintage in Nanny's trunk was a length of batiste with a white and blue design. Nanny had kept it for no fewer than thirty years, and I passionately wanted her to make me a gift of the material. She had refused and made herself a skirt out of it, but suddenly she changed her mind and gave me the skirt so that I could transform it into a dress.

Schmalhausen asked what my specialization was. I said that I was a geneticist. Genetics was plunging headlong toward its demise. Only a few months before our conversation the main destroyer of genetics, the semi-illiterate agronomist Trofim Denisovich Lysenko, a favorite of Stalin's, had been elected a full member of the USSR Academy of Sciences. The path for him was cleared by an article in *Pravda* that attacked two other candidates for the academy: my father and Nikolay Konstantinovich Koltsov. I'll have more to say about that later. Now Lysenko was not only an academician, but a member of the academy's Presidium as well. Koltsov lost his post as director of the institute that he had created. The Medico-Genetics Institute, the best of its kind in the world, according to Muller's statement, had already been closed three years earlier. Its director, S. G. Levit, and his co-workers, I. I. Agol and V. N. Slepkov, were arrested and never seen again. The International Genetics Congress that was supposed to be held in Moscow in 1937 was canceled. It was convened in Edinburgh in 1939. Forty geneticists, myself included, sent theses and submitted papers. Vavilov had been the vice-president at the previous, sixth International Congress, held in Ithaca, New York, and it was he who proposed Moscow as the site of the next Congress. Now he had been elected the Congress's president. But he, the great geneticist, an agronomist and explorer, didn't receive permission to go to the Congress. He was replaced by Lysenko as president of the Academy of Agricultural Sciences.

"Considering the catastrophic situation in genetics, do you want to change specializations?" Schmalhausen asked me.

"No," I said. "I want to continue my work in population genetics."

"Do you have any published works?" he asked. I handed him the offprints. He looked at them and said, "I have the honor of knowing your father. Although heredity isn't held in great honor now, the genes of *Nomogenesis** will be taken into account upon your entering our institution. I can offer you a postdoctorate position. You'll have to take an examination in Marxism. Hang on tight."

And I held on. The syllabus for Marxism-Leninism contained eighty-four recommended works. I crammed all summer long. In addition, there were examinations in two foreign languages.

I passed the language exams with "A's," though I nearly failed the Marxism-Leninism exam. No matter what I said, it was wrong. The syllabus listed works by Kant, Hegel, Feuerbach, and the French materialists. But in responding to a question about Kant, you were supposed to discuss Lenin's and Engels's pathetic musings on Kant's mistakes instead of *The Critique of Pure Reason*. To my good fortune, I was asked a question about Engels's book on Ludwig Feuerbach. It would be an exaggeration to claim that if you thrust a needle through the book, I could tell you which letter it had gone through on each page, but I could leaf through the book in my mind's eye and summarize everything in the right order. The examiners were delighted. The final question was on "The Role of the Geographical Environment in the Development of Society." I was prepared to talk about Rousseau and the French materialists, about Élisée Reclus and V. I. Vernadsky. But instead of that I said, "On such-and-such a page of *A Short History of the Bolshevik Party* it says so-and-so, and on such-and-such a page so-and-so." The *Short History* was published in 1938. You can't read it without complete and total disgust, but you had to know it by heart. While memorizing it by rote I didn't think about anything except the hope of going to Dilizhan to catch flies. I earned the highest marks and the philosophers' recommendation, and soon thereafter left for Dilizhan.

In comparing wild flies with laboratory strains, I had found that the wild ones were more diverse and that mutations arose among them more frequently. It would seem that life in captivity, standardized feed, and controlled conditions made the laboratory flies stable. It seemed to me, however, that the reason for the captives' loss of mutability were not the standardized laboratory conditions, but isolation itself. The high

* *Nomogenesis* is the famous anti-Darwinian book by my father published in 1922. It was condemned by official ideologists, Prezent among them.

mutability of flies in their natural habitats is the result of selection for flexibility. It's a process of group selection. If part of a fly population dies as a result of an infection or the use of insecticide, and a part survives, a selection of the more mutable groups will occur. The probability that among survivors there will be mutants possessing an increased hereditary resistance to infection or to the new insecticide is greater for groups having high mutability than for the groups displaying low mutability. These representatives of the more flexible groups that survived the fly epidemic will create a population of their own kind, and their progeny will colonize the deserted centers of reproduction. Repeating itself from year to year, this process of intergroup competition for mutability leads to its increase. In order to test that hypothesis, I needed to investigate the mutability of populations that were isolated in natural surroundings instead of in the laboratory. The population at Dilizhan was just such an isolated population. But that was only the half of it. The population at Dilizhan wasn't *one* of the populations that I needed to study in order to resolve the question, it was the only one of its kind.

The history of population genetics in Russia begins with a theoretical article that S. S. Chetverikov published in 1926. S. S. Chetverikov was one of the four founders of genetics in Russia. The three others were Yu. A. Filipchenko, who was anathemized; N. K. Koltsov, who was spat upon and ousted from his position of director; and N. I. Vavilov, who in 1939 was living out his last year before his arrest.

Chetverikov became a geneticist at Koltsov's urging. He lacked Koltsov's great gift for organization. If it hadn't been for Koltsov, Chetverikov would have spent his whole life as an entomologist, a specialist in butterflies. He was a great evolutionist, and one of the best works on the laws of evolution that I've ever read belongs to him. It's devoted to the advantages that a chitinous exoskeleton gives to insects. That was what I dreamed of studying—the technical and technological bases of progression and regression in the organic world.

Koltsov asked Chetverikov to give genetics and statistics courses at Moscow University's Department of Experimental Biology, which Koltsov had founded just after the Revolution and which he headed until 1930. In 1925, at Koltsov's suggestion, Chetverikov became the head of the Genetics Laboratory at the Institute of Experimental Biology.

The history of culture abounds in clusters of talent: Florence at the time of Lorenzo the Magnificent; Paris in the mid-nineteenth century, when its artists revealed a sparkling, multifaceted world to the viewer; the Mighty Five; the pleiade of poets that included Pushkin, and another pleiade of poets that included Tsvetayeva, Akhmatova, Pasternak, Mandelshtam, Mayakovsky, and Esenin. In 1925 the Genetics Laboratory at the Koltsov Institute featured just that sort of accumulation of talents. Every name was a major one: Nikolay Vladimirovich Timofeev-Resovsky, Petr Fomich Rokitsky, Dmitry Dmitryevich Romashov, Boris Lvovich Astaurov, and seven others—all of the distinguished scientists whom the Soviets had failed to crush at the dawn of their scientific activity. Sooner or later almost all of them were the victims of monstrous repression, with one exception. The exception was Sergey Mikhaylovich Gershenson. He could tell about a lot of things, but he won't. In Dantes' inferno he'd be one of those whose mouths are sewn shut.

Population genetics had existed before Chetverikov as well. Its subject matter was the observation of variability of separate traits in populations of some species and the analysis of the inheritance of those traits. Outstanding people had worked at creating it. One of them was W. Weinberg, a German Jew who never collaborated with the Nazi regime, and who was saved by a miracle from the extermination of the Jews in Nazi Germany. His studies of human blood groups, started long before Hitler came to power, and the principles formulated by him formed the basis for a new branch of science: population genetics. Chetverikov's experimental approach to that sprout of knowledge heralded a new era in population genetics and was the basis for experimental population genetics. But Chetverikov studied the genetic variability of a species rather than separate traits. He worked with the fruit fly, a single population of a single species. His procedure was just the opposite of the one that Mendel used to control the distribution of traits in the offspring of hybrids. Mendel forced pea plants, which normally reproduce through self-pollination, to produce offspring through cross-pollination. Unlike his predecessors, Mendel observed the offspring for a single, particular trait rather than for the differences in general. Chetverikov forced the fruit flies, which in their natural environment mate without concern for degree of blood relationship, to mate only with full siblings. He observed the offspring of these matings and kept track of the variations in all their traits, whether body coloration, eye shape, or the number and size of the bristles.

In his theoretical article, Chetverikov wrote that a species that reproduces through random matings is capable of accumulating hidden genetic defects, of soaking them up, he said, like a sponge. Further observations provided a brilliant confirmation of his prediction. A new branch of genetics—experimental population genetics—came into being.

Many important discoveries were made over the course of the four years that the laboratory existed. And it existed only those four years. No one knows who fabricated its denunciation or what lies that denunciation contained. In 1929, Chetverikov was arrested and exiled without trial or investigation. His laboratory was disbanded. In 1935, Chetverikov would be invited to the city of Gorky to head the Genetics Department at its university, but he never returned to population genetics.

Chetverikov's arrest was a blow for the Institute of Experimental Biology and a personal loss for Koltsov, its director. For Koltsov, 1929 was the beginning of the end. To say the least, his gift for diplomacy betrayed the great politician, born diplomat, and great thinker that he was. It was precisely that gift that got him into trouble. To fill the position left by the deposed Chetverikov, he invited Nikolay Petrovich Dubinin, a promotee among promotees, a very young man who had already distinguished himself in the arena of combat against hostile ideology. Among the scourges wielded by such chastisers as Prezent, his lash cut the air as well. He lashed out at the idealistic faults of Filipchenko, Serebrovsky, Levit, and everyone whom he envied and whose place he hoped to occupy.

Dubinin's social origins are lost in the mists of obscurity. From the first edition of his autobiography, a compilation of lies and boasts that was published by the State Political Publishing House in 1973 under the title *Permanent Movement*, we learn that the children's home where he was raised was under the aegis of the state security organs. He writes of their deeds with unflagging approbation. At the age of sixteen he entered Moscow University. His teachers were Koltsov, Chetverikov, and Serebrovsky. Serebrovsky was about to take him into his newly organized laboratory at the Timiryazev Institute of Biology, but quickly saw through him and tossed him out. That was a bold step. A few months later, Serebrovsky's laboratory was closed down. A fateful finger, not the Lord's, but all-powerful, indicated to Dubinin that he should be a professor. He became one at the age of twenty-five when he became the head of the Genetics Department at the Institute of Swine Breeding.

In 1932, Koltsov founded the *Journal of Biology* and became its main editor. He opened its pages to the battle with "idealistic perversions" in genetics. "Idealistic perversions" was a code name for political denunciations under the guise of scientific polemics. Two aims were pursued: to exterminate scientific rivals standing in the way of high administrative ranks, and to show proletarian vigilance to ideological watchdogs searching for victims among intellectuals. Open warfare was waged against Filipchenko and Serebrovsky. Dubinin had another weapon to use against Koltsov and Chetverikov.

In that same year, Dubinin became the head of the genetics branch of the Institute of Experimental Biology. The promotee was supposed to serve as a shield, and he did so until seven years later when it came time to reforge the shield into a sword. Then the sword's tip was aimed at Koltsov. In his autobiography, Dubinin relates that in 1939, the attacks on Koltsov became so intense that it became essential to save the Institute of Experimental Biology, and that in order to avert a disaster, Koltsov, its founder and director for twenty-two years, had to be ousted. The institute meeting at which Koltsov was "worked over" was conducted by Dubinin. The Presidium of the Academy of Sciences, one of whose members was Lysenko, removed Koltsov from his post. Soon after that, Koltsov died of a heart attack in a Leningrad hotel room. His wife committed suicide. Three co-workers from the former Koltsov Institute—B. L. Astaurov, V. V. Sakharov, and I. A. Rapoport—three fearless people—came to Leningrad to accompany their teacher and friend and his wife on their last trip to the institute.

From the frank admissions that Dubinin isn't ashamed to make, we learn that on the day after the meeting at which Koltsov was ousted from the institute, the institute's party organization nominated Dubinin for the post of director. But the Presidium of the Academy of Sciences, whose composition included Koltsov's institute under the new name of the Institute of Embryology, Histology, and Cytology, failed to confirm Dubinin as director. Dubinin writes that this was because of intrigues on the part of his enemies, Lysenko's followers. He is right, because Vavilov, Serebrovsky, Filipchenko, and Koltsov were not Lysenko's main enemies. Lysenko's main enemy was that bald, young, little promotee who thirsted for power, whose goal was to become a leader, to replace Lysenko. For Dubinin chairmanship at the Institute of Experimental Biology was a needed step toward much greater positions. Lysenko and Dubinin were both claimants to the same role. The social mission of both of these men was the same. They were levers of revolution—no, the blades of its guillotine, decapitating representatives of the bourgeois intelligentsia. In their unrestrained striving for power they got in each other's way. Each of their victories was a defeat for the rival.

It was the dialectics of nature in action. Only later would the headlong flight of time and the change of rulers unite Dubinin and Lysenko in a touching alliance. A follower of Lysenko became the director of the former Koltsov Institute.

Dubinin was nine years younger than Lysenko. But he turned out to be a representative of the first post-revolutionary formation of the intelligentsia, that new type of intelligentsia that was created at accelerated speed. Lysenko belonged to an even newer kind. Dubinin was the incarnation of the ideal Leninist intellectual, while Lysenko was a Stalinist fiend. Their skirmish in the battle for leadership is full of profound significance.

There are thirteen years between Lenin's writing *What Is to Be Done?* (1906) and his creating the program for the Bolshevik party. In 1919, when that program was being composed, the Revolution had triumphed. It was no longer necessary to call on all classes of society to unite in the struggle against autocracy and for democracy. Now that same intelligentsia who had grafted the idea of revolution onto the proletariat was declared by Lenin to be an enemy of the Revolution. The intelligentsia had to be forced at gunpoint to serve the new regime, he claimed, but one needed to learn from them, and in order to weaken their resistance, one should feed them a little bit. Lenin needed the bourgeois intelligentsia (and there was no name other than this abusive one for all the people who received an education before the Revolution) for the creation of a worker-peasant army of intellectuals capable of putting the achievements of Western science at the service of building communism. Dubinin was an intellectual of precisely this new type.

Lysenko was an intellectual of the newest type. In fact, he wasn't an intellectual at all, just as Caligula's horse wasn't a human member of the Senate in whose proceedings it participated. According to Stalin's policies, the professors who had been destroyed were supposed to be replaced by collective farm peasants and workers who incorporated into life everything that was to be incorporated. The practice of building communism, since it was the criterion for truth, was a science itself. Stalin's followers wanted to encroach on mathematics and physics too, but because this would jeopardize the war potential, Leninist policy in that sphere remained in force. In the struggle between Dubinin and Lysenko, Dubinin would win the ultimate victory. But the battle lasted thirty years.

In 1932, Dubinin created a new group of population geneticists who took up the comparative analysis of hereditary diversity among fly populations in their natural habitats. It was shown that not only mutations that change a fly's external appearance but also thoroughly harmful inherited diseases that kill fly embryos are present in concealed form in fruit fly populations. Differences between populations were disclosed. Different populations absorbed different mutations and in differing quantities, but those differences seemed to be a matter of chance. According to Dubinin, whether a mutation was absorbed or rejected by a population was purely accidental. The differences among the populations didn't prove anything other than the existence of these random processes. A population was even found in which harmful mutations were quite absent. That population had colonized Dilizhan.

By contrast, my experiments showed that in nature there was no such thing and could be no such thing as a population without concealed hereditary diversity. In 1938, my co-workers Galkovskaya, Alexandriyskaya, Brissenden, and I sent an article to the *Biological Journal*. Dubinin was not the head of the *Journal*'s editorial board, but the person next to the head. I should have foreseen that the article would not be published. Keep in mind in reading my book that you are a witness to a miracle, because miracles are needed for my writings to be published. In 1938, as in all other years of my life, I made one mistake after another in that respect.

That same year I gave a presentation for a seminar at the Institute of Experimental Biology. Dubinin was a chairman of the meeting. I spoke about the differences between normal genes in wild and domesticated fruit flies. Dominance is the ability of normal genes to create a normal trait with a mutant gene present in a concealed state. "Dominance is a perfidious thing," I said. It saves individuals having hidden diseases, but it puts the population in very disadvantageous circumstances. An enormous number of harmful genes accumulate in the population under the cover of normal genes. True, their influence is suppressed, and occasional family matings contribute to their being winnowed out. But the huge load of accumulated harmful mutations prevents the population from taking advantage of useful changes. And so it turns out that where mutations occur frequently, they are often manifested in the presence of the normal gene, and where they occur rarely, they manifest themselves infrequently and weakly. But if that is so, then the load of accumulated mutations must be the same in wild populations of flies and in the laboratory strain. In uncultivated ones, mutations occur frequently, but they're poorly defended. With domesticated flies everything is just the opposite.

At the time I was eager to study the fruit fly population in Dilizhan. My audience listened attentively. After the presentation, Dubinin called me aside and said, "I don't recommend that you go to Dilizhan. The bus route between Dilizhan and Erevan has been eliminated. You won't be able to get there. Anyhow, you won't find any flies. There are very few flies there."

"Don't worry," I replied, "even if I were told that the only way for me to get from Leningrad, let alone Erevan, to Dilizhan was to crawl on my hands and knees rolling a pea with my nose, I'd say 'Hurry up and give me the pea.' "

Dubinin didn't say anything. That was the end of the conversation.

In September 1939, I took the train to visit my flies in Dilizhan. A huge young man who looked like a foreigner turned up at the same table with me in the dining car. He addressed everyone at the table in three languages and asked someone to order for him. I offered my services. He was a Dane by the name of Ebbe Hageman. We spoke in German. One of the people in his compartment turned out to be a German. They were both on their way to Teheran via Baku. "Just what do the Germans have against the Jews?" asked the Dane. "They say that the War of 1914 is on their conscience," replied the representative of Nazi Germany. The Dane threw his arms up in the air, and as though pushing away someone who was attacking him, exclaimed, "Propaganda!" The Dane and I later carried on a written correspondence. His letters mysteriously disappeared from my briefcase in Moscow. When I think of that now, at a remove of forty years, I'm afraid for myself—for myself back then. But back then I wasn't afraid, although I ought to have been.

Dilizhan is located 4,500 feet above sea level amid low hills that are overgrown with oaks and hornbeams. The orchards there provide all of Armenia with wonderful apples. They make wine from the apples and also distill a liquor that resembles vodka. It's called *chacha,* and the barrels in which it ferments are fly heaven. While there is a multitude of flies, it's difficult to get to them. Everyone has home wine distilleries, but wine-making at home is forbidden by law, and an outsider, especially one from Russia, isn't allowed near the vats where the fruit ferments. On my first trip to Dilizhan, I was assisted by Misha, an agronomist assigned to help me find flies. He took me to a collective farm winery. There, underneath an awning, stood barrels full of apple mash. There were swarms of flies.

Misha helped me find a room near the winery. On the way to the collective farm, he would occupy me with pleasant conversations in the Armenian style. He told me how much Armenians liked Russians. At the agronomy station where he worked they were conducting experiments in accordance with Lysenko's instructions and ideas. "And do you have control groups?" I asked. "Of course we do," Misha said. "We sow two plots. The one where things grow well is the test group and where they grow badly, that's the control group."

The population at Dilizhan upheld my expectations. In theory, it should have been less mutable than the Uman and Nikita Gardens populations. It was isolated in a mountain valley and group selection was limited in its effects. Yellow flies turned up here, too, but less often than in the Crimea and the Ukraine. Mutations, including the yellow type, occurred, but with less frequency. The population's saturation with hidden harmful mutated genes was by no means lower than in the Uman and Nikita Gardens populations. The amount of hidden harmful mutations was twice as high as in Uman, and surpassed the amount found in Nikita Gardens.

It turned out that bus transportation between Dilizhan and Erevan had never been interrupted, and I rode to Erevan through the Semyonov Pass, past the beautiful Sevan, the highest lake in the world. In Erevan I caught flies at the famous Ararat wine factory. Flies lived in the factory basement where the wine aged in enormous barrels. The basement extended down into the earth for four stories. There were few flies that deep down. But yellow males turned up even there.

I left Erevan on a warm October day. Having arrived at the station early, I waited for my train in the park in front. I was drawing a delicate design on a little piece of paper with a pencil. Two little girls came up to me and watched me draw. "Do you like it?" I asked the older one, who was about eight. She nodded yes. "And what do you like most of all?" She pointed at the part of the design that was the most delicate. Then she pulled a handful of sunflower seeds from her pocket, handed them to me, and said, "Here." A blind musician's flute began to play as I accepted her gift gratefully.

An elderly Armenian came up. "Why are you sitting here? Let's go to my place," he said. I knew perfectly well how to deflect that sort of attack. "Tell me," I asked, "do you mean that it's customary among Armenians for a woman in the street to accept an invitation from strange men?"

"Why are you offended?" he said chivalrously. "What am I proposing that's so bad? Do you think I want to tear a chunk of flesh out of you? I only wanted to treat you to some breakfast—some tomatoes, some cheese. And right away a Maginot Line."

The war between France and Germany was already going on, but the French defenses, the Maginot Line, hadn't been broken through yet. Brilliance and chivalry, inherent in Armenians, brought into being this comparison of my rejection with the Maginot Line, which in the fall of 1939 was on everybody's lips.

In Moscow, I studied the progeny of the Dilizhan and Erevan flies. I was supposed to study Marxism-Leninism again, too. A young woman dropped into the laboratory and asked, "What classic work of Marxism-Leninism are you now studying? How many pages have you worked through?" She was ready to put my name down on her list. She wouldn't sit down at the table—she was planning to write standing up so as to be as quick as possible. There were more than a hundred co-workers and laboratory assistants in the institute. Her inventory was based on a political measure, the implanting of ideological conformity, rather than sociological research. "Have a seat," I told her, "I'll dictate the book and article titles to you."

"I don't have time to sit down—tell me as quickly as possible."

"But that will take about two hours," I told her. "I'm planning to give you at least eighty titles." She made a gesture of disgust and left. But nothing came of the official philosophy classes that were required of all doctoral candidates in all institutes. Our group consisted of Kushner, Babadzhanyan, and me. I was Schmalhausen's student, while Kushner and Babadzhanyan were from Vavilov's Institute of Genetics. Only three of us, but you couldn't imagine a more diverse crew. Kushner was a geneticist, a bold defender of genetics, and a future traitor who went over to Lysenko's side. Babadzhanyan was a follower of Lysenko, a foal born by Caligula's horse. Babadzhanyan and I clashed at the very first class, our views were irreconcilable, and no dialectic could reconcile them. Kushner kept silent. Our professor, Arnost Kolman, the editor-in-chief of a journal *Under the Banner of Marxism*, did not intervene. We had seen him only once. He was busy studying the genetics that he had been ordered to rout.

Kolman was told by the highest spheres to organize a debate on his own journal's initiative, pretending the initiative to belong to "the broad masses of workers." The official support of a prohibition of the false science, genetics, was guaranteed. "The broad masses of workers," of course, didn't know anything about the behind-the-scenes machinations involved in preparation for the rout, nor could they. And had it not been for an article by Vera Rich in the December 1977 issue of the magazine *Nature*, I would never have found out what had gone on in the Science Division of the Central Committee shortly before the debate whose outcome was supposed to put an end to that servant of capitalism, formal genetics. Arnost Kolman, Vera Rich reported, gave a speech—and not for a committee at some USA Institute of Sovietology, but . . . at the Biennale in Venice. This time the Biennale didn't feature the usual art exhibit. The 1977 Biennale was devoted to the dissident movement in Eastern Europe. In his speech of repentance, his *mea culpa,* Kolman gave away a secret: What was supposed to have taken place in 1939, was what actually happened nine years later, in 1948, when all genetics research came to an end and anyone who didn't take his (or her) place under Lysenko's banner lost his job. At any rate, the 1939 "debate" took place. Vavilov and Serebrovsky again spoke out in defense of genetics, and Lysenko and Prezent tried to crush it. The press was wholeheartedly on the latter's side, but no prohibition followed.

One can only guess why that didn't happen. World opinion was centered on the Soviet Union, and the attitude toward the land of victorious socialism was hardly favorable. For example, here are a few dates: August 1939, the treaty of friendship between the USSR and Germany; September, the German and Soviet invasion of Poland. In October, Latvia, Estonia, and Lithuania were occupied. They were included into the makeup of the Soviet Union a few months later. In October 1939, in accordance with an agreement with Hitler, the resettlement to Germany of Germans who had lived in those countries was begun. On November 30, 1939, the Soviet Union invaded Finland. At the same time that a debate on genetics that was supposed to be the final one was taking place in Moscow at the USSR Academy of Sciences Institute of Philosophy on Volhonsky Street, the Soviet Union was expelled from the League of Nations. There was no point in adding fuel to the fire. Vavilov remained free. His worldwide fame delayed the inevitable course of events for a moment.

In 1940, I studied fruit fly populations at the northern limits of the area in which the species could be found. Two small towns not far from Moscow—Kashira and Serpukhov—are surrounded by apple orchards. I caught flies at the nonalcoholic fruit-drink plants and at the fruit warehouses in these two towns. A laboratory was set up not far from Kashira at the Kropotovo Biological Station. The station was located in a small estate house that had been abandoned by its owner. It had been organized by Koltsov and belonged to his now renamed institute. Institute members conducted experiments there during the warm parts of the year. When Koltsov was ousted and the institute was named the Institute of Cytology, Embryology, and Hystology, the new institute retained the station.

The landlord had built his home on the outskirts of the village of Kropotovo, on the banks of the Oka River. If you were to ask me where paradise was, I would answer without the slightest hesitation that it was on the banks of the Oka. Steep banks overgrown with woods, fields and meadows, sandy beaches and shoals, pure streams running through areas that haven't been turned into swamps, a cool summer and a dry winter—these are the attributes of the paradise along the Oka. It was here that Levitan painted his *Golden Autumn,* and Kluyev, whose verses Herr Strizheshkovsky taught us in the German school, called this area a "birch paradise."

I had occasion to travel all over the gigantic empire and to come into contact with the workers of its various cities and various nationalities. I would come to a plant or a vegetable or fruit warehouse with my fly-catching apparatus and ask permission to use it. I would show my document with the Academy of Sciences seal at the main office. The workers would be presented with a disheveled woman scientist grasping a pump for catching flies. You'd have to admit that it was a rather strange sight. It gave birth to contradictory feelings—always the same ones, in all cities and all the republics—a mixture of mocking condescension and respect for science. I think that the feelings were actually all the same, but the commentary was varied.

The brightest people were in Kashira. The director of the vegetable warehouse told me: "Each vegetable and each fruit has its own fly. Your red-eyed little ones are on the apples. On tomatoes there are flies with the tiny little heads and the brown eyes, the littlest ones. On the pickles—you will not find any of those; the big dark ones are over there." He was right about everything. The girls were examining vials full of flies. "How did this one get so fat?" one of the girls asked. "A lot you know," said another, "that big one is another species. What do you think, that if you stuff and stuff a mouse, it'll turn into a rat?" A fly that resembled a wasp flew up. "A wasp," said one of them.

"It's not a wasp, it's a fly. It has two wings, and wasps have four."

"But it looks like a wasp."

"That's just mimicry."

Members of the former Koltsov Institute worked at the station, and they were the brightest of the bright. Sakharov was creating his strain of buckwheat. Astaurov was conducting his experiments on the silkworm.

And in the birch paradise on the banks of the Oka lived a Russian peasant with a red beard, Dmitry Petrovich Filatov.

*A red-head asked a red*
*What you dyed your beard with.*
*I didn't use dye*
*Or any pomade.*
*I lay out in the sun*
*And held my beard up.*

That's what Filatov, laughing at his own looks, wrote me when the war had separated us. Filatov lived in the same house where the station was located.

My friend and I were walking past his window one day. He was sitting in his room in front of the wide-open window and sewing. "That's Dmitry Petrovich Filatov," said my friend, "go up and give him your hand. He'll be glad to make your acquaintance. But he won't be able to get up. He'll be mending his pants, and they're the only ones he has." Later I told Dmitry Petrovich about the circumstances of our meeting. "How did he guess?" asked Filatov, referring to the mending. It wasn't difficult to guess. Even now I can hear the sound of his bare feet on the veranda where the station co-workers ate their meals together. Dmitry Petrovich wasn't a Tolstoyan, like my father. He didn't follow anyone or anything. He was a law unto himself—a hunter, an inhabitant of the woods, a man of the people.

"I went to the village and persuaded a woman to bring me milk," he told me once. "She asked me where I worked. I told her at the station. She wanted to know if I was a watchman. I told her I was a scientist, a professor."

Well, how could I help but tease him a little? "Why did you say you were a professor?" I asked.

"I just said that for her. Otherwise she might have thought I wouldn't be able to pay."

Or another conversation with him: "Do you want me to cook up a little feast for us?" I asked. "I'll fry a chicken." "No," he said, "I don't want you to stand among a bunch of chickens, point at the one you want, and say 'Slaughter that one.' " "But any housewife in the village will sell me a plucked and gutted chicken," I countered. "I don't care," he said, "I feel bad for the chicken." "But you're a hunter," I said. "But maybe I won't go hunting anymore," he responded. "All right," I decided. "I'll collect some mushrooms and fry them up." "But fry the white ones separately," he said. There was only one frying pan, so I put the white mushrooms on one side of the pan and the other mushrooms on the other side. "Oooh, look how she's scattered the white ones all around the pan," he said.

Dmitry Petrovich Filatov was a great scientist. He founded a new branch of science—experimental embryology. By rearranging the protogenic organs in a developing embryo, he (simultaneously with Spemann, but completely independent of him or of anyone else) discovered an important principle—the management of some parts of the embryo by others. To a certain extent he outdid all of his contemporaries. It was not enough just to discover the laws of interrelationships among the various parts of a developing embryo. It was essential to understand how those very laws change in the process of the progressive development of the organic world. It was necessary to link the new experimental method with the old comparative method of cognition.

Schmalhausen was also an embryologist. By comparing the growth of various organs in a baby chick before the chick hatched out of the egg, Schmalhausen had established the principle of the independence of the growth of certain organs in regard to others. Filatov showed that the principle of independent development was present in higher forms, and that the same processes that are independent in higher forms are extremely dependent in lower ones. He experimented on newts, frogs, and toads and studied the development of the sense organs in their embryos.

He transplanted an embryo newt's ear bladder under the skin of the tail and observed the formation of an ear capsule on a new place out of material intended for quite another purpose. He told me that the protogenic ear bladder seemed to attract the surrounding cells until it would be entirely covered with their interlacement "the way a dog's nose gets covered over with cobwebs in fall when it's on the trail of a wild animal," he said.

Filatov wasn't poor. He had rank and privilege, and he was an administrator. He headed the Laboratory of Experimental Embryology at the former Koltsov Institute, was on the faculty of Moscow University, and there was another laboratory too, with a large group of his students and admirers working under his leadership. His asceticism was hardly a denial of earthly joys or a return to nature; he had no need to return to nature since he had never left her. The voices of birds at dawn in a forest shot through with the rays of the sun stood higher in his hierarchy of values than the enjoyments granted by civilization. He lived by himself in the woods for weeks at a time. An admirer of his, the handsome sybarite Andrey Makarovich Emme, once said to him in my presence, "But how do you get along without a bath house?" Dmitry Petrovich explained that the human organism can easily be weaned from daily washing and will stay clean by itself.

His rejection of comfort was a sort of protest in the name of freedom, against compromise, and in support of service to higher ideals. He, and those like him, would not exchange immortality, contempt for fate, freedom of the spirit, and the right to have one's own opinion for a country house and a personal car with a chauffeur who is on the USSR Academy of Sciences payroll. The money that he had, too, was a guarantee of freedom—if his staying on at an institution were to become incompatible with the demands of his own conscience, he could survive on his savings. He foresaw the rainy day coming and insured himself. He was not at all a renegade or a dropout from society. His evaluations of individuals and human nature were quite realistic.

He told me: "Don't expect your work to be acknowledged by your laboratory colleagues. For recognition you must turn to the world. Publish." I understood him to mean the following: "There are few people in the world who are truly interested in the narrow field in which you work. It's unlikely that you'll find such people among your comrades at work. And the lack of interest will make you grow cold." Much later, disregarding his precepts, I became persuaded that I had misunderstood what he meant. It wasn't the lack of recognition that Filatov wanted to warn me about, but the sort of recognition that ends in plagiarism. His warning was in vain. But he was perfectly well aware of the poor quality of many representatives of the human race.

And he wasn't a dropout of the human population hierarchy either. An interesting incident demonstrated his ability to be subordinated and to subordinate. We were all sitting on the veranda after supper. Someone had said something funny. Andrey Makarovich Emme, laughing along with everyone, released a sound that was a cross between a yelp and a grunt. Everyone laughed even harder. Emme yelped again and everyone laughed even more loudly and uncontrollably. "Yes, let's have a laugh," Emme said and laughed as tears rolled down his beautifully shaped nose. Everyone was laughing except me. Filatov and Astaurov were laughing like children. "That's enough," I said in a quiet, firm voice. "Put an end to this psychosis." The laughter grew more intense. Suddenly Filatov ceased laughing, pounded his mighty fist on the table and said, "Stop." And everyone stopped at once.

Though he encouraged others to publish the results of their experiments, he himself published little. He said that it was difficult for him to express his thoughts. He had written literary works too. Unfortunately, I didn't get to read them. My friend Alexander Alexandrovich Malinovsky once recited to me by heart a fairy tale of Filatov's composition: Four baby bears were walking through the woods and came upon an extinguished campfire. They found a potato in the ashes. "What can this be?" asked one of them. "No point in wondering," said the second, "we wouldn't understand anyway." "We ought to take a look at the inside of it," said the third. "We should ask someone older," said the fourth. Just then a lynx walked up to them. The baby bears asked about the potato. The lynx blocked their way to the potato, bit it into two parts, and ate it all. Then it turned to the baby bears and said, "There wasn't anything in the ashes." And then it left. Each of Filatov's baby bears said one sentence each, and the phrases correspond exactly to the philosophical outlook of each of them, since what they'd said earlier had exposed that. The agnostic, I remember, said: "If we didn't have any means of knowing it when we had it, then we certainly won't find out now that it's gone." And the one who had suggested asking the lynx said that there really probably hadn't been anything in the ashes. And the four little bears continued their stroll.

During the war, the institute was evacuated to Alma-Ata. Filatov didn't want to be evacuated. His apartment house had been damaged by a bomb raid and he was moved into a room in the institute where there was a wood-burning stove. He had a supply of firewood too. I've been told that he kept the room frightfully cold, refusing to stoke the stove because he didn't want to expend institute firewood on himself. Any possibility for experimentation disappeared. Filatov wrote his last work—a tract on the morality of the future. He wrote me that he would read it to me when we met again. I returned to Moscow in November 1942, but Filatov didn't read anything. We drank tea. He used antique tongs to break the sugar into little pieces so that he could sip the tea through sugar held between his teeth. We talked about the war and Dmitry Petrovich predicted imminent victory. He said that the Germans would be defeated at Stalingrad and that that battle would be the turning point in the war. As is well-known, he turned out to be right. "You'll see," I said, "the war will end, and we'll recall how we drank tea through blocks of sugar." "No one knows whether we will or not," he said significantly, with the stress on "we." Here, too, he turned out to be right. A few days later, he was felled in the street by a stroke. A militia car took him to the hospital where he died without regaining consciousness. He had not lived to see the victory on the Stalingrad front. He was sixty-six.

Filatov's manuscript didn't disappear. For thirty-two years it lay in the archives under lock and key. The morality of the future, in Filatov's conception, didn't contradict Communist morality in the least. Filatov rejected the moral codes of all religions, just as he did faith in personal immortality. He didn't cite the classics of Marxism-Leninism, nor did he develop their founding theories. Instead, he developed his own. He dared to seek the roots of lofty human behavior in mankind's prehistory—the animal world—rather than in class warfare. He saw the reason for the suppression of egoism in a mother's love and in her readiness to sacrifice her own life in order to save her child's. A person who brought that sort of a manuscript to an editorial office would not only have been unsuccessful in getting it published but would have jeopardized his or her own career. This is still true today.

The question of the biological roots of altruism had been raised by another crazed and brave fellow, who was capable of throwing himself naked into battle and fighting with warriors girded in the armor of Marxism-Leninism. That crazed man was Efroimson, who had recently been released from a Stalinist death camp—Dzhezkazgan. He wrote an article about the biological roots of altruism, about the fact that man couldn't exist without mutual assistance, self-sacrifice, and the suppression of the instinct for self-preservation. His article was printed in the *New World.* The journal's editor, Tvardovsky, and Efroimson had let the genie out of the bottle, the same bottle in which Filatov's manuscript was sealed with a genie immured. Astaurov edited the manuscript and brought it to a finished state. It was published in the 1974 issue, number eleven, of the almanac *Paths into the Unknown.* The title of the piece is: "The Norms of Conduct, or the Morality of the Future from the Point of View of Natural History."

Filatov asserts in the article that in the future there will be no morality.

A code of instructions as such will disappear. Instead it will merge with the norms of behaviors and become one with them. Good will rule, and egoistic impulses will be suppressed, as will the instinct for self-preservation. Love of people will become part of love for every living thing. The man of the future will extract the highest pleasure from concern for the welfare of other people and from love for life in its entirety. Even now there exist people of this higher type. In the future they will conquer life. Their steadfastness in life is a pledge of that. "They emerge victorious from all of life's squabbles, that is, they retain their habitual relationship, the only one possible for them, to everything and everyone surrounding them."

Editing Filatov's article was one of Astaurov's last deeds. There will be a discussion of him and his death later. In the almanac the name of the editor and the author of the preface are framed in funereal black. In the preface to Filatov's article Astaurov writes: ". . . Filatov was a typical representative of the progressive, democratically inclined, Russian intelligentsia. Born into the family of a very wealthy landowner, he came into conflict with his own class, by giving away his land to the peasants [to be more accurate, he sold it for practically nothing—five kopecks a *desyatina* (2.7 acres)!]. This was the reason for his painful family conflict, the rupture of relations with his wife, his subsequent reclusiveness of sorts, and withdrawal to science and philosophy."

Flies were the joy and delight of my soul. Old hypotheses were confirmed, new ones created. In numerically smaller populations of fruit flies, mutations arose less frequently than in huge populations. Isolation, whether in the mountains or in a valley, had the same effect as isolation in the laboratory. I was able to mark out a strain of flies that was characterized by a higher incidence of mutations. Among them was the yellow one. Thanks to it, I was successful in developing that remarkable strain. It was called "Kashira-6." The study of the offspring of wild flies continued all winter.

My psychic equilibrium had been disturbed, however. I sensed disaster approaching. In December 1940, I was in Leningrad. One item of horrible news followed after another. Vavilov was arrested. Koltsov and his wife were dead. The university professors Karpechenko and Levitsky were arrested. Edna Brissenden, the most talented of my co-workers and co-authors, left the university as a sign of protest. I invited her to come and stay with me in Moscow, but I never saw her again; she and her mother were in danger of being arrested. The ignoramus Lysenko had become the director of the Academy of Sciences Institute of Genetics. Stalin and Hitler had begun carving up the world.

I was supposed to spend the fall of 1941 continuing my study of the populations of Kashira and Serpukhov. I was living in Kropotovo. My flies were at the Kropotovo Station. On July 22, 1941, we learned that the war had begun. I applied to the Military Comissariat many times, asking to be mobilized. I have an elementary medical education. They wouldn't take me then, and I needed to go on with my own work. The institute denied my request for a travel grant to go to the nonalcoholic drinks plant in Kashira—they said that I would arouse suspicion and would be picked up as a spy. I left without a travel grant. I had not got as far as the plant gate when they were already opening it for me! "Come to catch flies again?" they asked. I was not suspected of espionage for Germany.

I brought the flies I caught back to Moscow, but I had little chance to work with them. From the first day of the war the Germans bombed Moscow every night. With German punctuality the air attack began at 11:00 P.M., and continued until 5:00 A.M. People took cover in the subway and in bomb shelters; that is, those who weren't running out onto rooftops to catch incendiary bombs. The bombs didn't ignite at once. One could avert a fire by seizing the bomb with tongs and flinging it into a box of sand.

I was told that when the doors to the bomb shelter were opened, the first person to enter was always Turbin, a doctoral student who had moved into my dormitory. Later he would become an ardent follower of Lysenko. After him came women with nursing babies. That's a very important little detail for understanding the fate of genetics.

People who were deprived of sleep night after night came to work half-dead. But I was fit as a fiddle. The reason was to be found in an anomaly. It's a family trait that I have and share with my brother. Sim was posted with his detachment in Moscow at that time. Violating military discipline, which was quite unlike him, he didn't go to the bomb shelter either. "I love it when I'm sleepy and a demolition bomb lands nearby," he would say with a lively voice, as though he were praising a favorite dish. Love of danger? Courage? I don't know. Rather than fear, the air-alert siren produced in me an unconquerable desire to eat and sleep. I would rush to the kitchen to warm up some tea before the gas was turned off, then put on my nightshirt and go to bed.

Once, a demolition bomb landed on my building, on Malaya Bronnaya Street. It was in the summer of 1943, when bombardments had become rare. I was resting in bed, in the morning. The building shuddered. The room was filled by a white cloud. The whitewash had been shaken from the ceiling and walls. The bomb didn't explode. That happened fairly often as a result of sabotage at German factories. I would like to know, but I'll never find out where and by whom that bomb was made.

The bombardments had another curious effect upon me. The migraines that had plagued me for years suddenly stopped.

On October 16, 1941, I witnessed the panic produced by the most recent German victory near Moscow. I still can't understand how they could have broadcast over the radio the short announcement, terrifying by virtue of its laconicism, that I heard that morning: "The situation has worsened. The front has been broken through." That was a lie. Three lines of defense surrounded Moscow. As we later found out, only the first of them had been broken through and the battle still raged with undiminished vigor. But a panic had begun in Moscow.

There are four indicators of panic in a city that is awaiting the enemy: Stores are looted. Authorities use their privileges and flee in official limousines, private automobiles, trains, and carriages. The institution of money ceases to exist. And the blackout ceases to be observed. I saw and heard all four of these. The sidewalks in front of stores were littered with flour and granulated sugar. Two pianos wailed back to back on a speeding truck without any covering. The mournful wailing came from jolting. People talked about how a hospital director demanded a horse and carriage from the stables of the hospital. These were the only means to bring bread to the patients. The carter punched the director in the face and refused to give them to him. At the market—I heard it myself, a few days later—a male voice yelled vigorously: "Who wants milk for money? Milk for money!" A barter system ruled. That reckless fellow was the only person in the whole market who would take money. At night there were lights in the windows; sometimes a window was only half-curtained, on the diagonal. The militia, who had fined violators earlier and would again afterwards, didn't take any action.

It became known that the government had fled and that Kuibyshev had become its residence. The Academy of Sciences was seized by panic. Institute directors had been evacuated away to the rear even earlier. Their deputies were managing things. In our institute that was Khachatur Sadrakovich Koshtoyants, one of the people whose signature had stood beneath the libelous preelection lampoon directed against Koltsov and my father. When I had shown up two years earlier, he had made a stab at courting me. He said that he had not written the lampoon, he had only signed it. Nuzhdin, too, another werewolf, told me the same thing, under different circumstances. Now Koshtoyants ordered me to destroy our equipment, the microscopes, and the thermostat. I refused. I heard about how Nuzhdin and Dozortseva had commandeered an official car and fled. They were a happy married couple who had linked up in accordance with the time-honored principle that *chaque vilain trouve sa vilaine.* Formerly they had been colleagues of Vavilov's, but now they were Lysenko's underlings and their names stood beneath a slanderous pro-Lysenko article.

On October 16, 1941, we, the staff of the Schmalhausen's institute, were told to pack our most essential things in shoulder bags, fall into column formation, and follow the Kaluga Highway out of Moscow. I tried to pack my journals, but realized that I wouldn't be able to drag them even three miles, and decided not to go. I couldn't believe that there wouldn't be battles with the intruders at the barricades. Signs of panic were everywhere, but the determination to struggle is only revealed at critical moments.

Many people stayed. In a few days it was announced that the institute would be evacuated to Przhevalsk, in Kirgizia. I received an order from Schmalhausen to go to a hostel for academicians in Kazakhstan, where he himself was living, and to where my father had been evacuated from Leningrad. As Schmalhausen's doctoral student, I, of course, had no right to stay at a hostel for academicians. The unit in which Sim was serving was transferred from Moscow to Sverdlovsk at that time. I was to travel to Sverdlovsk with the institute employees who were going to Przhevalsk. I could see my brother in Sverdlovsk and go from there to the Kazakhstan refuge.

I traveled by train as far as Sverdlovsk together with other people from the institute. We had sleeping berths. I got the lower side bunk, a very narrow one. I was ill and remained in bed day and night. An old man sat at my feet. He was traveling without a ticket to Sverdlovsk, where his son was in a military hospital. He sat down with us somewhere along the way. The others hadn't wanted to let him in, but I insisted. He kept falling asleep and falling over on my knees. A doctoral student traveling with us quoted Pushkin in this regard: "Eugene, like one awaiting sentence, / Fell at her feet in wild repentance."

In Sverdlovsk, I was able to locate Sim through the commandant's office at the train station. I set off to find him.

As I was leaving Moscow I had taken some vials full of flies. My valuable, highly mutable strain, Kashira-6, was among them. I took food for the flies, too—raisins and agar to make marmalade. Procuring the raisins from the institute was quite an adventure. My request to have the warehouse issue them was denied. I made an agreement with the institute's fire patrol, and we broke into the warehouse at night, pried open the box containing the pressed raisins, knocked off a good-sized chunk of them—probably twenty pounds or so—and were about to nail the box shut when we were caught. It wasn't the manager, but one of her workers, who was on duty that night. I was guilty of theft with breaking and entering. The militia hadn't yet been called, and the issue of determining the exact gravity of my crime arose. The authorities brought the ledger, which indicated how many pounds of raisins had been at the warehouse before my pilferage. Well, how about weighing the chunk that I had nearly filched? But no. As luck would have it, they decided to weigh the remainder. And—miracle of miracles!—it turned out that there was exactly as much left as the ledger indicated. The raisins that I was stealing had been stolen, on paper at least, by warehouse employees before I even got near them. They hadn't caught me. I'd caught them. Everything had taken place in the presence of witnesses—the firemen standing there with their axes would have to sign the official statement. The box was nailed shut.

The raisins went with me into evacuation, as did my flies. The vials containing them rode in a wooden box tied to my chest with towels under my fur coat; otherwise the flies would have frozen to death. Then the unfortunate happened: I was walking around the train station looking for the military commandant, floating along in the current of crowds that overflowed the station. The door that I, along with many others, wanted to go through was slammed in my face by a militiaman. Actually, all would have been well if the door had slammed in my face; instead, while trying to drive people off, the militiaman shoved his fist into my chest and therefore into the box of flies. I was arrested and taken to the military commandant. I showed the soldiers and militiamen the flies and told them how similar the laws of transmission of hereditary traits from parents to offspring were in flies and humans. "Here are flies that have white eyes. Normal flies have red ones. Having white eye color is a disease. It's inherited in the same way hemophilia is. Tsarevich Alexey, Nikolay II's son, had it. He inherited it from his mother. She was a carrier of hemophilia, and it surfaced in him. Half of the Tsaritsa's sons, if she had had a number of them, would have suffered from hemophilia. But not a single one of the daughters. It's exactly like flies, when white-eyed crossbreed with the red-eyed." Impressed, they found my brother's address for me and suggested that I leave the flies at the commandant's office. It might be cold at my brother's apartment. And they told me the special military distribution center had felt boots—my brother should buy me some.

The flies stayed at the commandant's office. Sim bought me felt boots.

# *A Year on Olympus*

And so I found myself, in 1941, in a hostel for academicians. A horse milk (koumiss) treatment center for tubercular patients at the Borovoye Sanatorium had been turned over to them.* This was in western Siberia or northern Kazakhstan, depending on your point of view. The Siberian railway branches at Petropavlovsk and the southern spur lead to Karaganda. The Borovoye Sanatorium is located halfway between Petropavlovsk and Akmolinsk near the Shchuchye Station, and it is one of the most beautiful spots in the world. Academicians and associate members of the Academy of Sciences lived here with their families. There were single academicians, too, such as Prince Shcherbatsky; there were also academicians' families who had lost their breadwinner.

There was a sprinkling of other people as well. The mother-in-law of the writer Leonid Sobolev was there with Manichka, her elderly hunched-over maid. There was also Maria Feodorovna Andreyeva, an old Bolshevik, at one time an actress, and Gorky's mistress. She was known as Gorky's widow. In Moscow she had occupied the high post of director of the House of Scholars—a very lucrative position. Membership in that club gave one privileges. A closed restaurant fed the members inexpensively and well; Andreyeva headed a committee in Borovoye (I don't remember what it was called). The committee dispensed the good things in life: rooms, clothes, footwear, and foodstuffs other than those that went directly to the cafeteria kitchen. Everyone ate together as in a regular vacation home. Two members of the com-

* In Russia, koumiss is considered a remedy for tuberculosis.

mittee, besides its head, were elected by a general meeting of academicians. One of those unfortunates turned out to be my father. He and another member, Zernov, were noted for their asceticism and readiness to serve people.

Twice I have been swept along by currents of people who saved themselves or were saved from calamity. The first time, in 1941, the academicians and their families were saved; the second time, in 1974, it was the Jews. What could there be in common between the evacuation of scientists and the emigration of Jews? Both those who did the saving and those who were saved were different. And the people in each of the currents were different, of course.

Every one of the cities where evacuated academicians lived turned out to be in a zone near the front. If those cities had been captured by the Germans, the victors would have acquired for their disposal enormous capital in the form of scholars and scientists of all specialties. Evacuation, sometimes forced, was supposed to avert that calamity. But even in the case of decisive resistance to the occupiers, the life of every person in a front-line city hung on a slender thread. The 700,000 Leningraders who died from starvation and shelling are proof of that.

The reasons for the emigration of Jews and the evacuation of the scientific and scholarly elite are antithetical. The academicians were saved by the authorities because the authorities needed them. The Jews were saved from governmental persecution by their fellow Jews outside the country. The similarity in behavior of the two rescued groups is valuable material for a sociologist (but what do sociologists have to do with it?), for a political scientist, for anyone engaged in the restructuring of society on just principles—*just* ones, of course.

In both cases I found that the comportment of people deprived of the chance to take care of themselves brings forth the worst aspects of human nature. The dregs float to the surface. The very worst people set the fashion. Each of them strives at any cost to take as much as possible from the saviors and providers, to occupy a privileged position among those who have been saved and provided for, and to set themselves up as well as possible at their comrades' expense. The people from both of these waves or currents are or were prisoners of sorts, and the mores of prisoners have been described by Solzhenitsyn with supreme accuracy in *Gulag Archipelago.*

The Jews emigrating from the Soviet Union to the United States were directed to Rome to wait for an entry visa. Among the emigrants in

Rome, the only conversations that one heard were about the deals pulled off by his or her fellow émigrés. I felt shame and despair when I heard the stories about the thievery, extortion, and swindles carried out. Several times I became a victim of these swindles, true, but only over a minor matter.

One feels ashamed to belong to them, to be the object of legitimate suspicion on the part of the saviors; one despairs when a grand idea is compromised brazenly, unforgivably, for the sake of personal gain. *Noblesse oblige*—nobility has its obligations. It's caste honor. There's plenty of caste, but no honor. An émigré girl, an adolescent, told me about an appalling incident and then, right before my eyes, deposited a coin attached to a string in the elevator pay phone and then pulled it back out.

That awoke a recollection in me. I had encountered something like that in the distant past, more than thirty years earlier. During the war. In the hostel for academicians. It involved a daughter-in-law of one of the academicians. No one would have found out a thing if she hadn't come down with typhus. It's transmitted by lice. But from where had she gotten them? And how? It turned out that she had picked up the pest in a Kazakh yurt. Tea for academicians in the common dining hall was served from an enormous dark lilac-colored enameled teapot. The Olympus dweller was trading bottles of that tea for skeins of Kazakh woolen yarn. That superbly colored yarn is used by Kazakh women to make carpets.

The pain that I felt became dulled when I pondered the juxtaposition of such different yet such similar phenomena. The analogy, deriving a model, brought the joy of cognition. The fruit from the tree of the knowledge of good and evil was prickly. It's not food, but slops. But if you have to eat it, then at least one shouldn't do so blindly.

Wartime played with people and its games were quite insidious and ambiguous. The hostel for academicians was a miniature model of a Communist society. Not the sort that one pictures in one's imagination, but a real one, where, by order, everyone is equal and equalized, but does not really want to be equal. Equals enter into a desperate struggle for a place in a rat's hierarchy.

The hostel thoroughly deserved to be called a den of witches. But among that crowd of the unclean, there were the clean as well. And they were in the majority. And some of the unclean had clean wives, or none at all. But there were Xanthippes who amazed everyone. The clean held themselves apart and didn't get involved. The trash floated along the surface, the stormy surface of the human sea. But among this mass of the privileged there wasn't anyone who was absolutely clean. Climbing to the highest ranks in the Soviet hierarchy is incompatible with elementary human dignity. A clean person—and I call clean those who don't exchange spiritual freedom for privileges—has to give up any hope of reaching the peak. However, there are exceptions to every rule.

I divided the academicians who were living in the hostel into five categories, according to their comportment. On the basis of their conduct, I extrapolated their philosophy. The first category included Vernadsky and Mandelshtam, the physicist. They believed that people were wonderful and need not be improved. One did not serve people, but offered help whenever it was needed and to whoever needed it. The second category included people such as my father and Zernov. They believed that the world and people were wonderful, but if anything was not, it could and should be corrected. They believed that serving good was counteracting evil, averting it. The third category was very large, including figures such as Zelinsky, Favorsky, Krylov, Schmalhausen, and Bernshtein. They had no illusions about human nature—it was the way it was, and that was it. There was no point in trying to correct it. They were above the game. They didn't participate in the struggle for creature comforts or a high post in the hostel's hierarchy of rats. Thanks to their creative work, they were far removed from daily life. The fourth category was made up of lambs, for example, Boris Mikhaylovich Lyapunov, a Slavic specialist and relative of Filatov's, and the historian Tyumenev. They demanded nothing for themselves, were satisfied with everything and grateful for everything. They were not only above the game, but above any evaluation. The fifth category was composed of self-seekers.

I told Schmalhausen about my system of classification. Dissembling, I assigned his category the first place. "I think that your father and Zernov should be given the first place," he said. There was that same Filatovian, Tolstoyan, Karatayevian quality in Schmalhausen too, in that resident of Olympus to whom visionary dreams were alien.

The Revolution had brought about changes in the academy's charter. Before the Revolution, the charter read that only actual members comprised the academy. After the Revolution, that was broadened to include research institutes and all the institutions serving the academy—cafeterias, hospitals and clinics, garages and photo laboratories, learned societies and archives, libraries, kindergartens and nursery schools. Drivers, cooks, guards, yardmen, and employees of the fire brigades at these institutes and institutions found themselves under the aegis of the academy Presidium. There was one set of cafeterias and hospitals for the highest members and administrators; others for scientists, scholars, and lower-ranking bureaucrats; and still others for the hegemon of the Revolution—the proletariat, that is, for workers and bureaucratic small fry. The free garage, i.e., cars and drivers, was only for the elite. The children of the highest and lowest castes were not divided. It remains that way to this day. During the war, academicians had their grandchildren live with them. Evacuated children from the other castes lived in the children's home.

One day, the authorities sent a gift: an entire airplane full of apples from Alma-Ata, the capital of Kazakhstan. A member of the distribution committee, my father, gave instructions that the Alma-Ata apples be given to the residents of the nearby children's home. With my own ears I heard Lina Solomonovna Shtern, the only woman academician, protest that decision, saying, "We are the country's diamond reserve, but no one knows whether those children will amount to anything."

Stalin did not especially value the country's diamond reserve. He would make gifts, ever increasing the difference between the academicians and everyone else, and suddenly—bang!—he who had been somebody would become a nobody. First this one, then that one. Lina Solomonovna Shtern was among the favored, who received dachas or apartments, caviar, pheasants, or crab, who were entitled to free gasoline and a driver at the academy's expense, in addition to an official car. After returning to Moscow from Kazakhstan, she was exiled back to Kazakhstan, to the same place where she had arrived during the war with an enormous pile of elegant luggage. Having arrived at the same time she did, I was a witness to the Mont Blanc of her suitcases. The sight of her wealth sitting on the train platform at Shchuchye was farcical. Now was she sitting on the platform at the way station, a lonely, impoverished exile. That really happened.* What follows is based on rumors, but they seem likely.

A Kazakh fellow came up to her and asked, "Why are you sitting here? Where did you come from?" His wife had just died, leaving him

* Shtern was arrested in winter 1948–49 together with hundreds of Jews, all of them cultural workers. On July 18, 1952, twenty-five of the most honored of them were sentenced. Twenty-four were condemned to death; the sole woman, Lina Shtern, was sentenced to penal servitude for life. She was rehabilitated after Stalin died in 1953, and died in 1968.

with a large number of orphans. "Come live with me and look after my children." She went. Several years later, along with hundreds of thousands of other rehabilitated people, Lina Shtern too had her rights restored. She turned up at the settlement near Moscow where she'd had a dacha (given to her by Stalin), threw out the president of the academy, Nesmeyanov, to whom the dacha had been turned over by the academy Presidium, tracked down all her luxurious furniture, and had it returned. But it was difficult for her to part with her wards. She adopted one of them. And it was said that at a Kremlin reception, a high-ranking official offered her a seat, to which she responded, "Thank you, I've already sat long enough."*

I'll return to the year 1941. Among the academicians there was one person who wouldn't fit into any framework and who defied classification—and there was only one person like that—the Chinese specialist, Vasily Mikhaylovich Alexeyev. In essence, the academicians were divided into two main orders, two types—those who were willing to do with a minimum of creature comforts for the sake of a clean conscience, and those who were self-seekers. Alexeyev, a person of the highest morality, of a sort far superior to Christian morality with its cult of suffering, did not belong to either of the types. He said that he didn't want to be a springboard for a scoundrel. He expressed his opinions openly. He dressed elegantly. He liked to eat well. My father, it should be said, was a vegetarian. He didn't eat anything that resembled himself. And he evaluated the degree of self-resemblance according to the structure of the heart. He didn't eat anything that had a four-chambered heart like a human's. That includes all mammals and birds. But he ate fish. Whether the times were hungry or not, he followed his principle. I used to make fun of him and say that crocodile should be excluded from his diet, since it has a four-chambered heart. The crocodile tears that the highly developed reptile sheds as it eats a human added to the humor of the situation. "You can't even get full here on four-chambered food, and without four chambers it's absolutely impossible," Alexseyev said when he learned of my father's principle. He didn't want the privileged clique, with its claws on the supplies, to plunder him, along with all the other meek people. As we shall see, he became a springboard for a scoundrel sooner than anyone else.

* Translator's note: In colloquial Russian the verb "to sit" is a widespread euphemism for going to prison.

Alexeyev's contribution to learning is vast. He was a historian, linguist, philologist, and an authority on absolutely all branches of Chinese art, including those that have no analogues in Western art. He spent four years of his life in China and collected popular prints and examples of expressive calligraphy. He brought about a world revolution in Chinese studies. That's not my opinion, but the belated, posthumous evaluation of sinologue authorities. He came from the old St. Petersburg School of China scholars. Their interest was fixed on bureaucratic, ostentatious, official China and the Confucian exoticism. Alexeyev made China studies a branch of learning devoted to the folk culture of China. His great sympathy for the people's poverty, their drudgery, and their hungry existence provided Alexeyev with a key to understanding their folk pictures.

Speaking of the exhibit of popular Chinese prints that had been brought to Russia by Vladimir Leontyevich Komarov fifty years earlier, Alexeyev said: "Only someone who has lived in China understands in what terrible poverty the Chinese masses live and how these dreams can arise. All those earthly blessings that are conjured up in the hungry imagination of a poor man are depicted in the folk picture directly and through the conventional language of symbols. A Russian peasant, too, from the end of the last century or the beginning of this one, especially a barge-hauler, more likely would not have decorated his hut with a lithograph of Repin's *Barge Haulers*. He would have preferred a depiction of a tsarevich dashing off on a wolf."

I'll explain: Repin's *Barge Haulers* depicts a group of peasants. Straining, they are using ropes to haul a ship upstream. It's a symbol of tsarist Russia; an eternal symbol of Russia. The symbolism in Chinese pictures is quite different and considerably more complex than the allegories of Western art. The Chinese popular print is not just intended to be looked at. It is meant to be read; moreover, it is meant to be read by an illiterate. It is a special sort of rebus. "The Chinese language," Vasily Mikhaylovich said, "encourages the building of such rebuses. The same word can indicate different objects and concepts, depending on its context and its relation to other words. 'Bat' is *fu,* and 'happiness' is *fu.* An artist depicts the ominous figure of a spirit conjurer. Above him is a bat. One needn't be literate in order to read: 'Bring us joy! Give us happiness!' The pictorial symbolism that is based on those sorts of homonyms is known to every Chinese. The orchid is a symbol of the perfect person, whose regal scent is capable of influencing sensitive people; the lotus is a symbol of the nobility of a person who, though growing in dirt, doesn't become dirty; bamboo represents steadfastness, a person strong on the outside and wide open inside. A dragon, a tiger, a fish, an open pomegranate, a toad—everything has a meaning. Peonies in a vase and three lemons comprise a rebus: a distinguished, well-to-do flower and three items of fruit, i.e. living well, being distinguished, and having three things in abundance [many years of life, much money, many sons].

Vasily Mikhaylovich told about how he had managed to penetrate the meaning of this symbolism. When he lived in China he had a teacher—a Chinese newspaper journalist reporting news published in French, English, and Russian newspapers. Alexeyev asked him to explain the meaning of the pictures. The teacher said that he didn't know. Only women and even then only old ones still remembered what it all meant. Did the teacher know a woman who possessed such knowledge? Yes, his own mother-in-law. But she would refuse to talk to an alien male: Her feminine virtue wouldn't permit her to meet with a young man. How old was she? About eighty, but that didn't make any difference. A woman was a woman. Vasily Mikhaylovich spoke of this trait of Chinese life with great respect.

Vasily Mikhaylovich proposed a deal. Virtue incarnate would explain the rebuses to her son-in-law, and he would write down the explanations for Vasily Mikhaylovich. In return, Alexeyev undertook to summarize the foreign newspapers for him.

During the ongoing seminar at Borovoye where the academicians reported on the results of their research in a popularized format, Alexeyev contrasted Eastern and Western art. He spoke with great respect of the chasteness of Chinese art. Not a single work of art in any of its countless branches, extending as far as inscriptions decorating the outside walls of homes or roadside rocks and tablets, contains anything bawdy or obscene. Leonardo da Vinci's *Madonna Litta* with her bared breast was nothing but a pornographic picture if one viewed it through the eyes of a Chinese.

The ladies at the hostel were shocked. Sobolev's mother-in-law was indignant. Evaluating the canons of one art according to the canons of another struck her as blasphemy. "The *Madonna Litta* is a pornographic picture? Alexeyev and the Chinese should keep their opinions to themselves," she said.

Vasily Mikhaylovich did verse translations of Chinese poetry. He was a member not only of the Academy of Sciences, but of the Union of Writers as well. He organized a literary circle in Borovoye: It included himself, his two daughters, Lyusya and Musya, and me. Lyusya was nineteen and Musya twelve. We wrote essays, as Alexeyev called them. The topics were designated by him.

One of the topics was Leningrad, a city to which we might be fated never to return, a besieged city in which 700,000 people perished, where Nanny stayed to watch out for the apartment and perished. Vasily Mikhaylovich wrote something that was unimaginable for those times: St. Petersburg (Leningrad)—the city of palaces, its exquisite architectural ensembles, its elegant embankments and public gardens, had been flooded by a filthy wave of deprived, defamed, ruined people whom the Revolution had turned into common rabble. The rabble was now dying there heroically, without surrendering, but the authorities were responsible for the deaths. What's true is true. After all, before the war they had yelled that the border was under lock and key, that we would beat the enemy, with few losses, on his own territory. And they had depicted an enemy more frightening than Hitler—a coalition of all the countries in the world against the first country in the world in which socialism had triumphed. In reality, they had permitted Hitler's armies, after getting as far as Moscow, Leningrad, and Stalingrad, to drown at their doorsteps, choking on Russian blood.

If that piece by Alexeyev had caught the eye of the "Organs"—the widespread euphemism for the security police in any of its numerous metamorphoses—the great authority on China would have wound up in the clink.

I wrote that Leningrad was the people who were there: my friends, Nina Ryabinina, her thin fingers at her temples, a migraine, and Natasha Eltsina—Botticelli's Madonna. Nanny, who said that she would never die: She'd live to see the Second Coming, she said, when paradise would be established on earth. And she would describe colorfully and precisely what would precede its establishment. A great plague would destroy the sinners. Very few people would be left. A person would be glad to hear another human voice. An iron wire would gird the sky and iron birds would fly over it, carrying death. In the end Nanny died of starvation. Not a single bomb exploded near the apartment house that it was her duty to guard. The plant that made parts for ships and that was only steps away from the house did not suffer either. Was it Nanny's prayers that saved Father's residence and his thousand-volume library, if not her?

I didn't know anything about Nanny's fate when I was doing literary exercises with Vasily Mikhaylovich, Lyusya, and Musya. But my soul was there, in the starving, besieged city, with the people I loved and to whom I am now paying my pathetic tribute of gratitude.

Party-member Andreyeva, Maxim Gorky's widow, as she preferred to be identified, the chair of the commission for distributing creature comforts among evacuated academicians, was consumed with hatred for the great authority on China. His daughter Lyusya became her victim. When an order came to mobilize the hostel's young people for work in the Karaganda coal mines, Andreyeva took its fulfillment upon herself. She drew up a list of those to be mobilized, a list that included only one name—Lyusya. Other academicians had daughters too—big, healthy ones—and they weren't working. But they were not selected. I decided to meet with Andreyeva and dissuade her. The whole thing seemed to me to be a misunderstanding. I asked for an audience. She said that she herself would come to see me. She knocked at the door before the appointed time. I had Musya there. I hid her in the closet. I cited arguments that anyone would find incontestable. Lyusya was working and not subject to mobilization. Her health was poor (she had had tuberculosis as a child). Her mother was blind. She was the only person in the family capable of looking after her elderly father. They couldn't take her, and if they had to send someone, they could have me. A counterattack followed; Alexeyev was a monarchist and an alien element, so his daughter should work a while in the mines. I knew that Musya was weeping in the closet, so far silently. I had only been able to make limited use of a description of her family's sacrifices and calamities. I had as yet kept silent about the death of Alexeyev's son during the first days of the war—he had gone off as a volunteer. I had to resort to that torpedo, too. In vain. Then I said—as quickly as possible, before the wailing in the closet could grow audible—that there were two of us in the room, but only one of us was an old Bolshevik, and that it certainly wasn't she. The day after the conversation with Gorky's concubine, I received a note. At first I couldn't make head or tails of it. "I wanted to kiss you," it said. Who was it from? What sort of brazenness was this? Who was it that had taken it into his head to kiss me? The note was from Andreyeva. She left her decision in force. Lyusya left for Karanganda. When Vasily Mikhaylovich's friends obtained her release, she was on her last legs. She had been robbed and had given away everything she had. In the wintertime. In Karanganda. Without any warm clothes.

You might think that after the Chinese Communists took power in 1949, Vasily Mikhaylovich's fate changed and that he began to be published. In 1950, I asked him outright whether his works, the best testimony to Sino-Russian friendship, had seen the light of day. "There's no friendship. It's all false propaganda. Nobody ever needed me, and they don't need me now," he said.

When the interests of the two countries collided, we learned from the press—no, it leaked through into the press—that there really wasn't any friendship. And that the USSR's aiding Mao Tse-tung was an anti-American action rather than a pro-Chinese one didn't leak into print at all.

Many bad things happened in my life during that year when I was living at Borovoye. Many good things, too. I was able to live independently, without burdening my father and stepmother's budget. I was receiving my doctoral student stipend. I earned a little extra by drawing illustrations for academicians' reports. For the Orientalist Freyman, I made huge-scale drawings of some sort of extremely ancient characters from manuscript books. For this work I used the reverse side of sheets of wallpaper. What did the scribe write with in those remote times? The edges of the marks made by them were curved. It was as though the letters trembled. Reproducing those jagged markings while strictly observing the proper proportions gave me great pleasure. I did the illustrations for one of Vernadsky's reports, too. I had met Vernadsky very soon after my arrival. He came to me to find out what was going on in Moscow. With him was his secretary, Anna Dmitriyevna Shakhovskaya, the daughter of the historian Prince Shakhovskoy.* My father was Vernadsky's pupil, and Vernadsky said that I was his granddaughter. I was the first listener and illustrator for his report, "The Geological Strata and Geospheres of the Earth as a Planet," published in the *Bulletin of the Academy of Sciences* in 1944. One of the earth's strata as defined by Vernadsky was called "a free atmosphere." "Long live a free atmosphere!" by the way.

The sea, Rome, and Vernadsky evoked similar reactions in me. It was as though the limits of existence were expanded, as though one were joining immortality, a thoroughly delightful feeling. I used to go to Vernadsky to ask him silly questions. "Silly question" is a collective concept: It's either silly to ask the question or the answer can only be

* Shakhovskoy, the "beggar prince," as he was called, went to extremes of asceticism. He was a fighter for democracy and a member of the Duma. He was arrested in 1937 and disappeared from the face of the earth.

silly. People usually see through the cunning immediately and refuse to answer. But a genius doesn't suspect craftiness in his partner in conversation, doesn't see any tricks, and doesn't consider any questions silly. "Are people good or bad?" I would ask Vernadsky. "Who will win the war? How long will it last? Will there be other wars, or is this the last one?" Vernadsky said that people were good, and especially those in Borovoye. The war would soon end with a Russian victory, and the war would be the last. But then I asked whether viruses were alive or dead. Vernadsky responded angrily that people should have understood years before that there are mesomorphic, or intermediate, states between life and death. I wanted to extract a definition of life from Vernadsky, but I didn't manage that. Vernadsky also got angry when I said that I thought with calcium ions, not brains. But I had every reason to ascribe paramount significance in my brain's activity, to calcium ions. The truth was revealed to me as a result of the Stalinist tyranny.

In 1940, after Stalin had introduced very severe punishments for absenteeism and tardiness, I fell ill. It cost unbelievable effort and time to make it to work by the deadline. To get from Malaya Bronnaya Street to the Kaluga Highway, I had to take one bus and then transfer to another. One bus after another would go by, filled to overflowing and no longer stopping to pick up passengers. A half hour later it would have been easy to make it to work. Before Stalin's order, I had always come late and stayed until late at night, but now Schmalhausen asked me to come at the usual time for staff members, even though doctoral students followed their own schedule. "I'm afraid there might be trouble otherwise," he said. The punishment for being fifteen minutes late was prison. Genetics was on the road to destruction. And I was being watched, of course. I began having nightmarish migraines. I had to take sick leave. Ionization of the brain was prescribed. A current carrying calcium ions was passed through my head. The migraines remained just as they had been, but a downright miracle occurred. My brain began working as it never did before or afterwards. I wrote five articles in a month and a half. Soon the electrical treatment ended, and so did my animation.

Vernadsky and I had serious conversations, too. He was interested in the question of symmetry. He connected the development of life on earth with the development of asymmetry in the structure of protein molecules. I showed my black-and-white paintings to Vernadsky. I had continued to make these abstract patterns since my childhood, when regardless of my stepmother's anger Father gave me small pieces of good paper produced as far back as in tsarist times. Some pictures were done in Borovoye (the one decorating the dust jacket of this volume among them). I never tried to imitate the third dimension in my drawings. All of them were strictly two dimensional. Vernadsky liked them; he called them "asymmetry on a plane."

Vernadsky told me that radioactive tracers should be used to study the self-reproduction of chromosomes. His idea was fifteen years ahead of its time in a field that was not even his specialty. Russian geneticists did not have the chance to try out his idea.

In 1942, in Borovoye, my drawings were exhibited. There was an admission fee for everyone except schoolchildren and members of the Red Army. The exhibit's shaft horse was Zernova, a professional artist—Zernov's daughter. My drawings were the only abstract works, and they were not universally popular. Zinaida Pavlovna, the wife of Gleb Maximilyanovich Krzhizhanovsky, who I would later have much to do with, exclaimed, "Why, these are completely without content!" when she saw my works. I was standing right there, sick, irritated, and in a fur coat. I had hepatitis. It is carried by mice. I felt sorry for the mice, but I tried very hard to keep them away from my bread. Not hard enough, apparently. So there I was, an abstract artist, with bile in the blood, in wartime evacuation. I was nauseated all the time as it was—without verbal abuse. Certain wives of certain of the academicians excelled in insults.

But my art had its admirers. Many more even than the other constantly berated abstractionists had. The rejection of representationalism in art, abstractionism is usually combined with an innovative understanding of beauty. It is novelty raised to a certain power. It is something absolutely unacceptable to the unaccustomed eye. I am an abstractionist of the eighteenth century. My understanding of beauty has its origins in the wrought-iron fences of St. Petersburg, in the Swiss embroidery on my skirt that, according to my Nanny, "looked as though it had been shot up by the Germans." But abstractionism is abstractionism. It betrays the expectation and joy of recognizing something pictorial and thereby causes indignation. The viewer thinks that he is the victim of a mystification. I will point out, by the way, that he often *is* the victim of a mystification. But not in my case. The labor invested in my pictures is there for all to see. The most merciless critics of abstract artists are realists, however, and there's no point in even talking about those who possess the armor and weaponry of socialist realism. A lucky proprietor of that weaponry was Ekaterina Zernova. "You're not understood," she told me, "and if the people don't understand you, that means your work is bad."

But artists differ, too. In 1936, six years prior to the events being described, I visited Pavel Nikolayevich Filonov. When the greatest of great changes occurs in Russia and the chains of official ideology are cast off with contempt and indignation, the basement vaults of the Russian Museum will be opened wide, and Filonov's marvelous canvases will be shown to the nation for whom he painted. Filonov is an artist of the same rank as Van Gogh and Rembrandt, the brightest star in the constellation of the Russian Renaissance. The brother of Yuri Alexandrovich Filipchenko (who was banished from the university and who died soon afterwards), parasitologist Alexander Alexandrovich Filipchenko and his wife heard about my pictures and wanted to see them. After seeing them, they decided to show them to Filonov.

The walls of Filonov's rooms were hung with his pictures; here were fantastic cities populated by animals with human eyes and nude, sexless human couples dancing in the open air against a background of skyscrapers. "People don't understand me," Filonov complained. "I flail at vice, and they tell me it's pornography. I draw designs on the faces of humans and animals because I think that's beautiful, and people tell me that I don't know how to draw and that I resort to tricks." He pointed to a picture that was hung low on the wall. That extraordinary, pathetic work of art has been preserved in my memory for forty-four years. I call it pathetic because it had to serve as a proof of the great artist's ability to paint as everybody does it. Its colors haven't faded nor have its contours been erased by time in my memory. A bast basket and two eggs are on a bare table. One of the eggs is whitish, the other pinkish-yellow. The picture seemed to radiate light, and the sense of space and belief in its reality gave rise to an almost insurmountable desire to stretch out one's hand to make certain that the basket would hide it.

"They say that I haven't kept pace with life and with technological progress. But I've drawn a radio wave," he said. And the radio wave was not in Filonov's style, but it was marvelous. On an intensely colorful background, against a dark, moving rainbow, were drawn the faces of simple people, frozen in strained attention. One of them had a spoon to his mouth and stood just like that, listening. "They *will* understand me," Filonov said. "The day will come and Stalin will enter this room."

Filonov was pale, thin, and tall, with very dark eyes. He was dressed very poorly. His origins were in the common people and his goal was to create a native art. He not only created his masterpieces but wrote treatises on the art of the future. The only exhibit devoted exclusively to his works was held in Novosibirsk's Akademgorodok in 1967. His sisters brought his pictures there before turning them over to the storerooms of the Russian Museum. The writer Granin later boasted to me that he had seen them. "I ask the museum people why they don't display them," Granin told me. "They say that then they'd have to take down everything else. It would look very bad next to Filonov." The Russian Museum's storerooms are a sort of closed *raspredelitel,* a distribution point*—only the select few are allowed in. That is exactly the phantasmagoria that Filonov wanted to overthrow with his art.

In the House of Scientists at Akademgorodok, thoughtful, patterned animals looked at me with human eyes from the walls of the superb exhibition halls. There were nude, sexless couples as well, and several abstract paintings—streams of colorful geometric figures—music incarnate, dynamism, and calm merged into one. But the bast basket wasn't there, nor was the radio wave. The exhibit's organizer, Makarenko, gave Filonov the chance to be himself. The Filonov exhibit didn't pass without consequences for its organizer, however. Makarenko's crimes were countless—exhibits of Falk, Neizvestny, Chagall, Goya's exposé prints. Neizvestny's exhibit was closed before it had even opened. I was lucky enough to see that miracle of art. "We have to run," I was told by a young fellow who had run from the House of Scientists to get me, "the pictures have been hung, but the Regional Party Committee has ordered that the exhibit not be opened. But the pictures are still hanging." With Chagall, the matter never went any farther than an exchange of letters. Makarenko was tried on trumped-up charges and sentenced to seven years in the camps. He is in the West now.

A Filonov exhibit in 1968 suffered the same fate as Neizvestny's. A young artist on a flight from Novosibirsk to Leningrad told me that the Academy of Arts in Leningrad had organized an exhibit and lectures. The interdiction arrived after the audience had already gathered.

Filonov liked my pictures. He wanted me to become his student, a professional artist. He did not predict any fame for me. "If you were in Paris," he said, "you would be declared the Messiah of a new school of graphics, but here nothing will happen." He idealized Stalin. He idealized Paris and my art. I never saw Filonov or Alexander Alexan-

* The Russian word *raspredelitel* denotes a store for the Soviet elite. The average mortal is not allowed near these well-stocked preserves for the privileged.

drovich Filipchenko again. Alexander Alexandrovich was arrested in 1937, his wife exiled, and neither of them ever returned. Filonov died December 3, 1941 in besieged Leningrad. He waited for Stalin in vain.

My stay in Borovoye sabotaged my father's mission of serving people. It was assumed by those who elected him to the Distribution Committee that his asceticism would extend to the members of his family too. He couldn't take me into his room—my stepmother was there. He couldn't offer me a room—I lacked the necessary rank. Besides, he didn't have the rooms to offer. Try as he could, things wouldn't work out. People kept arriving, and there was nowhere to put them. The entomologist Kuznetsov, my father's co-worker at the Zoological Institute, was saved from besieged Leningrad, and he and his family were living in a tent. I was moved again and again. At one intermediary stage in my wanderings, I was living in a small room in a wooden wing. My door opened onto the hall. I was writing the last of a series of articles on fly populations. I still had my hepatitis, and my mood was a corresponding one. A woman suddenly entered, without knocking. "Is it all right to come in?" she said.

"You've already come in."

"No, I mean for my comrades."

"Of course." The door opened and a crowd filled my room and the hall. "Who are you?" I asked.

"We're cooks," they said. "We're taking courses here, raising our qualifications. We'll be working in military hospitals."

"What can I do for you?"

"We're finishing up, and we want to express our gratitude to our teachers. We need someone to write and draw."

"You've come to the wrong person," I told them. "Go to see Zernova. She'll draw you a cook in white apron and a pot with a leaping lid, and there will be steam coming out of the pot."

"No," they said, "we want you to do the drawing. We saw your pictures at the exhibit."

"You probably think that it's easy to draw designs like that," I said, "but work like that costs money."

"Any price," they said. I designed and executed three calligraphic addresses for them. Zernova was mistaken about the people and their understanding of art.

My stay at Borovoye was coming to an end. Anna Dmitriyevna Shakhovskaya had taken me in, and I lived with her and her mother in a single room. Anna Dmitriyevna was planning to go soon to Maloyaroslavets near Moscow to fetch her nephew. It was obvious, without words, that I would have to surrender my place before their arrival. I left before Anna Dmitriyevna's departure.

In Borovoye her mother, Princess Shakhovskaya, was writing her memoirs. At eleven o'clock the lights were turned off. The Princess railed at her fate: She didn't sleep at night, and if she had matches with which to light her kerosene lamp, she could write a bit. But matches were like everything else, rationed, and we didn't have any ration cards. The hostel provided us with all the essentials. Fortunately, the local store not only sold, i.e. "gave," but bought, i.e. "took" as well. That word "give" instead of "sell" is full of meaning and significance. It is the key word to understanding the socialist political economy. In exchange for a certain number of kilograms of mushrooms, cleaned and washed and ready for pickling—and no wormy ones, of course—one got from the store a bonus in addition to money: a box of matches. I gathered mushrooms and washed them in the lake. And we began to have something to light the kerosene lamp with.

In October 1942 I left Borovoye for Moscow and took my flies with me. They had survived evacuation in fine shape. A cook at the sanatorium prepared food for them. He spooned it into the vials. I had only to transfer the flies from the old vials to the new ones.

# *Stalin: The Creator of the GOELRO Plan*

It was the end of 1942. Moscow was cold and hungry. Filatov had died. The Battle of Stalingrad took place. And just as Dmitry Petrovich had predicted a few days before his death, the Red Army's victory at Stalingrad determined the outcome of the war.

It was growing more and more clear that Germany's forces had been irreparably undermined. The air defense was working superbly, and air raids became a rarity. People were returning from evacuation. The dormitory on Malaya Bronnaya Street was filling up with people, with their friendships and their squabbles. Scientists were given privileged supplies. Schmalhausen returned and brought a wolf with him; it had been a gift of hunters, and for some time it lived on Schmalhausen's balcony.

The articles that I had written at Borovoye were being published one after another. The findings based on the experimental research had been drawn up, and I got down to work on an introduction to my doctoral dissertation. I wrote a long book. I wasn't interested merely in populations. The concept of species as a whole occupied my imagination. Groups of similar organisms, taken together, no matter how far apart from each other they may be found, form a single species. In the sphere of interspecies competition, the species appears as a single entity. Foxes or chamomiles, fruit flies or octopi of a given species aren't crowds, but fighting legions. The competition is for the ability to preserve oneself. But not only for that: Sometimes it's necessary to remain oneself, to preserve the perfection that has been achieved, to cut off any novelty so as not to be wiped off the face of the earth; at other times inventiveness is the only thing that saves a species. Man perfects technology. Plants and animals perfect themselves. Change is a risk, and those who are willing to take it sometimes win. Species compete with each other in their ability to change. A harmonious transformation is the property of the victor. The very ability to change is improved.

In a game of chance where the players number in the millions, novelty is what's at stake. If you lose, you die out. If you win, you flourish. The hordes who thrive are those who are able to use the mutations arising from generation to generation to the greatest advantage. Millions of species have perished. A good dozen of millions have been preserved. If they survive, they have realized the wisdom of change. The interspecies competition from which they have emerged victorious is the harsh school of evolution. Nature's harmony is rooted in the rivalry among species. There are no conflicts more irreconcilable than those involving the interests of related species. They must either mark off spheres of influence or destroy each other. The relations among representatives of the same species are quite another matter. The range is enormous—from mortal combat and cannibalism where one's own children are consumed to self-sacrifice and total rejection of struggle. Group competition, with a fatal outcome for the defeated, occurs within a species as well. Populations compete in speed and quality of transformation. Interspecies selection consolidates a species. Intraspecies selection makes the population an integral system, directs the process of a change into a specific channel, and eliminates the element of chance.

My dissertation was called "Species as an Evolving System." It was never published; and I didn't manage to finish it before my time as a doctoral student ran out and I was to be taken off the rolls. That meant that I would be expelled from the dormitory and I would lose my ration cards and salary. It took some running around, but Schmalhausen arranged matters and I received two jobs—as a docent at Moscow University and as a senior scientific worker at the same Institute of Evolutionary Animal Morphology where I had been a doctoral student up until then.

In the winter, at the beginning of 1944, before I had yet been expelled from the program and during a period of desperate poverty, I decided to go to a vacation home near Moscow. This privileged place was called "Uzkoye." It was both an unfortunate and a fortunate idea. Unfortunate because I didn't have the money for the stay, and because vacations are absolutely inadvisable for me—the migraines start and there's no end to them. It was a fortunate idea because it was in Uzkoye that I met Gleb Maximilyanovich Krzhizhanovsky and his wife, Zinaida Pavlovna.

Ultimately I got the money. My hepatitis, instead of subsiding, had become chronic, and it prevented me from consuming the herring, vodka, and American canned pork which I had received with ration cards. The mail girl sold them for me at black-market prices. But I still didn't have enough money. I decided to borrow from Schmalhausen. "Lend me six hundred rubles," I said.

"I don't have any money," he said, "but we'll fix you up. Madame," he called to his wife, the beautiful and terribly kind Lidia Dmitriyevna, "do we have two bottles of vodka? Give them to her." Then he turned to me, "But please return the bottles, because at the distribution point to which we're attached, they demand the bottles be returned or they won't honor the vodka coupons!"

The mail girl (I'll be eternally grateful to her) brought me money—600 rubles—and the empty bottles. Six hundred rubles was equal to my doctoral stipend. A one-month stay at Uzkoye cost 1,800 rubles.

Gleb Maximilyanovich Krzhizhanovsky and Zinaida Pavlovona, my companions at Uzkoye, were marvelous storytellers. Zinaida Pavlovna told about how Stalin's daughter, Svetlana, enrolled at the university. She was told that she was too late, that exams were over.

"But my papa wants me to enroll in the History Department," she said.

"History is already overcrowded. We could make an exception and allow you to take the entrance examination for the Economics Department."

"But my papa wants me to enter the History Department."

"Papa, papa! Who is this papa of yours, after all, that he wants this and that and the other? . . . Stalin?! Well, you should have said so in the first place, and there wouldn't have been any questions asked."

Gleb Maximilyanovich was a poet and a lover of life. "I was born with a silver spoon in my mouth," he told me.

"What are you talking about?" I said. "You were exiled."

"Exile was sheer pleasure for me. I was with Lenin in exile. And Zinushka came to see me. Her traveling companions were the members of an assize. One of them presented himself to me and said, 'I'm your wife's boyfriend. What are we going to do now?' What'll we do?" I said to him. "We'll have to have a duel—to the barrier, old boy, to the barrier."

Zinaida Pavlovna would cut him off in an angry voice: "You've gotten completely carried away with your fibs, Gleb Maximych. What a liar you've become. And what a boyfriend you invented, Glebasya, Gleb Maximych, Gleb Maximilyanych!" (She selected various names, depending on the circumstances.)

Her words gave Gleb Maximilyanovich obvious pleasure. The only reason that he made up his stories was to tease Zinaida Pavlovna. But not everything that he said in her presence was a fiction. "She used to be a fatty," he said. "She fell under a horse-drawn carriage. The horse stepped across her, but she was run over by a wheel, and she wasn't hurt the slightest bit. She was just like a ball." Zinaida Pavlovna didn't dispute it.

But Gleb Maximilyanovich's friendship toward me involved a highly unpleasant incident, too. I told him that at first I had had superb neighbors at my table in the vacation home's dining hall—two women who were older than I, both of them lovely, one of them the beautiful Nadezhda Isayevna Mikhelson, the editor of the academy journal where my articles were published. But then Yakovlev, a court artist, the Kremlin's pet, who'd been Lenin's portraitist, turned up at our table and I became the target for his attacks. Before I could even say so much as a word, he pretended to know exactly what I was like—the embodiment of nihilistic ignorance, contemptuous of the older generation and the cultural values of the past. He laid down the law and then turned to me, "You don't agree with me, of course." And then he launched into an explanation of just why I didn't. The ladies maintained a polite, embarrassed silence. I tried to get in a word to the effect that he didn't possess sufficient information for such judgment about my character, but I only added oil to the fire and so fell silent. I began coming to meals later so as to avoid meeting him. He, too, began coming later. "What, didn't it work?" he said. "Truth is hard to swallow!"

But then I developed hideous migraines because of the good life—no air raids, four meals a day, the warmth and light—and since I couldn't eat, I didn't go to the dining hall at all. I told Gleb Maximilyanovich that I only survived in conditions of stress; that the good life was the death of me, and I told him about the harassment I had to suffer during each meal. He knew that I brought to Uzkoye a collection of my drawings, which he had admired so much at the exhibit in Borovoye. "Show Yakovlev your pictures," he advised. "Listen to what he has to say." He obviously expected that my craftsmanship would impel Yakovlev to stop the abuse. I predicted that Yakovlev would be outrageously rude, and that's just how it turned out.

But amid the torrent of abuse—timid imitation of so-and-so and so-and-so—amid all that mistreatment, there were words that I consider the highest praise ever directed my way. In a tone of contempt, Yakovlev said, "You know how to combine black and white, that's the only thing you know." In a letter, Pushkin once cited the opinion of one of his critics: "Pushkin's works don't possess the slightest merit; the only thing he has is the ability to use words in their true meaning." Pushkin exclaims that he couldn't dream of a more laudatory response.

A group of historians from the academy's Institute of History was vacationing in Uzkoye at that time. We had debates with them. Historians are deprived of the chance to have their own opinions. I was in the best position. As a scientist in another field, I not only could have my own opinion—a faulty one, of course—but I could predict their opinions faultlessly. More than two years earlier Stalin had said that "in half a year or a year Germany would crumble under the burden of her crimes." He had been very frightened then, the leader and father of all the peoples of the world. The nation that he had oppressed he addressed now not as comrades but as "brothers and sisters," and his voice shook. In January 1944, in answer to the question of how long the war would last, the historians said a half a year or a year and Germany would crumble under the burden of her crimes. I predicted the war would last four years. Gleb Maximilyanovich, an energy engineer, not a historian, was on my side. History confirmed that we were right.

In the summer of 1944, Gleb Maximilyanovich took me to his dacha, located in a pine wood on Nikolina Hill. The dacha wasn't his own property, but it was granted to him for lifetime use. It was surrounded by a pine wood, and without even going beyond the fence you could gather wild strawberries. Gleb Maximilyanovich plucked a flower and asked me what it was. Back then I didn't know, but now, having become a botanist by force of circumstances, in my mind's eye, I can label the plant that he showed me then. It was *Veronica longifolia*. "See, you don't know. But Bukharin did. He recognized birds by their calls," said Gleb Maximilyanovich. Both he and Zinaida Pavlovna, and her sister, Maria Pavlovna, and her brother, Pavel Pavlovich—all of them old Bolsheviks, had known and liked Bukharin. Gleb Maximilyanovich said that Lenin, too, liked Bukharin. "They'll ruin my Bukharchik," Lenin had said with regret.*

* Lenin used a diminutive form of Bukharin's last name and Gleb Maximilyanovich reproduced it. Nikolay Ivanovich Bukharin, one of the major figures in the Bolshevik Party, was Lenin's favorite. He was the moving force of the New Economic Policy reforms in the early twenties, and in 1936 the *de facto* author of the so-called Stalin Constitution with its articles guaranteeing freedom from arbitrary arrest, inviolability of the home and secrecy of correspondence, freedom of speech, of the press, of meetings, and of demonstrations. In 1938 he met his death at Stalin's hands. Khrushchev's rehabilitation enactments of the fifties did not restore Bukharin to Party favor.

This life was a little reminiscent of kings in exile—very close to the sun, yet at a remove. The most distinguished of the four residents at the dacha wasn't Gleb Maximilyanovich: It was Zinaida Pavlovna's sister, Maria Pavlovna. Her daughter was married to Malenkov. "Stalin's heir," Gleb Maximilyanovich said of Malenkov.

On the day after the attempt on Hitler's life, a magnificent Kremlin car, a new ZIM model, rolled up to the dacha, along with a driver who seemed to have been specially selected for aristocratic beauty and bearing, and through the car window he thrust at Gleb Maximilyanovich a newspaper with the important news about the attempted assassination. The car had been sent to fetch Maria Pavlovna.

All four of them were extremely kind. The royal favors, as much as possible, were sprinkled on those around them. If the cook's son had a toothache, Gleb Maximilyanovich summoned a car from the academy garage and no matter what the weather, took him to a dentist in Moscow.

One unforgettable day I was introduced to the Alliluyevs—the parents of Stalin's late wife. They lived nearby in a splendid vacation home; I had never in my life seen anything like the truly palatial luxury with which that vacation home was appointed. The vases that graced the living room belonged in the Hermitage. They had been expropriated from the Hermitage to decorate the vacation home of the Soviet elitist elite before the war, Krzhizhanovsky said. Krzhizhanovsky also told me that Mr. Alliluyev, a tall, wizened old man wearing extremely plain garb, was an ascetic revolutionary. He didn't say anything about Madame Alliluyeva. The Alliluyevs' granddaughter, Svetlana, had just married a Jew, and Mr. Alliluyev spoke of the usefulness of hybridization and even saw something useful, in this regard, as in the Tartar invasion.

The brunt of the conversation was carried on by Gleb Maximilyanovich and Mrs. Alliluyev. We were so seated that her back was directly in front of me. Her flesh overflowed her clothes and shoes like oversoured dough from a clay pot. She was concerned about whether the Krzhizhanovskys were being well supplied by the academy distribution centers. The Kremlin ones were significantly better; Krzhizhanovsky ought to talk to people about getting registered there.

"I've talked to them," said Gleb Maximilyanovich, looking me right in the eye.

"Well, and what did they say?"

"They said it's time for me to die, old fellow, and that I should be thinking about my soul instead of provisions."

She believed that typical arabesque of his. His directorial talent had worked flawlessly. "But you have to know who you're talking to," she continued.

"Well, I think that no matter whom you talk to, it's better to be silent," said the king-in-exile.

She was indignant about the fact that captured Germans had been led through Moscow. Tens of thousands of prisoners in convict's garb passed through the streets of Moscow at that time. The spectacle of humiliation is unbearable for me. But there were lots of them, and I went to take a look. The Germans had the chance to demonstrate their army's invincibility, to humiliate their conquerors. If they had marched in even columns at a brisk pace, the effect would have been stunning. Instead, they shuffled along like a tired, pathetic herd. Crowds of Muscovites observed them in silence. Even if the Germans had retained a fighting spirit, they wouldn't have been able to get organized. Flocks of birds in flight can observe strict discipline, herds of mammals are less organized, and man succumbs to organization only upon command.

"A lot of honor it does us to lead those vile creatures around Moscow," Stalin's mother-in-law said. "We should have poured kerosene on them and set them on fire."

"What on earth are you saying?" exclaimed Mr. Alliluyev.

"Why pour kerosene on them—why not insect powder?" Gleb Maximilyanovich asked, no longer looking me in the eyes.

"It's time to go home," said Zinaida Pavlovna.

Gleb Maximilyanovich was telling me about his life. "Lenin and I were walking along the banks of the river Yenisei. A barge carrying watermelons had sunk somewhere upstream. Watermelons were floating downstream. I said to Lenin: 'When the Revolution triumphs, the first thing we'll do is abolish capital punishment.' 'No,' said Lenin, 'you don't wear white gloves when you're making a revolution. We won't abolish the death sentence.' "

Gleb Maximilyanovich said that the entire group of political exiles banished together with Lenin and Krzhizhanovsky to Siberia tried to dispute Lenin. In their opinion, the victory of the Revolution would bring the people freedom and equality, and the revolutionary movement should celebrate their proclamation. Lenin argued for the maximum limitation of freedom, for inequality, and for dictatorship. The Revolution's banner, its goal, and its achievement should be the dictatorship of the proletariat. "He crushed our opposition," said Gleb Maximilyanovich.

Lenin liked Krzhizhanovsky, and appointed him chairman of the State

Planning Board. Gleb Maximilyanovich was removed from his post in 1929, at Stalin's order. He later defended victims of Stalinist terror even while he himself was in official disgrace. Sensible projects were labeled the work of wreckers, and those who criticized absurd projects were put in prison. He told me he had received a letter through illegal channels from people who were in prison for having opposed the construction of the Caspian-Aral canal. That outrageously absurd project was not brought to fruition. I don't know what prevented its realization: Gleb Maximilyanovich was out of favor by then.

"Illegal channels" sounded very strange to my ears. I hadn't even suspected that there were any other than the official ones. Samizdat, the struggle for human rights, underground groups—these are all spirits released from the bottle by Khrushchev. In Stalin's time, no one had ever heard of such things. There weren't even any jokes about Stalin, and up to this day there are very few of them. The Khrushchev era will go down in history as the era of the political joke. The Stalin era will be remembered as the time of love for the leader.

When the war ended, Gleb Maximilyanovich expected reforms that would give the people rights and freedom. "A government that cannot reward its people for the blood shed, for the monstrous wartime sufferings, for freeing the country from invaders, should go into retirement," he told me. "The time of freedom is approaching."

"What on earth are you talking about?" I said to him. "Stalin is attributing victory to himself and to him alone. Don't you know about the antiprofiteering detachments, about the chains of soldiers that executed those who retreated?* Don't you know that everyone who was a German captive, soldiers as well as peaceful inhabitants who were plucked up for slave labor, is being suspected of treason? They're being tried and exiled. There isn't any freedom and there won't be."

"The people have arms," he said.

"Don't worry," I parried, "measures against an armed people have

*On July 28, 1942 Stalin issued an order of the day, N227, which stopped the retreat. The name of the order was "Not a single step backwards." Troops had to stay where they were even if their positions were tactically disadvantageous. Behind their backs were placed the chains of soldiers, the units of NKVD (the name of the KGB during that time), with automatics to kill those who fled. The name of these chains, coined in the revolutionary jargon, was *zagradotryadi*, "antiprofiteer detachments." The word is in my issue of the Russian/English Dictionary (*Soviet Encyclopedia* Publishing House, Moscow, 1969, page 182: *Zagraditelny otryad*). I was surprised to see the term in the dictionary. An obvious contradiction exists between the official desire to forget the shameful reality and this punctiliousness.

been taken. The common soldiers have been demobilized. The officers have been left to serve in the army for a year. The lessons of the First World War haven't gone unnoticed, but the only one who can make use of them is Stalin and Stalin alone, not the people." I told Gleb Maximilyanovich about events known to everyone not dazzled by the ecstasy of victory, not blinded by a smoke screen of propaganda. During my expeditions I had met captives forcibly returned from Germany to be sentenced as traitors. And an old school friend had told me that the detachment he had served in was sent to force a crossing over the Oder, having a *zagradotryad* chain behind them.

By 1952, Gleb Maximilyanovich would no longer have any hopes for the people or for Stalin. I was on my way to visit him on Osipenko Street. I had left the house early. In order not to arrive before the appointed time, I dropped into a café. It was in the center of town; there were gigantic plate-glass windows and silk curtains. I thought I would sit a while and have some tea.

Because in 1952 I wasn't going on any expeditions—genetics had been completely forbidden—I was writing a book about my father's travel around the lakes of Siberia and Central Asia. I had grown quite out of touch with the life of the people. And now, right in that café, I saw them—the people. Alcoholic beverages weren't served, and there was absolutely nothing to eat. All the tables were occupied, however. The waitresses in lace aprons and caps were serving some sort of pathetic salads. The company was exclusively male. About eighty people filled the beautiful, well-lit interior. It was in the afternoon, around two o'clock. Everyone was drunk. There was loud, impudent swearing, noise, and shouts. One of the drunks, trying to prove something to his drunken companion, kept removing his cap with an excited gesture and then slapping it back on. The stream of mother-curses and the rhythm of his movements apparently served as important ingredients in his argumentation. At the coat check, I said to the elderly attendant: "I think this isn't quite the sort of place I was looking for." He understood. "But it's the same everywhere," he said. "It's only different in expensive restaurants."

"Who are those people?" I asked Gleb Maximilyanovich that afternoon. "Where is the proletariat that is realizing its dictatorship in order to build communism? The people that I saw just now are some sort of rabble, *lumpenproletariat*. . . ."

"What has happened to the country is worse than the Tartar invasion," said Gleb Maximilyanovich. "The country has had its head cut off. I myself survived by accident. The country needs a Lenin, but a hundred Lenins have perished in Stalin's torture chambers."

But it wasn't by accident that Gleb Maximilyanovich survived. He survived because he accepted Lenin's blood morality, and because he was ready to falsify history for the sake of the leader of all the world's peoples.

Long before that conversation in 1952 I had had a sad opportunity to learn how slippery Gleb Maximilyanovich's moral principles were. During one of our walks in Uzkoye he had told me a very old story, a drama that had unfolded in Warsaw, long before the Revolution. It was about revolutionaries escaping from a Warsaw prison. They had been condemned to death. Their comrades (I don't know whether Gleb Maximilyanovich was among them, but probably not) came to the prison dressed in high-ranking army uniforms and bearing documents counterfeited at a print shop. They pretended to take the prisoners off for execution. They hired a cab driver, and before the police realized what was up, the revolutionaries' tracks were already covered. "Just imagine how many people were involved," I said, "and no one betrayed them. Neither those who supplied the uniforms, nor the workers at the printing plant where the documents were run off, nor the driver." "Oh," said Gleb Maximilyanovich, "the driver was killed."

As for his falsifying history—that I heard with my own ears. I knew that he was the sole creator of that grandiose plan for the electrification of Russia, the famous GOELRO* plan. A centralized electrical network had been his goal since even before the Revolution. The owners of factories and plants realized with what danger they would be threatened if energy distribution were centralized: A worker's strike at a major generating station would paralyze a multitude of enterprises. But the Revolution provided the opportunity to bring the plan for centralized electrification to fruition.

In 1947, I was leaving Moscow, and I called Gleb Maximilyanovich to set up a farewell meeting. His niece told me that he'd left for the House of Scientists to give a lecture. I went to listen to him. Gleb Maximilyanovich was speaking on the plan for the electrification of Russia. The plan had been created, he said, by Stalin. I walked out without hearing the end of the talk and left Moscow without saying good-bye to Gleb Maximilyanovich.

Love, however, has its way. When I visited Moscow I began seeing

*Russian acronym for State Commission for the Electrification of Russia.

Gleb Maximilyanovich once again. Zinaida Pavlovna had died. Her brother Pavel Pavlovich had passed away, too. Gleb Maximilyanovich railed bitterly at fate. He had taken care of Pavel Pavlovich, but in the hospital they hadn't—they had let him catch cold. Now that his brother-in-law was dead Gleb Maximilyanovich was completely alone. There was no one left with whom he could perform his theater of the absurd. No one except the cook.

"I tried to hire a cook a few days ago," Gleb Maximilyanovich told me. "We just about came to an agreement. 'How old are you, my dear?' I asked. 'Sixty,' she said. 'And how old are you, Grandpa?' 'I'll soon be ninety,' I said. She crossed herself and said, 'Lord, have mercy.' 'Why are you so frightened?' I asked. 'The Lord,' she said, 'takes *good* people in proper time.' "

Then for some reason his cook turned up in the room just at that moment. "And when did you try to hire another cook, Gleb Maximovich?"—she used that abbreviation—"I don't seem to know anything about that. Surely I'd know if you'd tried to hire one," she said. Gleb Maximilyanovich beamed.

"Zinushka and I were staying at a vacation center in the winter," he told me on another occasion. "It was about twenty-five years ago. There was a skating rink for the vacationers. The young people were just skating around, but I was doing figures." Gleb Maximilyanovich used professional skating terminology to describe the figures he was doing. "There were women standing at the edge of the runway with brooms—they were cleaning ladies who'd come to sweep the snow off the ice. 'What are you doing?' they said to me. 'Those folks are young, after all, and you're an old man. You should be ashamed of yourself.' "

"What did you say to them?" I asked.

"I didn't say anything. I went just like this!" He took an issue of *The Light** from the magazine table and showed it to me. On the cover a young woman skater, leaping into the blue sky, was executing a full split.

At that moment the cook's skeptical voice rang out: "Why you've never had a pair of skates on in your whole life, Gleb Maximovich. What a storyteller you are!"

At my request, Gleb Maximilyanovich gave me that magazine and its cover lifted my spirits for many years.

*A large illustrated Soviet weekly magazine, *Ogonyok*.

My father and I visited Gleb Maximilyanovich shortly after the war, in 1946. I was still living in Moscow. Father would arrive from Leningrad, and in the absence of my stepmother, would go out of his way to give me pleasure. Once my choice was between seeing the Dresden Gallery and a visit to Krzhizhanovsky. The Dresden treasures were kept in a repository at the Pushkin Museum. They were shown to the elect. Father had been elected an academician in 1946, and we could have gone to take a look. I was smart! I chose Gleb Maximilyanovich. Before returning the pictures to Dresden, the Dresden Gallery was opened to the general public in the same museum eight years later. You had to stand in line in front of the building for at least four hours in order to get into the exhibit. But what were four hours in line when you were going to have the chance to see the Sistine Madonna, Rembrandt's self-portrait with Saskia sitting on his knees, and Van Gogh's pears?

On several other occasions Father obtained tickets for me, too. I heard the poet Pasternak twice. He was marvelous. At his first reading he stumbled twice, and both times the entire audience prompted him. They knew his poems by heart. They asked him whether he preferred Tolstoy or Dostoevsky. He had trouble answering, but in the final analysis decided that Dostoevsky was closer to him. He had in mind, I think, the inscrutable depths of the human soul. The treatment of social phenomena in *Doctor Zhivago* and *War and Peace* is identical.

At the second reading, Evgeniya Vladmirovna, Pasternak's first wife, and I were together. He began with a request not to ask him questions—his relatives had told him that his answers tended to sound silly. I have a feeling that it wasn't his relatives, but someone else who suggested that he restrict himself to reading what had already been published, lest he exceed the limits of the allowable. Pasternak, who had miraculously survived, was a sop thrown to freedom-lovers. Dostoevsky was under a prohibition. But Dostoevsky was hardly the issue. Pasternak could be dangerous if he were allowed to answer questions; he could expose many bloody secrets. In short, he knew too much. The destruction of the flower of the Russian intelligentsia had taken place right before his eyes. He knew personally many of the people who had rotted in the secret police's torture chambers and he knew that they were not guilty of anything. He knew why the poets Mayakovsky and Tsvetayeva, the crème de la crème, driven to suicide by the blind machine of terror, had taken their own lives. He could only puzzle over the fact that he had not only survived but was now appearing before an audience of 1,000. If it is true that his family advised him to limit himself to reading poems, one can be certain that they were guided by the fear that he would state a truth of some sort and would be imprisoned. He needed to be silent.

He read with rapture, with a joyous ecstacy, and the ecstasy had nothing to do with the content, which might be tragic and often was. In his voice one heard admiration for his own creation, an artist's wonder at what he has created. His voice seemed to be creating impeccable form, a new, a beautiful sonority, the easy breathing of his poems. There were unexpected, fresh, never-before-existing combinations of words that conveyed meaning perfectly. In a voice ringing with joy, he spoke not of love's exaltation, but of its demise. Without lowering his exultant intonations in the least, he summoned from nonexistence "the home where the bitter bromide of half-insomnias, half-somnolence is drunk like water, where the bread is like goose-foot."

He read one poem written for Evgeniya Vladimirovna, whom he had long since left, the beautiful artist with a sloping brow and potent smile who was sitting next to me. There wasn't a trace of the smile left. After the recital she introduced me to Pasternak. I said that I'd read his translation of Verlaine's poem in *New World,* "Music first and foremost of all . . ." He immediately recited the entire poem in French. When Evgeniya Vladimirovna and I left, she said, "He's a vile, cruel baby."

I wrote that Dostoevsky was under a prohibition. That's an unforgivable inaccuracy. No official prohibition is needed when there exist Jesuitical methods for implanting herd ideology: Dostoevsky, and all critical works about him, were excluded from school programs and from publishers' programs. Everything written by someone in disfavor is removed from the shelves of public libraries and transferred to special archives (a humane substitute for the Inquisition's and Hitler's bonfires), or else their work is quietly recycled. They aren't used to wrap herring—the customers bring newspapers for that, so long as they are pages without portraits of our leaders. And so even without an official prohibition, Dostoevsky was added to the very legions of those to whom during the time past, long ago, the poet addressed himself: "And your memory for future generations, is buried in the ground like a corpse."

To be under official prohibition was something different, however: It meant having books confiscated during a search, a prison sentence for reading them, an additional one for dissemination. Pasternak's own novel, *Doctor Zhivago*, met that fate. People weren't imprisoned for reading Dostoevsky, therefore he wasn't forbidden. After 1958, when Pasternak was awarded the Nobel Prize, reading Pasternak was considered sedition, but even then people weren't imprisoned for it. The great hypocrite Khrushchev, who replaced the great hypocrite Stalin, proclaimed that we had no political prisoners. An open statement that there were political prisoners in the country was a criminal act punishable by articles 70 and 190 of the Criminal Code both under Khrushchev and under Brezhnev. When the "organs" found out that someone was keeping, reading, and disseminating Pasternak's novel, i.e., was committing a political crime, they framed him or her on a criminal case—violation of the administrative regime, disorderly conduct, or offering resistance to the militia—and then imprisoned that person. And it's no trouble at all for the militia to concoct such a violation—it's as easy as spitting. I myself have been through the whole mess, and I came close to going to prison. Between the time that I was caught reading and disseminating Pasternak's novel and the moment when I wound up at the very edge of the abyss, two years passed. What humane mores reign in the land of victorious socialism! And I didn't go to prison not because of the regime's humaneness, but because of bureaucratic confusion and poor work on the part of the "organs." I'll explain all of this in detail later.

Krzhizhanovsky couldn't have wished for better accomplices for performing his "plays" than my father and me. The required qualities were absolute credulity until the arabesque's surrealism became apparent and an understanding of the allegory's meaning when the fantasy became obvious. We possessed them. We visited Gleb Maximilyanovich—and listened to his versions of the latest academy scandals—regularly. "For heaven's sake don't tell anyone about these conversations," my father would say when we were out on the street. The "cult of personality" was at its apogee. Gleb Maximilyanovich was walking along the edge of an abyss.

When I was already in the West, I learned where the expression "cult of personality" came from. It was created by Stalin, long before the war. The great hypocrite Stalin protested against the idolization of his own person. During the victory parade on Red Square in May 1945, however, flags from the state that had been trounced into the dust were thrown at his feet. I know that by rumor. The parade to show the nation's love was carefully prepared at the Academy of Sciences Institute of Evolutionary Morphology, where I was a senior research worker. Each person was assigned a row and number in a column. The right flank of each row was selected by the institute's party organization from among the most—how should I say this?—from among the most politically conscious, and therefore the most petted, the most privileged. The itinerary that the marching column was to follow was determined exactly. The column's right flank would be closest to the tribune on the mausoleum, where the genius and coryphaeus of all the sciences, the honorary academician Stalin would be. The party organization bore the responsibility of the generalissimo's valuable life. It guaranteed that there would be no plotters among the right flanks. I didn't go to the victory parade. Perhaps I would have gone, but I had a migraine.

The last time that I saw Gleb Maximilyanovich was in the summer of 1956 at his dacha. After the war, academicians were granted dachas. These six-room, one-and-a-half story structures were a contribution of the Finns, who had been defeated along with the Germans. For rural buildings these dachas were quite comfortable, with running water and tile stoves. They were heated with firewood or peat briquettes; later, telephones appeared. The dachas were given to the academicians as their personal property and could be willed and inherited. The dachas were standardized, and they all look exactly alike. They are laid out in three villages. One of them is near Leningrad and two near Moscow. One of the Moscow villages, Mozzhenka, is where Gleb Maximilyanovich lived. The dacha I inherited from my father was near Leningrad.

At that time—summer 1956—Stalin had passed away and Khrushchev's secret de-Stalinization speech had already resounded.*

On the day of my visit, Gleb Maximilyanovich was sitting by a television that was on the fritz. All the doors of his dacha stood wide open. He was waiting for the repairman. "Well, I listened to the measures being taken to overcome the cult of personality, and then the machine [the TV] refused to give away more information," he said. "Great changes are on the way."

"There will be changes, of course," I told him, "but they won't be great. Stalin's spirit is ineradicable; besides which, it's not in the authorities' interests to dismantle an irreproachable system that protects the authorities themselves. And this regime is beginning its liberating actions with decreed word combinations. The 'cult of personality' is one

* I call the speech secret because until recently the famous report of Khrushchev unmasking Stalin had not been published in the USSR. (See footnote page 225.)

of them. Gleb Maximilyanovich," I enquired, "tell me, what do you think the Revolution has given to the people?"

At precisely that moment a young man came in. He had come to fix the television. No one who had spent a significant part of his adult life in the Stalin era would express his views in the presence of an outsider. The very impossibility of continuing our conversation about the fate of the people in the presence of one of its representatives spoke of the results of the Revolution more eloquently than any words. Gleb Maximilyanovich responded: "The Revolution gave the overwhelming majority the chance for an education. There's an example right at hand." And he pointed at the young repairman. The question of whether that situation would have been any different had there not been a revolution stuck in my throat.

On March 31, 1959, *Pravda* announced Gleb Maximilyanovich's death.

I spent the end of the summer and the fall of 1944 at the Kropotovo's Biological Station studying the fruit fly population of Kashira. The flies had nearly disappeared. The soft-drink and juice plant that had made kvas* before the war, and where one could, with a bit of difficulty, capture hundreds of flies was now making noodles. Working with flies at the plant required the greatest discipline. Otherwise, you risked losing the respect of the plant personnel and your jars of fly food would be thrown out. I scraped together twenty-one flies. One of the females, unfortunately, turned out to be infertile. Twenty flies served for my experiments. The quantity of concealed mutations fell by ten times. The frequency of their occurrence remained the same.

Getting to the Kropotovo Station, and then from there to Kashira, cost a great deal of effort. A little steamer that traveled along the Oka River was the only transportation linking Kropotovo and Kashira, and it often broke down. Then I would walk—about nine miles. I had to get to the plant and back in a single day. Furthermore, there was no commuter train from Moscow to Kashira at that time, and getting onto a long-distance train wasn't so simple. Obtaining a ticket involved unbelievable labor and the patience of a prisoner. And a ticket didn't guarantee a journey. At the entrance to the track tickets, as well as documents, were checked. While militia officers fussed with business-

*Kvas is a sourish malt drink.

trip certificates, and their bearers, compressing themselves into an agitated crowd, fought with each other for a place in the check line, the train might leave. I once received permission to go out onto the tracks only to catch sight of the tail end of the departing train. Having lost the possibility of joining my flies and lost the money I'd paid for the ticket, I was standing among militia officers awaiting their next victims, who showed up in the form of women with sacks over their shoulders. A suburban train arrived. I'll never understand why you can enter by one door in a train station, while the one next to it is barricaded. The militia officers were rudely shoving the old women, who could hardly stand on their feet under the weight of their sacks.

"How do you dare to treat people like that?" I yelled; I was beside myself. "Brutes, fascists!" I screamed and rushed to protect the old women. The attack on the women ceased. I found myself the center of the militia's attention. They surrounded me, and a ring of women formed behind them. They were watching to see what would happen.

"Present your passport," said a militiaman wearing a uniform of high rank. He apparently was the head of the train station's militia section.

"Don't give it to him, don't give it to him," the ring of women cried. But I handed it over.

"Come with me," he said, and I was lead off.

The high-ranking officer did, in fact, turn out to be the section commander. He settled himself into an armchair and opened my passport. It showed my place of legal residence and bore the stamp of the Academy of Sciences Institute. "We'll notify your place of work about your conduct unbecoming to a Soviet citizen."

"And I'll notify the Central Militia director of your conduct," I said. "And I have nothing against your informing the institute about me. Please do."

"A senior research worker!" he exclaimed theatrically when I showed him my travel document. "And this senior research worker has insulted me so! Why, even from the president of the academy himself, Karpinsky, I've never heard anything but pleasantries."

"You've had occasion to deal with the president?!" I asked.

"Yes," the high-ranking officer answered. "I said to him: 'Comrade President, please allow me to have your autograph.' "

"Well, if you had asked me for my autograph, I would have been pleasant with you too," I said, dying from laughter and not concealing it. "But," I continued, "you not only insulted misfortunate people right before my eyes and engaged in physical brutality, you deprived me of the chance to leave for Kashira. Because of you, my ticket has gone for nothing. I'm going to complain."

I was going down the stairs to the square in front of the station when a voice rang out behind me: "Comrade Berg! Comrade Berg!" I turned around. "Come with me." The commander went with me to the ticket office, exchanged my ticket to Kashira for another one, and conducted me out onto the track. His farewell words were: "The militia knows how to punish. And it knows how to show mercy as well."

# The Eve of the Rout

It was 1945, the end of the war. The rejoicing of Muscovites was indescribable. On the night of May 9, thousands of people greeted each other on Red Square, strangers took each other by the hand or arm and walked together, kissing, laughing, and crying. A tent formed by the intersecting beams from searchlights covered the sky. Fireworks created fountains, sheaves, and gigantic chrysanthemums of light. The slow flocks of expiring stars sailed away from the spot where they had been kindled.

In 1945, my father and I reconciled. At the hostel for academicians, under my stepmother's vigilant and jealous gaze, Father had not only refused to talk to me, but would not even look in my direction. As a member of the board of governance, he kept moving me to shabbier and shabbier lodgings. I understood the complexity of his position and obeyed unquestioningly. When I would appeal to Father for help, he would never turn me down. And I appealed to him once a month, and every time with one and the same request. Every month I prepared an article for print. I was describing my research step by step. I had all my expedition journals with me, each and every page. (I still have them now, all of them, from the first to the last, but no longer in the form of expedition journals, but as microfilm secretly sent abroad by friends of mine shortly before I, too, left the country.) The articles had to be written in Russian and English, because at that time the journal, *Doklady Akademii Nauk* (*Reports of the Academy of Sciences*), was published in the two languages. I would translate my articles into English myself. Father was always happy to correct them.

It was at just such a time that my father's confession slipped out that, at my stepmother's urging, he had been the one years before who'd proposed to Mother that she take Sim and me. The impetus for such unexpected candor was a reproach of mine: I had just criticized Father for not inviting tutors to come to our home to teach me foreign languages. Subsequently I had only been able to acquire the most primitive knowledge of English at school, and now I was having a very difficult time.

He was terribly indignant, saying that he had sacrificed his life for my brother and me, had kept us with him, although my stepmother had agreed to become his wife only on the condition that he have no children with him. Mother had rejected us. I had only myself to blame for my ignorance. Tutors had been invited for me, but I had refused to study.

I replied that the German tutor, the woman who had prepared us for admission to the German school, had given me a book with the inscription *"Meiner fleissigsten Schülerin"* (to my most diligent pupil). And I didn't say anything more.

But father and I were at least headed toward reconciliation. A talk I'd had with Sim in 1943 provided the groundwork for the rapprochement. We talked, on Malaya Bronnaya Street, shortly after my return from Borovoye. "Do you love Father?" I asked Sim.

"Yes, I do. I feel sorry for him. He's an old man."

"But doesn't it seem strange to you that we have no reason to love a father who is adored by so many people?"

"No," said Sim, "devotion to science and paternal responsibility came into sharp conflict when Marmikha turned up. [Marmikha was the name we gave to our step-mother.] His only way out was to consider us unworthy of his care and attention. The highest social exploit—I gave birth to you, and I'll kill you as well—is very much in the spirit of Father's philosophy. It didn't come down to murder, but he could drive us out of his life without feeling any pangs of conscience. Clarity of spirit was essential for his work. And you shouldn't condemn Marmikha. Jealousy causes terrible suffering."

All of this was not spoken without humor. Our stepmother's usual sigh in response to Father's question, "Where's Raissa?," was reproduced dramatically and quite eloquently portrayed the whole depth of my moral fall.

In 1945, Father came to Moscow for the academy's anniversary session. The 220th anniversary was being commemorated. There were a multitude of foreign guests. The academy was supposed to show itself to its best advantage. There would be elegant ladies at the receptions in the Kremlin and academy. There were rumors among the women scientists that the Domestic Division of the academy Presidium, which was in charge of distributing the good things of life, would be giving women coupons with which to purchase clothes and shoes at the academy's distribution point, to which they didn't otherwise have access, since they weren't academicians. But the women scientists realized clearly that they had lost; the coupons would go to the ladies of the court. I had no hopes at all; I wasn't anxious to be at the banquets, and I didn't feel any spiritual uplift over the coming festivities. I was getting ready to go off on an expedition.

To finish constructing my theory, I needed one more fly population. I had enough isolated populations and strains; I needed to study one more large, unisolated population. I decided to go to the Transcaucasus, to the Rioni valley, and catch flies at the gigantic wine factories of the city of Kutaisi. Preparations for the expedition were proceeding well. All of my footwear had been taken to repair shops. At the university laboratory my friend, the young artist and poet, Igor Alexandrovich Nechayev, was preparing an exhibit of illustrations for my research. If foreigners came to visit the Department of Darwinism at Moscow University, we would have something to show them. So much for the banquets, I thought. But things turned out otherwise.

Julian Huxley, one of the most distinguished guests, wrote the Presidium that he wished to meet two scientists—Alexandra Alexeyevna Prokofyeva-Belgovskaya and me. And we showed up. We had our suits repaired and turned at the same seamstress's. She was a master at putting on patches and finding a way out of difficult situations in general. There were "bits" for all of that—it was her favorite word and was a broader concept than "patch." My canvas shoes, the only footwear I had that wasn't at the shoemaker's, went perfectly with everything else. My colleague and I examined each other's pathetic toilettes with merriment and mutual understanding, both of us realizing full well how they would stand out from the court ladies' finery. Alexandra Alexeyevna, for all her exceptional femininity, felt the same way about toilettes that I did. She liked to get a little dressed up, but everything in its place. If Huxley were to pay attention to our dress, damage would be inflicted on the academy's false front. But it was obvious from Huxley's own clothing that he was no dandy.

For the Bolshoy Theater, where the official opening of the session took place, one of my patrons provided the ticket. I don't recall who it was—Schmalhausen perhaps, or Orbeli or Krzhizhanovsky or perhaps it was Father.

Father was a corresponding member of the academy. "They gave me a ticket for the first row of the stalls," he'd said with surprise. "Why on earth did they do that?" After the performance of Tchaikovsky's *1812 Overture* with its cannon fire, "Marseillaise," and "God Save the Tsar," the curtains parted and it became obvious why. The stage was occupied by full members of the academy. Foreign guests in the fancy robes of their academies and the corresponding members of the USSR Academy of Sciences were seated in the stalls. You can't imagine anything more wicked than the farce presented by that stage. As though to emphasize the spectacle's tragicomicality, the gray hairs, wrinkles, and corpulence of some, the gauntness of others, the black caps covering the bald spots of a horde of immortals and the bald spots themselves, all were drowned in a sea of roses. Pink roses on one side, and white ones on the other.

Two of the high-ranking guests, Huxley and Ashby, visited the Department of Darwinism at Moscow University. Ashby, a professor at the University of Sydney, was on his way to London to occupy a high post in England's Ministry of Agriculture. The stairway along which the guests would have to make their way to the department was covered with a red carpet and two guards were assigned the task of maintaining the carpet's virginal purity. A separate entrance was opened for the rabble. We turned up. The foreigners were admitted with great reverence. "But you use the other door, please," the Cerberus whispered to me. "Don't bring shame upon yourselves," I said to him quietly, and pretended to the foreigners that nothing had happened.

Huxley and Ashby invited me to a lecture by Lysenko, specially organized in conjunction with the festivities. Lysenko, the director of the Academy of Sciences Institute of Genetics and a full member of three academies, spoke; Huxley and Ashby listened; and Eleonora Davidovna Manevich, who had a perfect command of both languages, translated. The hall was packed.

The hall of the Biology Division of the USSR Academy of Sciences, where the assembly took place, is a marvelous, well-lit amphitheater. Lysenko stood behind a table covered with a red cloth, and seated there were two academicians—Keller, the plant biologist and Gamaleya, the microbiologist. Neither of their very different faces reflected so much as a shade of awkwardness, much less shame. Nor interest: For you foreigners this is a novelty, but we have long since appreciated the significance and importance of the works of our academy colleague. . . .

Their interest was in what was happening on the part of their listeners, and they couldn't be bothered with facial expressions. Especially noteworthy was the hefty thoroughbred Gamaleya, with his slanting jaw, next to the gray-haired Keller. Lysenko looked amazingly like Hitler. Even the lock of straight hair falling on his forehead was the same. The ability to exercise a hypnotic effect is one thing, attractiveness quite another.

Lysenko showed sheaves and spoke in a hoarse, barking voice. I only recall a bit of one sentence: ". . . this characteristic is considered a dominant, and this one here is recessive, and unfortunately there are still people here who understand what that means. . . ." It was a threat to dissidents. After that I listened, but I didn't hear. Vavilov had been lying in a common grave for over two years already.

I don't remember Lysenko's lecture, but I remember very well the questions that Huxley put to him and Lysenko's answers. One of the questions was: "If there is no such thing as genes, how do you explain segregation?"

As is well-known, hybrids, as a rule, exhibit a similarity to only one of their parents. But one-quarter of their offspring manifest the trait of the other parent, the one suppressed in the first generation. It's the famous 3:1 ratio. Ratios are an important thing in science. We are obliged to them for the discovery of the elementary units of the universe—molecules, atoms, and genes. The difference among hybrids in the second generation is called "segregation."

Huxley asked Lysenko, "If there aren't genes, how do you explain segregation?"

"That's difficult to explain, but possible," said Lysenko. "You have to know my theory of fertilization. Fertilization is a process of mutual devouring. After swallowing, digestion takes place, but it's not complete, and the result is eructation. Segregation is nothing other than eructation." Eleonora Davidovna translated: "We know in our own persons, that digestion is not always complete. When that is so, what happens? We belch. Segregation is nature's belching: unassimilated hereditary material is belched out."*

After the lecture, the two self-restrained Englishmen no longer young looked at each other in embarrassment, then suddenly turned to each

* Cited by Huxley in his *Heredity, East and West: Lysenko and World Science* (New York, 1949), page 102.

other, threw their arms atop each other's shoulders, and roared with laughter. Act one of the performance was over.

The second act was an examination of the experimental plots at the Academy of Sciences Institute of Genetics. Lysenko left. Glushchenko, his trusty satrap, was in charge of the parade. He was supposed to show the influence of stock on graft, the mighty transforming effect of nutrition on heredity. But it turned out that there was absolutely nothing to show. There was not a single instance of the mutual influence of plants on each other, nothing that would demonstrate the validity of the doctrine preached by the institute director, the great transformer of nature, Lysenko. Large fields had been planted, and those plantings were a sorry spectacle.

The third act was to make up for the unsuccessful first two: a banquet. Not only were the participants strictly chosen, but the places at the table were as well. There were no more than twenty-five guests. There was food enough for a good hundred gluttons. Huxley's worldwide fame prevented the banquet's hosts from showing me the door; he had insisted I accompany him. In proposing a toast to the health of our distinguished guest, Glushchenko said: "Let's drink, comrades, to Julian Huxley, the grandson of Thomas Huxley, Darwin's great brother-in-arms!"

"I'm a grandfather myself," Huxley said, seemingly to himself.

Eight years had passed since the day that another director of the same institute had greeted a no less distinguished guest. The director's driver had sat with us, and we drank tea, and ate chocolate and white bread with smoked fish. Vavilov. Muller . . .

I knew where it all came from—all those hams and pâtés, those countless riches under the weight of which the table groaned. One of the people responsible for it was my friend, Igor Alexandrovich Nechayev. Honesty had prevented him from receiving a higher education. He couldn't rattle off things he didn't believe in with an easy conscience at examinations on political subjects like I did. He worked as a lab assistant, and in addition to those duties cleaned the pigpens and fed the pigs at one of the institute's subsidiary farms. The hams and pâtés came from that very farm. The worst parts of the hog carcasses were distributed among the institute staff.

At the time I had an admirer. Yuri Arkadyevich Vasilyev had been one of I. P. Pavlov's co-workers, was an outstanding scientist, and was an executive secretary to Leon Abgarovich Orbeli, the vice-president of the Academy of Sciences. Yuri Arkadyevich's manner of courtship was purely maternal. He had the unconquerable desire to feed me. This desire was combined with stinginess in the most implausible and fantastic way. He was married, and his wife was just as stingy as he was. He brought me a half-pound of chocolate candy: "Take exactly half of it." "Thank you," I said, "but if you don't mind, I'll just take two pieces." He did mind: "Exactly half of it or not a single piece." "But why?" It turned out that his wife would weigh the remaining candy, and if it came up at all short, she'd consider that he had been short-weighed and would send him back to the store to have it out with the director.

And so, at a time when I myself was getting more than enough food, thanks to my privileged coupons, and to superb dinners with my father at the House of Scientists, Yuri Arkadyevich would waltz into the department with pressed caviar sandwiches for me. "Oh thank you, thank you so much!" I'd exclaim. "We'll give these sandwiches to Igor Alexandrovich."

"Nothing doing! The sandwiches are for you and no one else."

"Then what's the point of the gift?" I asked. "Do you want to give me pleasure? Let me offer the sandwiches to Igor Alexandrovich. That will give me great pleasure." But it would only give Yuri Arkadyevich pleasure if I were to eat them myself. "I don't want your sandwiches. Get them out of here! And may the fish that they get caviar from become extinct and may the waters where they breed turn into a desert."

"Well, if that's the way things are, take the sandwiches and do what you want with them."

All of this was said in Igor Alexandrovich's presence. From the very beginning of the conversation his face expressed great pleasure, shot through and through with irony. Pleasure was replaced by delight when it got to the part about the waters turning into a desert. I handed him the sandwiches and he wolfed them down immediately, without the slightest resistance. When Yuri Arkadyevich left, Igor Alexandrovich commented with great praise and merriment on my biblical curses.

The next day he brought a glass jar full of stewed bits of pork and potatoes. We had a hot plate in the laboratory, behind a partition. There the staff heated up food; the smell of canned American pork wafted constantly from behind the partition. Igor Alexandrovich went to warm up his stew when Yuri Arkadyevich turned up. "Where's Igor Alexandrovich?" he asked.

"Behind the partition."

"Go away," he told my friend. "I'll warm up your stew myself. I brought you something." He had brought a half-kilo of black bread and a large piece of lard—that was a fortune for those times. We were conquered.

I was setting off for an expedition to the Transcaucasia. Both the university and the academy were supporting me. Three students were going with me—two girls and a young man to whom I was to teach population genetics. Their expenses were covered by the university. The academy provided food and medicine from its own supply room. The academy obtained the tickets. "Why are you studying something so disgusting as flies?" asked the department's deputy dean, a woman ready to deny me funding. "There's nothing on earth more beautiful than a fruit fly," I told her. "If you had ever seen them under a microscope you wouldn't have said such an unjust thing. Their eyes are like a burning pomegranate chandelier." She signed the funding order. I received a great deal of food from the academy storehouse, including a bucket of rendered butter. But I did something unforgivably stupid. I sold off the butter at black-market prices, thinking that if I had money, I could buy butter in Kutaisi, where I was headed. As it turned out, there was neither milk nor butter, nor meat, nor fish at the market in Kutaisi. We made millet gruel with condensed milk. The fruit was first-class and abundant, as were the vegetables. We were protected against scurvy then, but only against scurvy. After the expedition, I came down with furunculosis (a disease that results from the lack of vitamins A and D in the diet; it manifests itself in abscesses all over the body). I was in a very bad way. Ultraviolet rays and cod liver oil cured me.

Getting to Kutaisi in 1945 was not easy. We had to change trains three times—at Rostov-on-Don, Baku, and Tbilisi. Validating our tickets required standing in line for hours. There were multitudes of wretched people standing in these lines, and getting a place in the ticket line was a struggle. "First they get fat on the easy life, and now they want tickets!" yelled a woman as she tried to squeeze me out of line. I was extremely thin, but she was infinitely gaunt.

Boarding the train itself presented mortal dangers. I was on my way to do research, not to die en route. There were students with me for whose lives and health I was responsible. The girls could take care of themselves, but the boy was weak and sickly.

The dangers threatening my expeditions were on the same scale as the risks facing the Polar explorers of centuries past. The lack of scales for weighing the component parts of fly food, the scarcity of light bulbs, primus stoves and needles for them, and the lack of many, many other things could foil the whole plan. The raging elements that had to be subdued on the way to the cherished goal were the economic problems of socialism under construction. The war wrought chaos and destruction. But even at the end of the 1950s, one didn't think of setting out on an expedition, and not to the desert but to a city, even a Republic capital, without first playing out in one's imagination every possible calamitous situation. I especially recall the catastrophic dilemma in which we found ourselves when the last light bulb that we had brought with us burned out. Without influential connections it was impossible to get a bulb either from the storeroom of the Pedagogical Institute where we were staying or in any of the stores in Kutaisi. I don't remember which saved us in the end, a bribe to the storeroom guard or an appeal to the institute director to unscrew the bulb from the lamp on his desk.

I had foreseen the battles in the train stations, and measures for avoiding mortal danger were taken. Bottles of 192-proof grain alcohol are an essential ingredient in expeditionary equipment: A bottle of alcohol is payment for a mercenary. He, the station porter, occupies the places in the train by force. The victory in Baku, bought at a high price, turned out to be only partial.

I was lying just under the ceiling on the third baggage berth in a hard-class coach. The shelf opposite was occupied by two Azerbaijanians. One of them was sitting on the edge of the berth. He thrust his legs across the passageway and his bare feet landed on my stomach. "Take your feet off of me!" I said. I thought that he ought to be able to position himself in some other way.

He was horribly offended. "A lot I need you. Why, even if you paid me money I wouldn't sleep with you," he shouted for the whole coach to hear.

"We'll postpone discussion of that question until better times come along," I said, "but in the meantime take your feet off of me."

He gave in. He was cutting a melon in an amazing pose. You should have seen him doing it! Only natural selection over the course of hundreds of generations could create the perfection and beauty of his motions. He held the melon between the soles of his bare feet and went to work with the knife. Neither his feet nor hands touched the melon's flesh. Amid the dust and dirt, the part that he was going to eat remained sterile. He offered me a slice of melon, extending it on his knife. "Eat!" Peace.

The train's departure was delayed. I watched out the window like someone bewitched. I'll never forget it: A thickset little old Azerbaijanian, squatting under the weight of his burden, fantastically bent over under its crushing pressure, was carrying a huge plaster of Paris bust of Stalin. I remember everything: the road, the old man's clothes, his gray hair, and the bust of the leader, grayish-white in the blinding light of the Baku sun.

The Kutaisi population ought to have been highly mutable. The frequency of occurrence of mutations turned out to be low. I decided that I had been mistaken in considering a population subdivided into separate, relatively isolated breeding areas. Kutaisi is a hurricane city. And they rage precisely during the wine-making season. Each winery is a fly isolation area, an island in the ocean. Going outside its boundaries is the equivalent of death. Selection is on the side of the stay-at-homes. The Kutaisi population, I decided, was an isolated one. The low incidence of mutation was the result of isolation.

Schmalhausen foamed and raged. I should cease my research and present my doctoral dissertation. A doctoral student who doesn't turn in his (or her) dissertation in time casts a shadow on his adviser. But I couldn't settle down to writing. I needed a large population, a superpopulation subdivided into fighting battalions that were engaging in mortal, though bloodless, combat with each other. The competition for hereditary flexibility could not be limited by restrictions or hurricanes.

What good fortune it was that I didn't heed Schmalhausen. The year 1946 was the last one of my research, which was then interrupted for a decade by circumstances of life and the August Session of the VASKHNIL (V. I. Lenin All-Union Academy of Agricultural Sciences, LAAAS), when Lysenkoism won its decisive victory and genetics was liquidated.

Had I listened to Schmalhausen, the director of the institute, head of the laboratory, and my adviser, I would never have seen global fluctuations in mutability. I knew that spacial differences existed. In some populations mutations arise frequently, in others—rarely. It wasn't for nothing that Nanny, comparing my personality to my brother's gentleness, used to say: "Raisska is mulish."

In 1946, I discovered temporal differences. My geographical investigations became geographico-historical.

The time leading up to my fly expedition of fall 1946 was rich in events. After spending New Year's Eve on the Moscow-Leningrad train, I arrived in a city that was slowly recovering from the siege. The contrast between Moscow and Leningrad was staggering. Postwar Moscow was being shown to foreigners by July 1945. In January 1946, the canals of Leningrad were blocked by heaps of trash and sewage. Only the very center of town had been cleaned up, and the Krushtein Canal, which curved along Count Bobrinsky's once-magnificent palace, was swamped in gigantic, jutting mountains of refuse. Palaces alternated with ruins. The walls of buildings were all chipped, as if they had had smallpox. Even in the seventies, one of the granite columns of St. Isaac's Cathedral on the side facing Konnogvardeysky Boulevard retained the scars from that smallpox epidemic of artillery fire.

People in bread lines in post-war Leningrad told stories about eating the corpses of people who had died on the street. I won't relate those stories.

I began my research into the evolution of the insect wing at the Academy of Sciences Institute of Zoology, under the guidance of A. A. Shtakelberg, the great authority on fly taxonomy. The insect wing is the brilliant invention of a brilliant aerodynamic engineer. My childhood dream of discovering the laws of evolution came to pass. Mutability, selection, the differences among species—everything was right before my eyes. I measured the wings of flies of different gender and species, compared the wing measurements with the arrangement of veins on the wing plane, and was happy.

In August 1946, Marina Pomerantseva, Sofya Lvovna, and I went to Tiraspol via Odessa. Marina had been with me on the previous year's expedition. Sofya Lvovna, an elderly woman who was called "Sofochka" behind her back, was a lab assistant for the Department of Darwinism. Sofochka was a product of the system. As she understood it, working meant receiving the pathetic crumbs that the state paid and extracting maximum profit from the place of work, minimizing actual labor as much as possible. She had no duties at the department. She was listed as a laboratory assistant, but did not do a stitch of work. Instead, I hired a lab student at my own expense. A zealous student also helped Abram Lvovich Zelikman, an instructor at the department. Zelikman was studying the advantages of low fertility under starvation conditions. His subjects were tiny cyclops crayfish. He worked from morning until night. Sofochka didn't even help him. Zelman Isaaskovich Berman, a research worker at the department, hired lab assistants at his own expense, just as I did. The names of the department's personnel may have surprised you. Its head, Schmalhausen, was a German. All the departmental graduate students and co-workers were Jewish.

I had been warned about taking Sofochka on the expedition. I hoped for the best. I forewarned her that we often caught flies far from the laboratory and that we would have to do a lot of walking along roads that would most likely be bad. We would need appropriate shoes. She didn't own the right kind of shoes. I gave her some money. When she finally showed up in Tiraspol—very late, because she had spent a long time visiting relatives in Odessa—work was in full swing. Go catch flies? She didn't have any shoes. She'd had to give away the shoe money to her relatives. Her sister was sick. And she had to sell the bottle of grain alcohol that was intended to help her procure a ticket. That's why she was late. She didn't do a thing, living as a parasite on the expedition's budget. The day and hour came when my patience was exhausted. But extraordinary events were required for that.

Moldavia is a land of wonderful people. The Moldavians' honesty, willingness to oblige, and kindness are legendary. The house where the laboratory was set up was located at the intersection of five roads. Pavan, the chairman of the collective farm, a man who reaped such harvests and had built such an irrigation system that he was written up in *Pravda*, gave us the house. He let me have a cook for the fly food at the collective farm's expense. He said that the doors had no locks because there wasn't any need for them. Thievery simply didn't exist. We rented rooms in the nearby village, but we only slept there. We ate and worked in the house at the five corners. I made arrangements with a Moldavian couple to have milk delivered. They would put earthenware pots on the windowsill, and we would put their money in the empty pot that was being returned to them. The system worked without a hitch. But once the husband and wife came and asked where the money was. I had given Sofochka the money and asked her to pay for the milk. Her tenure on the expedition staff came to an end. What was she counting on? Leaving at an appropriate moment before her machinations were uncovered, and then accusing the milkman and me of lying?

We returned to Moscow. Sofochka checked in the expedition equipment borrowed from the department. There were a few things missing. The oil stove had broken and I gave it to the cook. Sofochka threatened to take me to court. A merry life. But then she got caught embezzling—in a big way. She presented spurious documents claiming that the department had spent a large sum of money on repairing microscopes, and pocketed the money. I asked Zelman Isaakovich Berman, "Do you think they'll fire her?" He laughed. "There's not a chance in heaven. They'll fire everyone, including Schmalhausen, but Sofya Lvovna will stay." But as monstrously unlikely as that prophecy seemed, it came true. And it was not in the theater of the absurd, not on a stage that the drama unfolded, but in real life, absolutely according to the rules of socialist realism. When it comes down to real life, socialist realism turns into surrealism and life begins to resemble the theater of the absurd. Less than two years after Berman's statement had been made, in 1948, every one of the employees of the Department of Darwinism was ousted, including the department's head, Schmalhausen. Sofochka alone remained.

Unfortunately, I had occasion to meet with thievery in Moldavia. Aunty Motya, a Ukrainian, cooked for the flies and for us. When we ran out of food that we'd brought from Moscow—and we ran out much earlier than we should have, the reason for that being Aunty Motya's thievish ways—she left us. Her morals, like those of the Moldavians, were legendary, but with an inverse sign. She hated Jews. "Soon we'll be polishing our roofs with their blood," she used to say. She was replaced by the cook to whom I had given the broken oil stove.

We were only in Kishinev, the capital of Moldavia, long enough to change trains. The destruction greatly exceeded anything that I'd seen before. The shells of houses, without roofs, without frames and glass in the windows, stood everywhere.

In a suburb of Kishinev, a tiny, tiny village, a stork fell in love with me. A black-and-white ribbon dangled from the wide brim of my white straw hat and reached to below my waist. It was the same color as the stork. He flew from roof to roof, following me, not taking his eyes off of me. I pretended not to notice him. Had I been wearing that hat a year earlier, I could not only have flirted with a stork, I could have walked down the Academy's red carpet along with Huxley and Ashby without encountering the slightest obstacle.

At first it seemed that the flies at Tiraspol were exactly what I needed. There were many mutants among the wild males. Yellow males turned up with the same frequency as in Uman nine years before. But then I began studying the frequency of occurrence of mutations. It turned out to be low. Since mutants were turning up among wild flies, that meant that the frequency of their occurrence had decreased *very* recently. The mutational process was pumping hereditary variation into the population, and we were seeing the fresh traces of its halted, no, diminished activity.

I needed to fly to Uman. Marina was sent off to Moscow. As it was, she came close to being expelled from the university for failure to appear at classes by the proper date. I went to Uman by myself. In some incomprehensible way, I managed to requisition an airplane through the government of the Moldavian Republic. It was given to me for my personal use. The itinerary was Kishinev-Kiev. There was no landing field at Uman.

For the first time in my life I saw an airplane—or what at the time we called an airplane—up close. This two-seater dragonfly had an open-air cabin. Cables no thicker than my finger were stretched between the upper and lower pair of wings. It was the antediluvian model U-2.

I thought that this dragonfly would take off and—*crack!*—fall apart. Moreover, the seat was broken and there were no seat belts. The pilot was fine, though—a hero of the Soviet Union, Pryakhin. I needn't fear an accident. But I was frightened of something else. We would be flying at the height of snow-covered mountaintops. I didn't have any winter clothes. I wrapped myself up in a shaggy towel and put on all the stockings that I had.

"Comrade Pryakhin, I need to go to Uman, not Kiev. After all, we aren't following a train track. How about it, let's fly to Uman," I said to the pilot.

"The itinerary is Kishinev-Kiev," he said. "I don't dare disobey."

We took off. It turned out that the travel sickness that had tormented me on the buses from Simferopol to the Nikita Gardens, especially when the highway would begin to twist between mountains, was much easier to tolerate in the open air. The contents of your stomach escapes from you into the sky as easily as breathing and is dissipated in the atmosphere without reaching the topographic map under you. You had to look into the distance. Nothing must rock before your eyes. I had just about adapted myself when Pryakhin handed me a note: "Where do you want to land in Uman?" "At the train station," I wrote. My motion sickness intensified. The U-2 is also called a "cornball." We landed not far from the train station, in a corn field amid sparse dried-up stalks. Pryakhin shook my hand good-bye.

One by one I dragged the boxes of Tiraspol files, empty test tubes, and stereo microscope to the road. A man came walking along, his boots on a stick swung across his shoulder. "Help me carry these things to the check room at the train station," I asked. He got booted slowly, tied my boxes to his stick, and swung the stick across his shoulder. As we walked, I tried making conversation, but he was silent. "Thank you," I said at the check room. "I'll pay you for your help. Wait! Where are you going? The money comes out of my expense account." "There's no reason to pay me. You're quite welcome." Very kind, very polite, and he left.

The Uman Pedagogical Institute put me up. I settled in with my stereo microscope at the Geography Department. As a senior scientific worker from the Academy of Sciences, Berg's daughter, the geographers treated me royally. They invited me to attend lectures that were given not in Ukrainian but in Russian especially for me. I felt terrible. And then things grew even more complicated. Having run out of money, I was starving to death. I had sent a telegram to Moscow requesting an extension of my official trip and money. But there I was without a kopeck, and still no answer from Moscow. Moscow was in no hurry. If it weren't for my colleagues treating me like visiting royalty, I'd have gone to the market, sold one of my dresses, and been done with it. But I could clearly imagine the reaction of anyone who recognized an Academy of Sciences senior research assistant, Berg's daughter, dealing at the rag market. And I couldn't appeal to the director of the Pedagogical Institute. He would have taken me for an adventuress. As for sending a telegram to Father, he would have sent money, but I really didn't want to do that. I was already seeing black when I decided I would have to go to the rag market in any case. With a dress to sell under my arm, I arrived at the post office to check to see whether the money had arrived. It had! My honor was saved.

I'll never forget the taste of the *ryazhenka* that I bought at the market, where I headed from the post office. Ryazhenka is a sourish drink made from baked milk, a piece of golden skin stirred up during preparation that is caught firmly in a delightful bit of marvelous food. Nor will I ever forget the taste of the borscht that I had bought in the cafeteria two days before with my last ruble. I was very careful in bringing myself back from starvation, ate very little, but nonetheless I couldn't avoid terrible diarrhea.

Uman didn't differ in the least from Tiraspol. Mutants were caught among wild males with exactly the same frequency as nine years ago, in 1937. But the frequency of occurrence of the yellow mutation proved to be low. The mutability had been lowered here as well just as in Tiraspol, and the diminution had also occurred quite recently. The changes in mutability in the fly populations of Uman and Tiraspol had occurred simultaneously.

---

I could study flies from Uman and Tiraspol at the Department of Darwinism in Moscow. In order to grasp the laws of innovative engineering, using the fly wing as an example, I needed to visit the collection at the Zoological Museum in Leningrad.

The evolution of the wing of all first-class flying animals is accompanied by a simplification of the network of veins that rig the wing. A wing is a sail capable of giving a stream of air the necessary direction. A wing differs from sail in the same way that a poet does from a prose writer. A wing is a propeller, but a hundred times better made than one created by human hands. The rear part of its flexible blades bends in flight, throwing the air forward, not backward, so that it carries the flier toward his goal. If there's no wind, wind can be created by flapping the wings—a favorable wind. Birds and insects don't push the air aside; instead, they glide along in the currents that are created by flapping their wings. The veins on an insect wing are a guarantee of solidity and flexibility. The diversity of their networks is an eloquent tale of the diversity of technical achievements, acquired step by step, by means of cooperation between mutation and selection, on the way toward perfection. The networks become simplified in the process of perfection.

Meanwhile, the thread of my fate was becoming torn and tangled. There was only one geneticist at Moscow University's Department of Darwinism and one geneticist at the Institute of Evolutionary Morphology, and I was both of them. Geneticists were being driven out of the research institutions whose purpose was genetics research. Yu. Ya. Kerkis, expelled from the Academy of Sciences Institute of Genetics when Lysenko replaced Vavilov, was employed briefly at the Zoological Institute in Leningrad, but then driven out of there as well. A world-famous geneticist, he provided bread for his family by working as the director of a sheep-breeding state farm in Tadzhikistan. A. A. Prokofyeva-Belgovskaya, expelled from the academy, became a microbiologist.

During one of my visits to the museum in Leningrad I sent a letter to Moscow asking to have my official trip extended so that I could keep measuring fly wings. Only Schmalhausen's permission was needed to extend the term of the official trip. But this time the departmental authorities got involved. The telegram that I received in reply read that I was to return, otherwise I would be dismissed. I returned immediately, only to find out that I'd already been dismissed from the university. At the Institute of Evolutionary Morphology, the party organization demanded that I give an accounting of my work on the evolutionary morphology of the animals. My population research did not correspond to the institute's goals and directions. But my lecture on the evolutionary morphology of the wing made an impression, so I was reinstated. Schmalhausen was out of town. Everything was done behind his back.

I wrote an article on the evolution of wing venation discussing technological problems, and sent it to the *Academy of Science News*, where the beautiful Nadezhda Isayevna Mikhelson was still the editor. My article did not see the light of day. After the rout of genetics in 1948, the editorial board returned the article to me along with a rejection letter. The article could not be published, since it treated an object lacking economic significance. Mendel's object of study, as we all know, was the pea. The pea, and not some quite harmless old fruit fly! The indisputable economic value of his object didn't save the deceased genius from being repudiated. What chance then, did I have?

By that time, Nadezhda Isayevna must have been thrown off the journal's editorial board, or else they were forced to change the publishing profile, as the official euphemism put it.

But on the other hand, another article of mine, written together with Marina Pomerantseva at Moscow University's Department of Darwinism, was lucky. It appeared just before the curtain came down, in the last issue of the *Journal of General Biology*, before its profile was changed and academician Schmalhausen, its main editor, driven off the editorial board.

The appearance of my article, the rout of genetics, the dismissal of all geneticists—including me—from their posts, the official victory of quackery, and the coronation of Rasputinism all occurred in 1948. Nicholas II retained Rasputin as a healer for the Tsarevich Alexis, but at least the tsar didn't make his healer the Minister of Health. Lysenko's quackery was elevated to the level of state doctrine, and he himself was made a member of government.

However, at the time of my short dismissal from the university we were still in 1947.

No matter how Schmalhausen raged at me for not writing my doctoral dissertation, no matter how often he pointed out to me the danger of my being expelled for failure to fulfill the production plan (and the plan comprises labor, not truth—useless labor spent in putting together a typed book which is not intended for publication and is only suitable for the endless bureaucratic procedure that is crowned by the receipt of the degree of Doctor of Sciences), I had no intention of writing my dissertation. Even my father and my friends urged me to hurry up and earn my degree (Father foresaw the coming catastrophe).

The flies that I had been studying up until that time were essentially domestic animals, man's inseparable companions. I wanted to study real wild animals, the inhabitants of the fruit forests of the Fergan Valley of the Tien Shan Mountains. I needed to go to Central Asia, to a small town named Shakhrisabz. Schmalhausen approved the plan. Ultimately he could not resist Raisska's mulishness (this is how Nanny described my disposition, to distinguish it from that of the heavenly angel Simochka). During one of our visits with him my father asked him, "Does she listen to you?" "When I do what she wants," was the answer.

But I never got the chance to see the Fergana valley, with its light-blue chains of mountains fading off toward the horizon, growing smaller, overlapping each other. Nor did I get to see the wild flies of the fruit forests. The reason was in order to follow one's calling unimpeded, one needs to maintain the freedom of choice at every stage of one's life. I had lost that freedom of choice two years earlier, in 1945. In December of that year I got married. Step by step, marriage, and then the appearance in the world of my daughters, determined my life, pointed out the way for me, and that way was far from the wild flies in the Fergana valley fruit orchards.

My husband, Valentin Sergeyevich Kirpichnikov, was demobilized from the army and received a job at the Fish Institute—pardon me, I mean the RSFSR Ministry of Food Industry Institute of Fish Husbandry—in Leningrad.

I moved to Leningrad. I wouldn't have moved there for anything: We just would have lived in different cities if it hadn't been for two *extenuating* circumstances: My elder daughter, Liza, was on the way, and I had lost my living quarters. The latter fact means that I was thrown out of the dormitory on Malaya Bronnaya Street. That happened in 1945 while I was studying the flies at the giant wineries in Kutaisi. A demobilized academy staff member moved into my room and my books and possessions were moved out into the hall in my absence. Bagrat's Temple had saved me from despair then. It had lain in ruins for a thousand years. High reliefs with depictions of lions and grape leaves had once decorated its walls. Its noble stones lay scattered in the grass. God, deprived of a shelter, whispered words of consolation to me at the ruins of His temple.

I rented a room from actors who were going on tour. I held onto

Moscow and to the Institute of Evolutionary Animal Morphology for dear life, realizing full well that in regard to work, nothing good awaited me in Leningrad. But Liza was already kicking, the actors would be back very shortly, and Lobashev, the former promotee who had expelled Roza Andreyevna Mazing and me from the department that he chaired, agreed to take me back. I moved to Leningrad.

# *Communal Apartments*

Liza was born in July. At first, until November 1947, we lived in my father's apartment. Then we received a tiny little room in a communal apartment in the building across the street. A communal apartment in the capital city—rather, in a former capital city—in a semipalatial building across from a grand duke's palace and the Dutch embassy, is a historical phenomenon. The building, which had survived the destruction of the Civil War, the bomb raids and artillery attacks, and again the insidious destruction of the Blockade, was a building covered with pockmarks, yet it retained traces of former luxury. There was a statue of Hermes and a crystal shade on the lamp in the vestibule, bronze plates over the slits in the mailboxes, and huge stained-glass windows on the main stairway. I'm only listing those things that were fated to disappear.

The first thing that a barbarian does when he sees a Greek statue of a youth is to knock off the sexual organ. Our Hermes was disfigured in another way—people snuffed out their cigarettes against his genitals. The second thing that a barbarian does on entering a palace is to write a sacramental four-letter word on the wall and add his name.

In the thirtieth year of the Great October Revolution, the communal apartment where I was to live looked like a nightmarish monstrosity. It wasn't barbarians living atop the ruins of civilization, but prehistoric cavemen. No, if prehistorical cavepeople had been as lacking in a social instinct as were the residents of apartment no. 6 in apartment house no. 1 on Maklin Prospect, man would never have arisen. The manager who had told us about the existence of an unoccupied room in the apartment warned us that it would be impossible to get into the apartment. There was no bell and no one answered if you knocked. The residents had keys. If anyone came to see or visit a resident, the resident brought the person in themselves. The manager was poorly informed about the simplicity of customs in the primordial cave. In addition to the front stairs, there was also a back stairway, which people used to bring up bundles of firewood. It led to the kitchen. That entrance was never locked. There weren't even any keys to it. There wasn't a thing in the kitchen to tempt a thief—not even matches, let alone pots and pans. All the doors of the rooms, each inhabited by a family—there were five of them when we moved in—were locked from inside. People protected themselves not from thieves, but from each other.

There was no light in the hall, kitchen, or bathroom. The Venetian glass shades with edges curled in the shape of rose petals, the color and texture of a mouse's skin under thirty years' worth of dust, were still hanging on cords. But there wasn't a single light bulb. Everyone had light bulbs in their rooms. But there was only one electric meter. The common electric bill was subdivided and paid proportionally, according to the number of residents. They couldn't come to any agreement about paying for lights in shared space. They went out to the kitchen with candles or kerosene lanterns.

They used primus stoves and oil stoves instead of the woodstove in the kitchen. All the kitchen utensils, and the kerosene as well, had to be kept in one's room. Anything left unguarded in the kitchen disappeared immediately. All I had to do was wash the Venetian glass in the hall lampshades and they disappeared on the spot.

The kitchen is the mirror of a communal apartment's soul. It would be impossible to imagine anything that cried out more about human nature than the kitchen in apartment no. 6. The luxurious kitchen, formerly an aristocrat's, had a huge window looking out onto a vast stretch of water, the confluence of three rivers—the Neva, Pryazhka, and the Moyka. The glass of this oak-framed window was covered by a layer of dust. The kitchen floor was inlaid with red-and-white slabs so solid that even though people chopped wood on them they were only a little nicked in one spot. Lyon velvet is renowned for its blackness. The kitchen ceiling was just that color. The blackness on the eleven-foot-high ceiling had been created by the accumulation of layers of soot over the course of thirty years. Long cobwebs, thick from the soot that clung to them, hung down. The sources of the soot changed—two epochs of iron stoves had alternated with two epochs of primus stoves and oil stoves. The first epoch of iron stoves, the time of war communism, came as a result of World War I, Civil War, and Revolution; the second epoch was a creation of World War II and of the terrible siege of Leningrad. Prosperity coming after each of the disasters was heralded by primus and oil stoves.

Besides the common electric bill there is another point of inevitable contact among the inhabitants of a communal apartment—cleaning the areas of common use. A professor taking his turn cleaning the toilet is a very pleasant spectacle for a janitor. It was obvious that the residents of apartment no. 6 in house no. 1. on Maklin Prospect, formerly English Prospect, not only had no disdain for this source of mutual humiliation but made use of it with subtle inventiveness.

The kitchen spoke for the derisive character of the possible tasks required of the on-duty person. If all the copper parts had to be made sparkling clean—handles and faucets—that wouldn't be anything out of the ordinary. But there was a reservoir built into the stove, the stove that was never used. Yet the copper lid on the reservoir gleamed. O, mighty language of things! The contrast between the layer of dust on the windowpane, the cobweb hanging down from the asp-black ceiling, and the warm glow of the lid that no one needed said that there was a legislator here, a strictly observed hierarchy. A rat king ruled here, and all were equal before him.

Subordination obviously excluded protest against an absurdity that indicated such a strict social order. The radiance of the kitchen lid gave me impetus to predict the presence of a dominant person in an otherwise atomized population of the communal-apartment inhabitants. What followed showed that I was not wrong in my guess.

Much later I was lucky to observe one—positive on this occasion—sign of relations between Maklin no. 6 population members. The tsarist griddle was replaced by gas ovens. The house management, ordered by the government, made the replacement, free. In the kitchen appeared an aluminum ladle with a broken handle, full of used matches. The ten burners of the new ovens were distributed among the tenants. A burnt match had some value: if you entered the kitchen at a time when one of the burners of your neighbors was lit, you didn't need a match from your own matchbox to light your burner. You could take a used match from the ladle to get light from a burner already lit. The goal of economizing was achieved. The match stubs became . . . socialized! The broken ladle, full of used matches, belonging to the society as a whole, spoke of the best side of the vaunted, broad Russian nature.

Who had moved into this apartment? The whole four-story building was abandoned by its residents in 1917. I don't know how it was repopulated.

Workers weren't given lodging in the former apartments of aristocrats. The first invaders, who moved in immediately after the Revolution in 1917, were nurses from the psychiatric hospital, the "Nicholas-the-Wonderworker Hospital" as it was called before the Revolution. (The rain kept washing away the red paint on the sign and "Nicholas-the-Wonderworker" showed through again and again, in black and white, until the sign was taken down.) Also former servants from the court of the Grand Duke, janitors and janitresses from his palace, located in front of our building, took up residence in the apartments from which the aristocracy had fled. Auxiliary workers from outlying factories, new townsfolk who flooded into town from the villages during collectivization had added to the number of residents in my new dwelling. There were hardly any men. Widows, children, single women abandoned by their husbands or who never had been married, with children, without children—these were the residents.

The room that we got was the only small one in a former general's seven-room apartment. That narrow lodging between the kitchen and the main stairway had once been the cook's quarters. We had to take it because we couldn't expect to find anything better, but we were taking it in order to exchange it for a larger room later. The black market in housing exists to this day and has its own laws, which are strictly delimited by the state. The black market isn't forbidden, because the state profits by its existence.

When I walked into that narrow little room and saw before me a huge window looking out onto the water from the third story, the same view as from the kitchen window, I realized that I would be living in that tiny little room. And I wasn't afraid of those barbarians living on the ruins of civilization. I knew that they were significantly better than those members of the intelligentsia who resided at the Academy of Sciences graduate dormitory on Malaya Bronnaya Street in Moscow, where I'd lived for almost five years. Each story of the five-story building on Malaya Bronnaya Street comprised a communal apartment. Passions raged. The nobility and baseness of human nature poured forth in full measure. I had had experience in dealing with people who found themselves together by chance, crowded together within narrow confines.

We moved in.

I don't like movies in color. Color removes a part of the abstractness from a picture and lowers its artistic quality. The late autumn projected a black-and-white film outside the window of my room. A timber float was being driven toward a wood warehouse on the bank of the Moyka. The float's trapeziums, becoming progressively smaller, faded away into the distance. They were covered by snow. The water was black, the sky gray. There were gray willows dusted with snow, a gray fence around an insane asylum, the black hulk of the ship factory. Spring didn't paint the landscape in any gayer colors. It brightened it just a bit—the water and sky grew a little bluer, and the willows grew a pale greenish yellow, as though the paint were placed onto the canvas with a dry brush. Dobuzhinsky's graphics were replaced by the gentle decoration on Danish porcelain.

The five other families who inhabited the apartment building consisted of five older women, six children, and one man. By some miracle that man had survived the war, the siege, and the deathly famine. We had only just moved in, and I hadn't even seen him, when we learned that he had perished. He got drunk, went for a swim, and drowned. His wife and her two daughters took our narrow little room. She was comfortable with a modest recompense, and with the much smaller rent she now had to pay for a smaller dwelling space. We moved into two luxurious rooms with a balcony. Then a marvelous view from the windows of the corner room was revealed to us: the Dutch embassy, the wing of the grand duke's palace, and the uncurtained windows of my father's study, through which one could see his gray head.

In 1948, Anna Ivanovna, who came to wash the diapers of my newly born daughter, Masha, looking at that head eternally bent over his desk and illuminated by a green lamp, exclaimed, "I wouldn't just do the wash for a man like that, I would starch his linen!" The happy periods of one's life fly past quickly, but in our recollection they seem long. You only need one thing for happiness—a bench on which to sit at God's feet. We found those luxurious rooms in 1948. In 1950 my father died. A memorial plaque appeared on his house.

I came to adore my neighbors despite their behavior. The comparison with the intelligentsia and nonintelligentsia inhabitants of the graduate dormitory on Malaya Bronnaya Street beautified my new neighbors beyond measure. They didn't even suspect how beautiful they were.

The five women belonged to three strata. The lowest one consisted of auxiliary workers, guards, warehouse workers, and laundry attendants.

They went from one of these jobs to another without rising any higher. There were three such women. Three "stone women," all of them drunkards and thieves. They wore scarves.

The second stratum had only one representative—the rat king, at whose initiative the lid on the kitchen reservoir glowed. This dominating personality, Vera Alexeyevna, belonged to the worker bureaucracy. She was some sort of manager at a factory cafeteria and was a member of the factory's ruling committee. She wasn't the least bit different from the stone women in her education, manners, or the way she looked, but she wore a hat.

The third stratum, too, had only one representative. The unfortunate Anastasia Sergeyevna, a cashier at the train station, considered herself part of the intelligentsia. She wasn't a stone woman and she wore a hat.

In the Land of Soviets, a hat is an object of extraordinary importance. On one side are men in caps and women wearing scarves tied under their chins, and on the other side are people wearing hats. For city dwellers a cap and scarf are symbols of modesty, simplicity, of belonging to the masses, to the proletariat. A hat symbolizes high rank, education, or the pretense of such. Modest, simple workers in caps and scarves hate people who wear hats. Squabbles in packed trams, buses, and trolleys do not transpire without the mention of a hat. The phrase "He put on his hat" is an overture to abuse and foul language.

Lenin knew what he was doing when he wore his famous cap. And the wags of the third post-revolutionary generation know what the point is when they tell jokes about Lenin. Here is one of them. It is the unveiling of a monument to Lenin. The covering is removed. The leader stands there wearing a cap. His right arm is thrust out, and his right hand is gripping another cap. Khrushchev wore a hat. The hat belonging to the rat king in apartment no. 6 and Khrushchev's hat have something in common.

In apartment no. 6 passions were not influenced by whether a neighbor woman wore a hat or a scarf. Everyone hated everybody—and was afraid of everyone. Everyone was involved in shady dealings. Anastasia Sergeyevna sewed and didn't report her earnings. The stone women brought home stolen odds and ends from factories—light bulbs, rope, and burlap. The rat king had a mythical person registered in her room. He had never existed. I even remember that phantom's name: Ivanov. I found that out by accident when the notices inviting us to vote were delivered. The king needed Ivanov in order not to have to pay for excess living space. Before the elections, Vera Alexeyevna took an absentee ballot for the phantom, as though he would be out of town. It was a phantom elector at phantom elections, since in actuality there are no elections and never were. We participated in the staging of them under the strictest system of checks and controls.

The residents of the apartment did not say hello to one another. And when they referred to each other, they used uncomplimentary nicknames. It sounded approximately like this: "We're not on speaking terms with Toska. We fell out over a kopeck," Elena Kirillovna, with whom we had exchanged rooms, told me, hoping for my sympathy.

"You're speaking figuratively when you say 'over a kopeck'?" I asked.

"Not at all figuratively—it was over a kopeck. I was figuring out who owed what on the electric bill. I added an extra kopeck to her share. She has an electric doorbell, while mine is hand-operated."

The reader has already realized that bells had appeared in the apartment. There were buttons by the main door. Next to each button stood a name. Actually, I had set up the hand-operated bell, for use by everyone. Supreme respect for private property was manifested in the fact that no one used it. One day I heard a terrible clatter. Someone was pounding on the front door with her shoes. I opened up—it was the rat king—she'd forgotten her keys. "Why don't you use the bell?" I asked. "The bell is yours," she said.

I civilized the apartment. Its external appearance changed, although the mores remained the same or perhaps even grew worse. Before I moved in no one had used the bathtub. Everyone went to the public baths. I replaced the broken sink. Now the tub and sink were used by many, if not by everyone. The responsibilities for the person whose turn it was to clean did not include the tub and sink. The sink was mine—and I was supposed to clean it. But no one asked me anymore for permission to use it. I never had the right to install it without their approval. All of this, just like many other things, was implied and understood without words. The bespattered sink and the shoe banging on the front door were the language that we spoke. Stalin was undoubtedly mistaken when, in a presentation at a "free discussion of linguistic problems," he asserted that there was no thought without words.

It occasionally happened that I heard conversations in the kitchen: "I yelled at them and they took off—they realized what might happen." That was Vera Alexeyevna's single note. She was describing her stay at Carlsbad, in Czechoslovakia, where the factory committee had sent her for a rest cure.

Anastasia Sergeyevna, hearing Pasternak being abused on the radio (on that day they had broadcast his humble request not to be expelled from the Motherland), expressed her joy: "They made him tuck in his tail—now let him crawl. . . ." Up to that time she had never even heard his name and couldn't even have had the occasion to hear it.

Elena Kirillovna was indignant over abstractionists. She had heard Khrushchev speak over the radio—he had said that abstractionists painted trees upside down.

I always kept quiet. Always. But this time I not only spoke up, but even dissembled: "You lucky people have seen abstract art—I've never seen it at all."

"What about the foxes that I saw in your room?" thundered Elena Kirillovna. That was the extent of her acquaintance with abstract art.

Someone had given me a single porcelain fox—a plain old everyday fox, a stylization that bowed to the canons of socialist realism and did not exceed the limits of the allowable. But the other fox, a plastic one made out of stereometric figures, was a delight. The artist who did it was abused in the local evening newspaper and a great work of art—the tail was a rotated semi-ellipse, a marvelous piece—was reproduced in the paper. Everyone had to see the depths to which the sculptor had fallen. I think that in reproducing the fox, the author of the exposé article was dissembling in the same way that I was. To Elena Kirillovna both the foxes seemed like trees drawn upside down.

Do you think that Elena Kirillovna, a warehouse worker at the shipyards, tasted the blessings of socialism and was a loyal supporter of the government? She washed paint reservoirs. Either her gloves didn't provide protection or she didn't have any gloves—and her hands were covered with sores. When she came home from the plant for her dinner bread, you wanted to weep when you saw her food. At the plant, "behind the employee entrance," was a closed store where good foodstuffs were sold to the workers at the plant. Her salary wasn't large enough to allow her to buy them, yet she sometimes bought chickens for me there.

She even told political jokes. There are two types of political jokes in Russia: One type is composed by members of the Union of Writers—anonymously, of course. And the other is created by the people. The Union of Writers produced jokes in the form of questions. What is freedom of speech? The recognized necessity of keeping silent. Or: What is freedom? Recognized necessity. The words *freedom* and *necessity* were pronounced with a Czech accent and made clear the joke's meaning: mockery of the invasion of Czechoslovakia. Elena Kirillovna told a joke about a mother who gave birth to twins. The boy, Nikita, was growing fat. The girl, Rodina,* was growing sickly. The time of Nikita Khrushchev was the time of the political joke.

Elena Kirillovna was irritated with the government, but her displeasure would have been a hundred times greater had there not been flights into the cosmos and abuse of the intelligentsia. The radio replaced the bonfires of the Inquisition. In the 1960s, televisions made their appearance in apartment no. 6.

You may wonder why I liked this rabble—rabble in the most literal sense of the word. But they were angels in contrast with the intelligentsia in the dormitory on Malaya Bronnaya Street. There the inhabitants played venomously dirty tricks on each other, and they knew how to spread gossip with devilish calculation. And that was all done out of envy, but at the same time disinterestedly. The intellectuals extracted no material gain from these pranks.

Let me tell you about one of their actions. Directly across from my door lived a family—husband, wife, and two children. The husband was Russian, the wife Jewish. During the war they had been evacuated. The war was still going on when the dormitory on Malaya Bronnaya began to be filled again. First the husband returned, and then his wife and the children. The young woman learned from the newspapers that the Germans had killed her parents in Kiev. She literally wailed. Just at that time a woman doctor, the wife of the Jewish war invalid and graduate student Gilechka Fridman, told the grief-crazed woman that during her absence her husband had betrayed her, that he had brought women home. I knew she was probably lying, because I had no occasion to see my neighbor in the company of a woman. The daughter of the slaughtered victims went insane. She sang and danced in the hall and made her children sing and dance.

One couldn't even imagine anything like that among the barbarians on English Prospect. Things did not go beyond the contemptuous ignoring of each other and petty theft. In comparison, you could like them. The mores of the residents of apartment no. 6 became unbearable when the stone women got old and poor Anastasia Sergeyevna came down with an overt form of tuberculosis, and when the rat king rose up the social ladder to unattainable heights and took upon herself the function of my ideologue, constantly threatening me with denunciations.

Keeping track of people in the Soviet Union is a process veiled with a fig leaf. You cannot arrest a person and raise accusations against him

* Rodina means "Motherland."

on the basis of information derived from reading his letters. So, when the KGB learns about the existence of sedition from private correspondence, it organizes a search. Letters found during the search may be used as evidence. O! The great land of triumphant socialism! The freest country in the entire world! O! The great reforms of the defender of civil rights—Khrushchev! Under Stalin such ceremony was not observed.

I only once caught Vera Alexeyevna listening at my door. I came into the hall from the street. She was wearing an overcoat, hat, and shoes and standing outside my door and eavesdropping. None of us was at home. The radio was blaring. They were broadcasting a play. A scandalous family scene was in full swing. When she caught sight of me, Vera Alexeyevna hurriedly concealed herself in the bathroom. She hadn't suspected that I had a radio. I never played it, because it irritated me. The children had left it turned on in my absence. If I had not turned up in the hallway at that moment, the whole scandalous dialogue that the rat king had overheard would have entered her treasury of knowledge about my private life.

In the good old days, back at the time when our apartment was still in its original barbarous state, I managed to get a telephone. No one wanted to have it and it was installed in my room and I alone payed for it. In my absence, when for five years, from 1963 till 1968, I was living in Novosibirsk, it was moved to the entrance hall. On my return it started to serve as a source of private information for Vera Alexeyevna. A telephone conversation is rather like a letter confiscated during a search. Any telephone conversation is samizdat. She caught me.

The way it happened was this. Just before New Year's I was flying from Leningrad to Novosibirsk. I bought a fancy layer cake at the airport restaurant. The inscription on the box celebrated the anniversary of Great October. As I was walking up the stairs to the airplane a soldier behind me said, "It's New Year's, but your cake is for the anniversary of October." The rampant celebration of great dates—the October Revolution, Lenin's birthday—has often been captured in jokes. One of the jokes is Armenian, to judge by the character's name. As soon as I entered the plane, I told the joke to the soldier who had started the conversation with me: "Do you know why Karapet has stopped shaving? He's afraid that if he turns on his electric razor, he'll hear about the hundredth anniversary of Lenin's birth." And now, over the phone, in the hallway of my communal apartment in Leningrad, I related the whole episode with the cake, the young soldier, and with Karapet and his refusal to listen to anything about the hundredth anniversary of Lenin's birth. I think that Vera Alexeyevna saved up information that she considered suitable for my denunciation, not with the aim of denouncing me, but rather to keep me in terror. When passing me in the kitchen or the entrance hall, she would say, "The radio irritates them. They don't want to hear about Lenin." In the twenty years that had passed since I had exchanged rooms with Elena Kirillovna, Vera Alexeyevna was still threatening to denounce us for our illegal deal. The deal was legal, but the extra payment that Elena Kirillovna received was not so much my crime as hers. Under Stalin a false denunciation could result in death. Under Khrushchev and Brezhnev the crumbs of information that Vera Alexeyevna had at her disposal would only be added to my file. They couldn't serve as a reason for initiating criminal proceedings. My dissidence was a secret for her.

When Anastasia Sergeyevna came down with an overt form of tuberculosis, they forbade her to use the bathtub. "Let her wash in the kitchen. They're afraid of infection." Animals, let alone cavemen, would have acted differently. I wanted to propose that the bathtub be given over for Anastasia Sergeyevna's exclusive use, and that all the rest of us use the kitchen sink. Anastasia Sergeyevna used her common sense and talked me out of it. "Let them have it their way," she said. She doubted that my proposal would be accepted and it could only cause unpleasantness both for her and me.

Elena Kirillovna lost her mind. She and Anastasia Sergeyevna both died of cancer. The apartment began to be populated by the intelligentsia of the new type. Then life became impossible. Maria Ivanovna Sorokina, a jurist with higher education, and her husband Lev Makarovich Sorokin, who looked like a Gypsy and was a fireman, knew much better than I did how to lay down the lines of coexistence with one's neighbors in a communal apartment. The Sorokins simply terrorized the others. I tried not to have contact with them.

They ask me here in America why I left Russia. "Because I lived in a communal apartment," I answer. That's a joke which, like any joke, is one-quarter serious.

Maria Ivanovna combined in herself the vices of the worst elements of Malaya Bronnaya Street together with the worst elements of the rabble. She played devilishly dirty tricks, while Lev Makarovich was as simple as one could be. "Professor, mop up the floor," he said to me. I replied with a joke. That was the bright dawn of our bitter relations. At their apogee he would walk into the entranceway from the stairs or come out from his room (his huge room where formerly Anastasia Sergeyevna had lived and where he now lived with Maria Ivanovna and their son, a university student, and daughter, who was still in grade school), and if he saw me talking on the phone he would depress the lever and say, "This is what you get, kike." The first time that he did that I told him that it was wrong to persecute a member of the intelligentsia who hadn't done him the least bit of harm, that the Constitution and laws forbade such actions—his wife could testify to that, and that his children were members of the Communist Youth League and he should be ashamed for their sake. His son, who was handicapped—his right arm had been amputated at the shoulder—walked past us in his overcoat and headed for the outside door, apparently signaling in this way his protest against his father's conduct. I assume as much because his parents, coatless, ran after him and tried to restrain him. I didn't see the end of that scene. The second time that the Gypsy-looking fireman cut me off in the middle of a phone conversation with the words "I'll make mincemeat out of you, kike," I told him that I'd heard that for the last time. The next time I'd slap him right across his mug, without warning. "But I'll make a puddle out of you!" he exclaimed. "Yes," I said, "you'll make a puddle out of me because you're a man and I'm an old woman, but just the same you'll get it on the mug." The outrages ceased on his part, but Maria Ivanovna continued staging her devilish vaudevilles right up until my departure.

They knew that I knew that Lev Makarovich had been assigned to spy on me. My elder daughter, Liza, had caught him red-handed. It turned out that he had wired my doorbell into his room.

I didn't know about that in 1972 when I told Andrey Dmitriyevich Sakharov over the phone, "But be sure to press the right bell when you come to our apartment." I not only didn't know that my doorbell rang in the fireman's room as well, but I didn't know to whom I was speaking on the phone. I'd been told that there was someone who needed my help and that he would be calling. I wasn't told his name. I realized that it would be best for my future acquaintance not to make a mistake when he rang the bell of our communal apartment. And I didn't find out who was coming until my bell rang late at night, when there was no one in the entranceway or the kitchen, and I opened the door and Sakharov and Elena Georgyevna Bonner came into our rooms and only then introduced themselves.

The bright dawn of my relations with Maria Ivanovna was marked by her yelling at me through the bathroom door, "You've washed your gynecology long enough!"

A former graduate student of mine, Lyusya Kolosova, was staying with me at that time. She had come from Novosibirsk and couldn't get over how beautiful Leningrad was. "What beautiful latticework," she said as we passed the Summer Garden. We were taking a taxi along the Neva embankment. It was the first time that she had seen golden spires and cupolas—"gilt roofs," as she called them—and she liked them best of all.

When she heard the way Maria Ivanovna yelled at me, Lyusya Kolosova said to me, crying: "Leave, leave right away! Go anywhere, but leave!" But gynecology was nothing compared to what was yet to come from Maria Ivanovna's bag of ugly pranks. Try as I might, none of my attempts to exchange rooms into another building met with success. Everyone who came to look at my rooms with an eye to an exchange understood the language of things superbly. The comfort and amenities of the apartment testified to the fact that I was fleeing from my neighbors. That was all potential exchangees needed to know.

I have described a perfectly typical communal apartment, and ours was far from the worst—we didn't have any fistfights, let alone murders. Yet that was not the case in other communal apartments. We had never heard any swearing until the fireman moved in and Elena Kirillovna lost her mind. Lev Makarovich swore at me with true artistry and Elena Kirillovna used the polite form of the pronoun "you," which clearly indicated her mental disorder. But can it really be that there were no good communal apartments? No; there were. There was, in fact, two of them among the multitude of apartments I had occasion to encounter.

My friend and colleague Alexandra Alexeyevna Prokofyeva-Belgovskaya once was visiting my home and happened to enter the room where Liza was hanging upside down on gym rings. Alexandra Alexeyevna rushed to save the child from falling, but I calmed her and assured her that my children were like monkeys and that it was quite natural for them to hang upside down. And it was then that we started talking about communal apartments. "That you can't leave an egg, an onion, or a frying pan full of cutlets in the kitchen—that's to be expected. But to have peace and quiet and no scandals—that was in the good old days," I said. "Well, we have refrigerators in the kitchen," said Alexandra Alexeyevna. "Each person can take anything he likes from any refrigerator and eat it, without asking the owner's permission. If the owner arrives and everything has been eaten up and there's nothing more to eat, he knocks at all the doors, and people bring out some food." I thought to myself that Alexandra Alexeyvna was a little pretentious. But I was soon persuaded on the basis of my own experience that what Alexandra Alexeyevna had told me wasn't a fairy tale. The Belgovskys had a dinner party. There were two guests—Boris Lvovich Astaurov and myself.

I was washing my hands in the bathroom when I developed a nosebleed. One of their neighbors, a woman, saw me and cried, "Just a second! I'll bring a chair and a newspaper. You sit down, lean your head back, press the wet newspaper to your nose and forehead, and it will pass in a moment." It was no sooner said than done. My nose stopped bleeding. I went to the kitchen. There was another neighbor woman there. I asked her where the Belgovskys' trash container was, because I needed to throw away the newspaper. "We have the same trash container for everyone," she said, "and each person watches to see that it doesn't get too full." And the container had to be emptied into a trash can in the yard. The apartment was on the fifth floor; there was no elevator; and the stairs were steep.

Or another example of that apartment building: I was in Moscow, where I'd been delayed and had run out of money. I called Alexandra Alexeyevna and asked her to loan me 200 rubles. That was before the reform.* "Since you don't have time tomorrow to come to our office, just go to our apartment. There's always somebody home. They'll let you in. Ring all the bells. The door to our rooms isn't locked. Go on in, and right across from the door is a desk. You'll find the 200 rubles on it." One of the neighbor women let me in. "You're from Leningrad, aren't you?" she asked. "Come in and have breakfast with me." I thanked her, but told her that I couldn't eat because my stomach was bothering me. "Then come in, and I'll treat you."

A miracle? No miracle at all. In 1934, when the Academy of Sciences was moved from Leningrad to Moscow, Vavilov had arranged to have that very apartment on Pyatnitskaya Street given to his co-workers at the Institute of Genetics. Now more than twenty years had passed, and they were all still living together, and everything that Alexandra Alexeyevna had told me was absolutely true. So the inhabitants of this

* Under Khrushchev there was a currency reform. Old rubles were exchanged for new ones at a rate of ten to one.

apartment hadn't ended up together by accident. The spirit of the deceased Vavilov hovered over them. Traitors had long since lived in separate apartments.

The inhabitants of the second legendary communal apartment found themselves together by accident. The inhabitants of the former luxury apartments in the house on the tenth line of Vasilyevsky Island had their per-person living space reduced. The new people who moved in, unlike the residents of our apartment, were neither expropriators nor members of families of those who, responding to Lenin's call, plundered what had been stolen by bourgeoisie while exploiting the proletariat. They were simple people installed by the ruling force, who took part of the living space from an old aristocratic family. A communal apartment resulted.

I made the acquaintance of an aristocratic family—descendants of a famous Italian military commander. Elizaveta Iosifovna gave English lessons at home. Margarita Vladimirovna taught Russian at school. They were both widowed, and raising their utterly charming granddaughter and daughter, Marina, who was in grade school. Though they had English, Swedish, and Russian ancestry, all three of them were the very image of beautiful Italians. Neither English nor Swedish nor Russian genes showed through their pitch black hair and eyes. Margarita Vladimirovna's features were just a bit heavy. The beauty of the grandmother and granddaughter was impeccable.

The events that changed the family's living conditions occurred in the German Democratic Republic (GDR). Walter Ulbricht, then the general secretary of the East German Communist party, and his wife asked their daughter Beata what she would like as a birthday gift. The willful child wished to study in the Soviet Union. Her wish was law. The choice of an educational institution for the German princess fell on the boarding school whose inspector was Margarita Vladimirovna, and the choice of a family with whom the princess should spend Sundays and holidays fell on Margarita Vladimirovna's. Three generations of beautiful Italian females found themselves at the center of the construction of a Potemkin village. A representative of the Central Committee came from Moscow to give instructions to the Regional Party Committee as to how the princess's life was to be fitted out. The house and apartment were renovated a little. No, that wouldn't do at all. Margarita Vladimirovna was offered a three-room apartment: two rooms for the three women and one for the princess. I called the old communal apartment when events were at their thickest. Someone picked up the receiver, but no one answered the call. I asked for Margarita Vladimirovna. Wailing burst out at the other end of the line. All that I could make out was, "They just moved." This is how I learned that at some communal apartments, deep affection could develop among the inhabitants.

With that I will allow myself to conclude my description of communal apartments. I have to add a few words about one of their inhabitants.

Marina was no good as a resident of a Potemkin village. "What does your *History of the Party* say about the GDR?" asked Beata.

"Eighty percent of the peasants have been collectivized," Marina answered.

"That's nonsense. Don't dare to answer that at the examination."

"But how do you know? Maybe it's the truth," said Marina.

Beata called her papa. He confirmed that there were only a few test collective farms in the GDR. "So are you still going to answer at the exam that eighty percent of the farms in GDR have been collectivized?" she asked Marina.

"Yes, I will."

"But that's not true."

"But the test isn't to find out the truth—it's to see whether I know what the textbook says." That was at the end of the 1950s.

In the summer of 1959 Marina visited Beata and her parents at Levanda. Once upon a time, Levanda had been Nicholas II's summer residence. Now Khrushchev was living there. We learned a few things about the people who were ruling us from Marina, who said that if she was invited back for the next year, she wouldn't go for anything in the world. There was barbarous luxury—such as barrels of wine standing on the lawn with decorative ladles floating in them—uninterrupted security checks, a thousand prohibitions, and terror that information about how the people's servants lived might become public knowledge. Marina gave us marvelous descriptions of daily life among the elite. She rather liked Khrushchev. Their acquaintance took place in the water and under circumstances that nearly ended with Marina's demise. The beach was covered with Turkoman carpets that went right out into the water. Anchored rafts gave the swimmers an opportunity to rest. Marina was sitting on a raft. It suddenly lifted and a mighty lever catapulted Marina into the sky and then the sea. When she fell she hit her face on a rock. It was a miracle that her face wasn't injured. Only two of her front teeth, two white lambs, as in King Solomon's "Song of Songs," were knocked out. Marina had been catapulted by Khrushchev when he hauled his incredibly corpulent self up onto the raft. And the teeth, biblical in their beauty, were immediately replaced by new ones that were indistinguishable from the original ones.

Hunting is the most royal of pastimes. Khrushchev was a good shot. Standing in a thicket of bushes and trees, a servant tossed white clay pigeons into the air. Khrushchev hit twelve of them in a row. "Leave it at that, Nikita Sergeyevich," his son-in-law Adzhubey said. "You'll miss the next one." "We're not fatalists," said Khrushchev. A young man with a notebook who was standing near Khrushchev, at the ready (those young men were called "companions"), jotted down: "We're not fatalists." When something went wrong with Ulbricht's rifle, Khrushchev said: "These fascists always screw things up." Marina didn't say whether the companion noted down that phrase as well.

For my husband and me, two homeless geneticists from Moscow, a tiny room in a communal apartment in a wonderful building on Maklin Prospect was at that time God's green acres. We were full of bright hopes. My husband, Kirpichnikov, a great specialist in the genetics of fish, had easily received the position of head of the laboratory of genetics and selection of fish at the Fish Institute in Leningrad. That institute had grown out of a tiny section of the Institute of Experimental Agronomy. The section was founded at the very beginning of the 1920s by my father and for many years consisted of a single employee: my father. Things at work were coming along very nicely for Kirpichnikov.

I, on the other hand, was counting firmly on Lobashev's promise to open the doors of Leningrad University wide to me. But Lobashev did not take me back into the department. I ended up as a half-time assistant professor at the Herzen Pedagogical Institute and lectured on Darwinism at the branch where elementary school teachers were trained. Nevertheless, my flies were held in honor at Leningrad University's Department of Genetics. The famous Kashira-6, which had survived the war, was among them. But I couldn't be bothered with flies. I couldn't be bothered with flies! Feminists of the world, note this! I couldn't be bothered with flies. I was bringing children into the world.

In the summer of 1947, Kirpichnikov and I went to the state fish farm, an experimental base of the Fish Institute, located not far from the city of Valdai and near the village of Yazhelbitsy, which is equidistant from Leningrad and Moscow. For four years in a row, from 1947 to 1950, we had gone to the fish ponds amid the meadows and forests of the Valdai Hills. From that moment, when, in the eighth month of my pregnancy, I for the first time stepped on state farm soil and until

September 1950, when I left Valdai forever, I was busy with child-bearing, clucking over my tiny brood, and learning about the life of the people.

The state farm where carp is raised for market was surrounded by collective farms. Serfdom grew firmer before our eyes. The Soviet system's organizational inventiveness is limitless. It is aimed at keeping tighter check on the people so that it will be more convenient to plunder them. The dismantling of farmsteads, the merging of small collective farms to produce big ones, the limiting of private plots—one decree followed another. We had occasion to observe the conduct of the chairman of an enlarged collective farm before and after it was enlarged. His behavior revealed an absolute scoundrel.

Aunty Nyusha, an individual peasant farmer, lived near us with her impoverished friend Aunty Sasha. By some miracle Aunty Nyusha had avoided collectivization. She sewed clothes that the collective farmers ordered from her, and had a pig and a small, marvelously well-kept-up kitchen garden. Aunty Sasha took care of the pig and the kitchen garden. It was obvious that the piglet had found a place under the sun, though violating the laws for the existence of individual farming. It never left the pigsty. But its oinking, hearty enough for a pig of grandiose size, rang out triumphantly, providing food for denunciations. Evidently Aunty Nyusha wasn't afraid of denunciations. Everyone knew that in addition to sewing, she used hired labor for raising "illegal commodities"—her only piglet. Aunty Nyusha obviously made money.

The chairman of the collective farm was forever at her door, requesting or perhaps extorting money for vodka. Have you ever seen a person with delirium tremens walking down the road? On more than one occasion I saw the chairman having difficulty stepping across nonexistent puddles. A hallucinatory spectacle.

The poverty of the surrounding villages was flagrant. I had written a book about my father's travels around the lakes of Siberia and Central Asia describing how difficult his travels had been. But with a certain slyness I thought to myself that a trip to Lake Balkhash at the beginning of the century, thousands of kilometers from the nearest resort center, was actually much easier than my feats of derring-do. You couldn't buy a thing at the collective farm shop. You had to bring absolutely everything from Leningrad—groats, sugar, and butter. It was the ration system. It was impossible to lay in provisions. You could get milk in the country, but only with difficulty. There was no point in even thinking about meat. Once in a while it turned up by accident.

On June 23, 1947, we set out so that I could have my child at the Leningrad Clinic. In order to get from Valdai to Leningrad, to cover the 200 miles that separate the two places, one needs to travel for at least twelve hours. The state farm truck takes you to Valdai. By train you get to the big train station at Bologoye. All trains that go from Moscow to Leningrad stop at Bologoye. It takes the passenger train six hours to crawl to Leningrad. Labor began on the train, when we were going from Valdai to Bologoye. At Bologoye I informed my husband of the approach of the blessed event. "Hang on until Leningrad," was his reaction, one quite shameful for a biologist. To slow down or speed up the appearance of a human being in the world is only in God's power, and the willpower of the mother has nothing to do with anything.

We went to the railway hospital and, less than an hour later, there appeared a lovely creature with white hair and gray eyes quite like those of the commonfolk, and she cried just as she was supposed to, no more and no less. The happy father set off for Leningrad to get baby things.

The poverty of the maternity section of the railway hospital at Bologoye defies description. I knew that the hospital's unantiseptic conditions threatened me with death. But I couldn't do anything.

Kirpichnikov soon brought me to Leningrad in a condition that did not offer hope of a favorable outcome. Yet septic infection and mastitis did not take me to the grave. Penicillin, which had just appeared then, and the surgeon's scalpel, delivered me from the claws of death. But my life hung on a slender thread. Grandmother was already coming to get me in her black coat and black hat, and in my delirium, I was ready to follow her.

In the summer of 1948, we went to Yazhelbitsy. Four of us traveled together. Liza was the third. The fourth was our nanny, Maria Nikolayevna. I was pregnant again and was very glad about the approaching addition to our family. As an assistant professor of the Pedagogical Institute I would have a maternity leave, but the term did not begin yet. Though I was close to my due date, they did not spare me at the institute. My "pedagogical load," as that enormous number of hours that a lecturer is supposed to work is called, included summer courses for teachers raising their qualifications, so I had to leave my hearth in Yazhelbitsy. I lived in Leningrad alone and had great difficulty making it up the steep stairs to my apartment. And the lecture hall was on the fourth floor of an old building. The architecture was just fine, but there was no hint of an elevator.

As my time approached, I went back to Yazhelbitsy to gather up Liza and Maria Nikolayevna and take them to Leningrad, where I could give birth in a clean hospital. But I didn't have time to take them to Leningrad. Masha was born in Valdai—with the greatest of ease again—on August 20, 1948. In contrast to the railway hospital, my husband and I, together with Maria Nicolayevna and a young midwife, created in a village hut such a perfectly sanitary environment that I managed to escape both sepsis and mastitis.

Liza was thirteen months old. One morning, when she was lifted up to the carriage where new-born Masha was sleeping, she exclaimed: "Baby!" The greatest amazement could be heard in her voice.

We had every reason to be satisfied with our lives: My husband and I were both working, the tiny room on Maklina was exchanged for two luxury rooms; the superb Maria Nikolayevna was with us; we had two healthy, charming daughters.

But our happiness was not to be. Both of us, my husband and I, and along with us my father and many of our friends and acquaintances, many of the people whom we loved and who loved us, became the victims of an unprecedented ideological slaughter.

# *The Negative Equivalent of Fearlessness*

In 1948, instruction and research in genetics ceased. The purge occurred at the historical August Session of the Lenin All-Union Academy of Agricultural Sciences (LAAAS).

In 1723, Jonathan Swift described Gulliver's visit to Laputa and to the academy in Lagado. Swift has earned the reputation of a wicked satirist. If Swift had described the USSR Academy of Sciences Institute of Genetics and the Department of Genetics and Selection at Leningrad University for the period beginning with the time of the August Session of the LAAAS in 1948, and if he rejected fantastic allegory as a device, he would have earned the same estimation. The energetic work of laboratories under the guidance of Lysenko, Turbin, Stoletov, Babadzhanyan, M. M. Lebedev, and many others who had hastened to take up Lysenko's banner, along with the stenographic record of the August Session of the LAAAS, produce the deceptive impression of a satire written by an unrestrained fanatic.

The parallels between the projects that Swift depicts and the practical recommendations of the Lysenkoists are amazing. The methodologies coincide completely. Vice isn't a vice as long as it's secret. Shamelessness, or more accurately, shame that has been overcome, is the negative equivalent of fearlessness, i.e., fear that has been overcome. Shamelessness is a variety of fearlessness, the ecstasy produced by one's boldness in the face of virtue and the law. I purposely use the word *ecstasy* to remind you of Pushkin's verse: "There is an ecstasy amid the furious fight."

The stenographic record of the August Session of the LAAAS has been translated into English. The cardboard cover of the English edition, published in 1949 by the Foreign Languages Publishing House in Moscow, reads, in bold letters: *The Situation in Biological Science, Proceedings of the Lenin Academy of Agricultural Sciences of the USSR Session July 31–August 7, 1948, Verbatim Report.** The translation really is verbatim, but the original does not contain the whole truth. It doesn't include Rapoport's remark. He interrupted Prezent, saying, "You're lying, you stinking jackal."

I should say a few words about Rapoport. Courage is the quintessence of that amazing man. He spent the entire war at the front, taking part in the most dangerous diversionary operations. His efficiency and valor have been discussed in print. As a front-line soldier he entered the party, which added to its ranks a large group of members of the intelligentsia who became party members during the war. They did not get the party membership card for the sake of the privileges that it granted in peacetime, but, rather, at the risk of death: The Germans killed captured Communists.

One needed insane courage to expose the all-powerful Prezent publicly as a liar. Wartime honors did not protect Rapoport from punishment. Immediately after the plenary session he was expelled from the institute where, after the war, he had continued the experiments that Koltsov had begun and which had brought him, Rapoport, worldwide fame. He was also expelled from the party. At the end of the 1950s, under Khrushchev, when Dubinin was organizing a genetics laboratory under the aegis of physics, he agreed to take on Rapoport on the condition that Rapoport undertake to have himself reinstated in the party. Rapoport refused. One needs to know the mores and customs that reign in the Soviet Union's ruling party, the conditions for acceptance in the party, the conditions for expulsion and reinstatement, in order to understand that this refusal was the most extreme, the boldest act of his life.

The transcript of the LAAAS plenary session makes it easier to compare the two academies—the one described by Swift and the LAAAS. Let us begin with methodology.

In Swift's Laputa there is an Academy in Lagado of project designers. There the professors contrive new rules and methods for agriculture. As a result, "all the fruits of the earth shall come to maturity at whatever season they think to choose, and increase an hundredfold more than

* Pages given in parentheses after each quotation refer to this issue.

they do at present. . . ."* There are innumerable other happy proposals: "The only inconvenience is that none of these projects has yet been brought to perfection, and in the meantime the whole country lies miserably wasted. . . ." The academicians, however, "instead of being discouraged are fifty times more violently bent upon prosecuting their schemes, driven equally on by hope and despair." Those who do not wish to follow the pernicious recommendations of the project designers "are looked on with an eye of contempt and ill-will, as enemies to art, ignorant, and ill commonwealth's men, preferring their own ease and sloth before the general improvement of their country" (pp. 175–76). It wasn't real help for agriculture or the well-being of the nation that were important to the rulers of Laputa, but rather the struggle against the bearers of a hostile ideology.

The August Session of the LAAAS was pursued "under the banner of Michurinism." The late gardener Michurin was Lenin's protégé. The most pitiful and obscure of his principles, and Lysenko's obscurant "teaching," have a remote resemblance. In accordance with the general Soviet rule to bestow infallibility on some of those who are dead, Lysenko gave Michurin's name to his own creation. He thus stressed his patriotic devotion to Russia's great scientific heritage and facilitated the gain of governmental support.

We turn to the "Verbatim Report." The strength and significance of the Michurinist biology lies in how it smashes hostile ideology. At the August Session of LAAAS the great ideologist of Lysenkoism, the great Laputian of the twentieth century, was Turbin. "You want to change an organism's inherited traits," spoke Turbin. "Change the manner of its feeding. Graft a tomato plant that yields red fruit to one that gives yellow fruit. The grafted plant is now taking nourishment with the help of the roots from the plant that bears yellow fruit. Its fruit is yellow, and plants bearing yellow fruit will grow from its seeds. Academician Lysenko," said Turbin, "does not intend to use these graftings to produce new types of cultured plants. He will use these facts in the struggle with enemy ideology. That is the great meaning of the experiments which have been described. If feed is capable of changing an organism's inherited traits, the theory of the gene ought to be abandoned" (p. 480).

I will note parenthetically that a red tomato plant, whether you graft it to a yellow one or allow it to feed itself through its own roots, yields

* Jonathan Swift, *Gulliver's Travels and Other Writings* (New York, London, Toronto: Bantam, 1971). Pages given in parentheses after quotations refer to this edition.

fruit, and that fruit is red. Among thousands of grafted plants there will occasionally be one with yellow fruit, and the same thing will also happen among ungrafted plants, and the incidence of occurrence of exceptional fruit does not depend in the least on their nourishment. My yellow flies were also born from bronze ones, and that had absolutely nothing at all to do with fodder.

Let us move on to practical measures for increasing harvests. We begin with methods of working the soil. Gulliver says: "In another apartment I was highly pleased with a projector who had found a device of plowing the ground with hogs, to save the charges of plows, cattle, and labor" (p. 178). In the Verbatim Report, G. P. Vysokos, the director of the Siberian Research Institute of Grain Husbandry, states: "In 1942 Academician Lysenko made a momentous scientific discovery—namely, that winter wheat can overwinter in the steppe part of Siberia sown in the entirely unploughed stubble of spring crops. This year the Ministry of Agriculture, taking into consideration the favorable results of our experiments, included in the collective farmers' plan in the Omsk Region the sowing of winter wheat in stubble on an area of several thousand hectares" (pp. 205–206).

I'll explain. There is spring wheat and winter wheat. They are different kinds of wheat. Spring wheat is sown in the spring and yields a harvest at the end of the summer. Winter wheat is sown in the fall. The young sprouts—green grass—spend the winter under the snow and the following summer, winter wheat yields a harvest. No kind of winter wheat can survive in the conditions of a Siberian winter with its spells of −40°C. temperatures. Lysenko's proposal amounted to a kind of reverse sowing. He proposed that spring wheat be grown, harvested, and the unplowed field planted with a variety of winter wheat. The remains of the sown spring wheat—the stubble—would protect the sprouts from freezing. And you wouldn't have to plow. Just as in Logado.

Now for a moment put yourself in the shoes of the Soviet chairman of the collective farm who is ordered by the plan to sow in stubble. What does he do? He sows and reaps magar. I know that from firsthand information. Magar is a crop whose seeds resemble millet but differ from it by their bitter taste. Because of its inferior quality, the government does not take it from collective farms; the farm can keep it. Siloed, it is used as green food for farm animals. Prisoners are fed with porridge made from it. Chairmen of collective farms prefer magar to any other crop because it is for sure better to have something than nothing. What did the collective farmers of the Leningrad region, with its climate absolutely unfit for corn, do when they were ordered to deliver a rich harvest of corn? They delivered a rich harvest indeed. They raised potatoes, took them to the Ukraine, sold them, bought corn, and turned it in as though it had been raised on their own native soil. I'm certain that it was exactly in the same way that the cameramen shot the newsreel that shows fields of corn, if one believes the announcer's heartfelt words, in the Leningrad region. That film is run before every feature, every day, in every theater. It was only in a movie theater that I ever saw those corpulent fields of Leningrad corn.

I return to Lagado and the methods used there for creating highly productive breeds of agricultural animals. A scientist assures Gulliver that spiders are much more useful than silkworms, because they not only spin but weave. Gulliver says: ". . . Whereof I was fully convinced when he showed me a vast number of flies most beautifully colored, wherewith he fed his spiders, assuring us that the webs would take a tincture from them; and as he had them of all hues, he hoped to fit everybody's fancy, as soon as he could find proper food for the flies, of certain gums, oils, and other glutinous matter, to give a strength and consistence to the threads" (p. 179). It was exactly the same at the USSR Academy of Sciences during the years of Lysenko's dominance.

Khilya Fayvalovich Kushner, formerly one of Vavilov's doctoral students and later a permanent co-worker and proselyte of Lysenko's, worked with chickens there. Just before the war, he worked out a means of raising enormous baby chicks in a very short time. His chicks were hybrids. He crossed purebred hens of one stock with purebred roosters of another. The new stock was not intended to produce progeny, but it served the aims of gastronomy splendidly. Having adopted Lysenko's banner, Kushner didn't give up his hybridization of chickens. Ideology changed and so did the methods for achieving hybrids. In place of sexual hybridization came vegetative hybridization. Now nature was to be changed with the help of feed. Before the chick hatched out of the egg, Kushner removed the egg white and replaced it with the white of an egg laid by a hen of a different stock. A chick hatched from an egg in which the yolk, i.e., the embryo and one part of the nutrient material, and the white, another part of the nutrient material, belonged to different stocks—that's precisely what a vegetative hybrid is.

The matter has nothing to do with economically valuable aims. It's a question of principle. A trait transmitted to the offspring via the egg white is sheer ornamentation, i.e., coloration. Yet the egg white of a white chicken was replaced by the egg white from an egg laid by a black hen. Since in Kushner's point of view food was the carrier of an organism's hereditary traits, it was expected that the "hybrid" chick would inherit the black coloration from the black hen. Nature didn't yield immediately to the Laputians, nor did it to the Lysenkoists either. The black hens kept them waiting a while. But finally a black chick was hatched from an egg laid by a white hen who had had no contact with any rooster other than a white one of the same stock. The egg white had been transfused. If Kushner had had a control group, he would have realized that he had before him a mutant chick. Mutation process and selection, carried out by the researcher, had created what the hapless laborer attributed to his own efforts. Black mutants turn up among purebred white chickens according to the same laws that yellow flies turn up among normal bronze-colored ones, and according to the same laws by which a cutting of a red tomato plant grafted to a yellow tomato yielded yellow fruit.

An outstanding American geneticist and a great specialist in selection among chickens, Michael Lerner, used the pages of an international journal to invite Kushner to bring his lab assistants and conduct the same work using Lerner's breed of chickens. According to his observations, black mutants were a great rarity among his white leghorns.

I met Kushner, and since I knew of the invitation, asked him: "Are you going to America? Did you agree?" "It doesn't make any difference whether I agree or not, the decision is the Central Committee's, not mine" was the answer. Kushner did not have to go to the United States.

The ace of trumps for *Michurinist* biology was its assertion that if a chiffchaff (a warbler) were to stuff itself on fuzzy caterpillars by mistake, it would give birth to a baby cuckoo. The birth of one species from another would occur as the result of a change in the diet. The students at Riga University, hearing about the cuckoo, made a preparation of a dissected cuckoo—you could see the ovaries with nearly mature eggs. They sent the preparation to Lysenko along with a tag that read: "What Morganist-Mendelist-Weismannist stuck cuckoo eggs in this cuckoo?" The students hadn't taken the bait. They realized that a cuckoo born of a chiffchaff fits in very neatly with the most advanced teaching in the world, but not with the established fact that cuckoos are born by cuckoos.

But I made a fool of myself with the cuckoo. When I heard that a baby cuckoo was born of a chiffchaff instead of a cuckoo, and that this was a model for oats producing a weed, wild oats, I decided that I was hearing a joke. That was in 1951, at the USSR Academy of Sciences

Institute of Botany. People were standing by tables there in the vestibule—a book store had organized a display of new books. Two elderly botanists were speaking animatedly. (I didn't know them then, but made their acquaintance later. One of them was Alexander Innokentyevich Tolmachev.) One of them was telling the other about the cuckoo. The same story that I just told—about a chiffchaff giving birth to a cuckoo. I lent an ear. They noticed and stepped away—it was the Stalin era. A while later, at Leon Abgarovich Orbeli's dacha (he was a great physiologist, who had been removed from his position as deputy president of the Academy of Sciences because he supported genetic research), I told him that jokes were predictable, that the one about the cuckoo could have been foreseen. He looked at me sadly and didn't say a thing. On another occasion he told about Lysenko's presentation, when the illiterate agronomist, the graduate of the Uman Institute of Agriculture, was expounding his notions of general parasitology. The huge hall wasn't large enough for all those who wanted to get in, and loudspeakers were set up in all the corridors of the building. But all the academicians, of course, were in the hall. Leon Abgarovich especially remembered how Evgeny Nikanorovich Pavlovsky, a full member of two academies—the Academy of Sciences and the Academy of Medical Sciences—a zoologist and parasitologist, and the director of the USSR Academy of Sciences Institute of Zoology, applauded.

As for the cuckoo and where it comes from and how it lands in the chiffchaff's nest—even little children know that one. Vitaly Bianki, the ornithologist and children's writer, described such an incident in a little book for preschoolers. Darwin writes about the cuckoo in detail in *Origin of Species*, and there is a special book devoted to the topic of nest parasitism in birds, and the genetic and ecological bases for the cuckoo's choice of a nest for its progeny have been studied.

But for Lysenko science wasn't a ukase. He said that a chiffchaff gave birth to a cuckoo if the chiffchaff itself had been raised on rations that were unusual for it—if it was fed fuzzy caterpillars by its parents. I guess you don't think that research was done about the cuckoo and chiffchaff. No research was needed. Lysenko had spoken, and that was enough.

E. N. Pavlovsky took on academician I. I. Schmalhausen at his institute when the latter, the director of the USSR Academy of Sciences Institute of Evolutionary Morphology of Animals and the head of the Department of Darwinism at Moscow University, had lost his job; Schmalhausen remained on the staff of that institute until the end of his days, in 1963, fifteen years after the August Session of the LAAAS. They still would not allow him to teach. In order to take on the anathematized academician in 1951, Pavlovsky applauded Lysenko.

A cuckoo's eggs resemble the eggs of those birds to whom a given variety of cuckoo entrusts the care of its offspring. Otherwise the foster parents throw the eggs out. Similarity is called mimicry. Pavlovsky's applause was related precisely to that category of phenomena. He, of course, would have preferred to withhold applause, but he was too much on view.

And now the August Session of the LAAAS. It was separated from Lagado by two-and-a-quarter centuries. On opposite sides of an ocean, the ocean of time, there are Swift's fantasy country and a real country that inevitably and predictably gave rise to Lysenkoism. I will sight the speech by V. A. Shaumyan from the fifth meeting. He is a director of the Kostroma state cattle-breeding station. Before getting down to the business of praising the Kostroma breed of cattle, he called for the destruction of the enemy.

Unmasking the hostile ideology lurking behind the chromosomal theory of heredity was far more important than raising crop capacity or the productivity of farm animals. A Soviet person had to realize that the enemy wasn't dozing. Vigilance was more important than a full stomach. It had to be impressed upon the Soviet citizen that the people responsible for his hungry, empty life were intellectuals: Mendelists-Morganists-Weismannists. They were purposely lowering the harvests and, by working for the enemy, they were forcing the wise leadership to raise the Soviet Union's military potential. The August Session was carried out precisely under the slogan: "Vigilance Reinforced by State Ideology."

Shaumyan's speech said: "It is high time to realize that today our Morganist-Mendelists are in effect making common cause with, and objectively—and in the case of some even subjectively—are forming a bloc with the international reactionary force of the bourgeois apologists not only of the immutability of genes but also of the immutability of the capitalist system" (LAAAS Report, p. 252). After that comes the praise of the cattle breed: "What is our great success due to? The first and basic condition of success in the formation of breeds is abundant and skillful feeding. . . . 'Fodder and feeding,' M. F. Ivanov used to say, 'exert much greater influence on the organism of an animal than breed and origin' " (pp. 253–54).

Shaumyan concludes his speech in a manner so typical for Laputa that I cannot resist reproducing that passage in its entirety:

"The formal geneticists have done us tremendous harm; they are trying to disarm millions of the foremost agriculturists who with utmost devotion work indefatigably and creatively day and night to increase the wealth of our country. We must now finally and irrevocably take this unscientific and reactionary theory down from its pedestal. Unless we intensify our 'external action' upon the minds of our opponents and create for them the 'proper environmental conditions,' we shall of course be unable to remake them. I am fully convinced that if we guide ourselves by the only correct theory, the theory of Marx, Engels, Lenin, and Stalin, and take advantage of the tremendous care and attention which the genius of Stalin bestows upon men of Science, we shall undoubtedly be able to cope with this task" (p. 262).

If you take into consideration that "the proper environmental conditions" which in Shaumyan's opinion ought to be created for the representatives of "formalist genetics" are the Gulag Archipelago, the Swiftian satire will grow pale right before your eyes.

The stenographic report of the August Session of the LAAAS takes up 631 pages. Fifty speakers called for a massacre. The opening speech and closing remarks by Lysenko are a drum roll, the ritual music of public executions. Only in the closing remarks, pretending to be answering an anonymous question that had been handed to him, did he inform the session's participants that his report had been approved by Stalin. No, not Stalin alone, but by the Central Committee of the party. The audience's reaction was: "Stormy applause. Ovation. All rise." And the concluding words of Lysenko's concluding remarks had the same effect: "Glory to the great friend and protagonist of science, our leader and teacher, Comrade Stalin!" The performance didn't end with that. Three of the eight defenders of real genetics delivered penitential speeches. A resolution was accepted and a letter written to Stalin, the great leader and teacher. No more voting was needed. The letter ends with the words: "Glory to the great Stalin, the leader of the people and coryphaeus of progressive science!" The audience reaction: "Stormy, prolonged, and mounting applause and cheers. All rise."

Don't forget that Swift's satire is satire. The Verbatim Report was issued with all seriousness. It is a model to be imitated.

# *At the Edge of the Abyss*

We were still in Yazhelbitsy occupied with my husband's fish and our own hatches when we learned from the newspapers that genetics had been crushed. At first it seemed to be just the usual discussion, like the ones that had been played out earlier upon instructions from on high—there had been such discussions in 1936 and again in 1939. It became even clearer that an end had come to genetics when Lysenko announced that his report had been approved by the Central Committee of the party—or what amounted to the same thing—Stalin, and when *Pravda* had published in full Lysenko's closing remarks.

After we returned to Leningrad I found out what had transpired at the Pedogogical Institute's Department of Zoology and Darwinism, where I was an assistant professor. The entire staff, including myself, was fired, along with the chairman, Yuri Ivanovich Polyansky, a brilliant scientist and lecturer. Lobashev, by then not only the head of the Leningrad University's Department of Animal Genetics, but the Dean of the university's Biology division as well, was ousted from both positions by Turbin, Lysenko's trusty satrap. The extremely rich fund of fruit fly strains kept at the Genetics Department was destroyed. My flies were destroyed as well. Kashira-6 was no more.

The art of ruling includes the ability to break the law on a legal basis. They didn't have the right to fire me. It was claimed I was fired because I held more than one position; not long before, a law had been issued forbidding the holding of multiple positions. I had no job other than the one at the Pedagogical Institute's Department of Zoology and Darwinism. But since I was part-time, it was assumed I held multiple jobs. I could have protested the administration's decision and I probably would have been successful. But I had no intention of making a case. If an attempt to protest my firing had been successful, I would have found myself forced to teach Michurinist biology under the command of an "appropriate" person when Yuri Ivanovich Polyansky, a brilliant scientist and lecturer, became inappropriate. I submitted a request that the formulation of the reasons for my dismissal be changed and asked to be released from my duties because of family circumstances. I was released.

I was no longer a geneticist now. The resolution of the August Session of LAAAS calling for the eradication of servants of imperialism, parasitical members of the bourgeoisie, and racist Mendelists-Morganists-Weismannists, multiplied a thousand times by resolutions having to do with biology of the Academy of Sciences, by resolutions of all research institutes, all universities and pedagogical institutes, all publishing houses and museums was carried out irreproachably. I was subject to eradication. No institution had the right to offer me work not in my field. The academic degree I had and the position had to correspond to each other. My rights as a Ph.D. were protected sacredly by the work legislation of the Soviet state. I was doomed to starvation on legal bases.

But I was a daughter, and my father, elected two years earlier as a full member of USSR Academy of Sciences Division of Geography, was not only a geographer, but the greatest authority on fish that the country had known. My husband, consequently, was his son-in-law. The Fish Institute where he worked was my father's brainchild. Kirpichnikov wasn't expelled from the institute and room was made for me there.

I have already spoken about and presented persuasive examples of my father's violent opposition to protectionism and nepotism, those invariable concomitants to and indicators of the moral degeneration of a society. If Kirpichnikov had been expelled, if I had been denied the position of staff member, Father would not have protested. He would have supported us financially. But the matter was transpiring in a society where protectionism had long since entered into the flesh and blood of social morality and not a single living soul believed in the existence of all those Tolstoyan or, if you want, Christian virtues. The institute acted in accord with Father's nonexistent desires, they protected me. The same mechanism thanks to which I had been retained in graduate school and later accepted in a doctoral program had worked here. My position, one long left vacant and forgotten by God and man, paid 550 rubles a month (that was before the monetary reform which meant the ten-time reduction of the ruble's value).

At the end of the war, privileges were showered on scientists as though from a horn of plenty. The pay for scientific staff members, both junior and especially senior, was raised in 1944 several times. Junior staff assistants were given from 1,200 to 1,600 rubles a month, depending on their experience and the presence or absence of a Candidate of Science degree. On that salary even a person without a family could hardly make ends meet. Senior assistants began receiving from 2,400 to 4,000, depending on experience and the absence or presence of a Doctor of Science degree. A person without a Candidate degree could not receive a position as a senior assistant. This Doctor of Science degree, which does not exist in the USA, in the Soviet Union gives one the right to occupy the position of professor and to receive the passionately desired 4,000 rubles. Earning that much, one could now make ends meet and even raise a child. People on top looked after the junior and senior assistants. But the plain old staff members were forgotten. Their salaries remained the same as before. Positions went vacant.

I worked almost a year at the Fish Institute. I like to feed people and animals, including flies. I adored feeding fish. Thanks to me, fish began to be fed during the winter season. Until then it had been considered that fish didn't need food in cold weather. I had been occupied with fish husbandry in 1948 as well. In my husband's absence I conducted control nettings at his ponds, directing old fishermen and young auxiliary workers. It was during one of those test nettings that the labor pains began that announced Masha's imminent arrival in the world. Labor pains last several hours, and I had time to finish the test netting. Masha was absolutely certain that children are caught in ponds, where they begin their lives as fry. In 1949, I worked at the ponds as a scientific staff member. I left the institute when the head of the institute's department where I was appointed, a very elderly man, said to me apropos of some insignificant matter—I don't even recall what had happened: "You and I will have to talk in another place." "Another place" could mean only one thing in Stalin's time. He would complain about me either to the party organization or the Special Department or to the Personnel Department. Then and now all these departments are arms of the KGB. I left the service.

It was better for me to sit at home. I decided to become a zoologist, a specialist in the classification of flies. Alexander Alexandrovich Shtakelberg placed the Zoology Museum's collection at my disposal and appointed me a specialist in the tiny flies *Agromyzidae.* The Academy Zoology Institute is obliged not only to describe all the species of animals that populate the USSR but also to draw up the characteristic features of the animals in each group. There had been no specialist in that group in Russia since the creation of the world, while at the same time those little flies have economic significance—in a negative way, of course. Their larvae live in the pulp of leaves. The white threads of their tunnels grow thicker as the larva grows and moves away, looping, from the point where it hatched out of the egg. Each species feeds on the leaves of a specific species of plant. I had to become not just an entomologist, but a botanist as well.

It was the summer of 1950, the last summer in Yazhelbitsy, on the Valdai. The sky in which the thunder clouds of my bitter fate were gathering seemed cloudless to me. My father was supporting me. My husband's salary, of course, was insufficient for the needs of an expanded family. Not long before leaving for Yazhelbitsy I received an invitation from my father to go see him. "Maria Mikhaylovna is wailing and complaining that you're always asking for money. I've been giving you 400 rubles a week and don't ask for any more." My stepmother was caught red-handed. She controlled the finances and had been giving me only 200 rubles.

I really did constantly ask my stepmother for money. I needed to buy firewood, order little wooden beds so that in the winter the children could sleep during the day on the balcony, I needed to sew sleeping bags . . . 200 rubles wasn't even enough for food. I left my father with his mistaken notions about my extortions. To my stepmother I said, "Father has promised to give me 400 from now on." She submitted. Now there was enough to pay Maria Nikolayevna and to buy tickets to go to the Valdai. Father was categorically opposed to my wish to receive a job at the Zoological Institute where he worked, and where I had earlier measured fly wings and was now studying *Agromyzidae.* He felt two members of the same family shouldn't receive such huge salaries. He had enough money to be able to give me 400 rubles a week.

And so arrived my last summer in Yazhelbitsy. There were six of us in 1950—my husband and I, Liza, Masha, Maria Nikolayevna, and Marusya, our maid, who worked for us for three years, since 1949. Maria Nikolayevna was often quite sick and we hired Marusya to help her.

In Yazhelbitsy I rented a wooden hut near Aunty Nyusha's. Toward the end of the summer, from off the Moscow-Leningrad highway and away from the village on a slightly sloping green incline you could see an unusually vivid unbroken carpet of flowers grown by me in the yard outside the hut.

Maria Nikolayevna did not stay long with us in Yazhelbitsy. She left for Krasnoyarsk, where her elder son was serving his term of exile. Of her five children only two survived—two sons. Both were punished by law. The elder evoked the wrath of Nemesis by being captured by the Germans. The younger failed to guard public property; the military base that he managed had been robbed. Maria Nikolayevna's husband had perished under a German artillery attack on the road of life along which the half-dead residents of the blockaded city were evacuated. Both of Maria Nikolayevna's sons were workers. The younger one was preparing to become an engineer. In her name I wrote petitions in defense of both of them, begging for pardon in consideration of their father's services to the Revolution. His party card had been preserved at the Museum of the Revolution (though he himself had very early seen through the meaning of the phrase "dictatorship of the proletariat" and had quietly left the party). We sent books to the younger son.

And for the duration of my vacation, Marusya went to her village, Plenishnik. Every year she harnessed herself to hard work there to help her collective farmer parents. She would gather red bilberries, which grew in abundance in the forests of the Vologda district. The berries would be taken to the collective farm market in Vologda or Leningrad and sold for a high price. According to an agreement we had made, I had to grant Marusya a vacation when the bilberries of the Vologda District ripened.

With the consent of Aunty Nyusha, we replaced Maria Nikolayevna with Aunty Sasha. The pious Aunty Sasha had never been married and she reminded me of my and Sim's Nanny, although there was less severity in Sasha. Aunty Sasha adored Masha, and Masha had turned out to be like my brother Sim. Masha's name Maria was transformed by Aunty Sasha according to its peasant usage into a very soft and caressing-sounding Manka. Aunty Sasha said of Masha as she was washing her in the bathhouse: "You could tear my Manka into pieces and she wouldn't even cheep." I caught flies, hatched *Agromyzidae* from their cocoons, and studied plants.

As long as it was my task to acquaint myself with the flora of the Valdai range, in order to be able to identify the plant by the fly and the fly by the plant, I thought that I might as well work out the subject that had occupied me when I was writing my introduction to the doctoral dissertation: the mutual relations among species of a single genus within a community. If species crowd each other out in an implacable struggle for the sources of existence and for space, then the older communities are different from the young ones. Each genus in a young community will turn out to be represented by several species, while in an older one by only a single species. If you divide the number of genuses by the number of species, the result is a fraction, which the old Russian specialists in forestry called the "coefficient of genus." As the community grows older, the significance of the fraction increases. The number of species diminishes until it becomes identical with the number of genuses. The significance of the coefficient of genus strives for 1:1.

The meadow on the gently sloping incline near our house had not been plowed for a long time. It was covered by rich plant growth. For a comparison with this mature herb system I marked off the young communities. I decided to put off studying them until the next year. The list of species inhabiting that luxuriant meadow comprised a good hundred names. Possessing the coefficient of genus for the old ecosystem that had been formed long ago, and being able to identify species quickly, I counted on being able to finish up my work the next year without any difficulty. I was frolicking at the edge of the abyss.

The gypsies enlivened life greatly. A significant and beneficial shift had occurred in their life. Marvelous purebred cart horses turned up among them. How had this miracle transpired? In the Baltic regions that had just been liberated from the German yoke, collectivization had begun, and the wealthy peasants sold off all their possessions before going into exile. Some of those possessions wound up in the hands of the gypsies. One young gypsy woman with a tiny little girl asked me for some medicine. Their horse had developed a sore from the harness. I gave her some disinfectant lotion. In a fit of gratitude and in spite of my assurances that I was glad to be able to help the horse, she insisted on telling my fortune. She blurted out her prophecy to me: "Your father will soon die, and your husband already has another family and will abandon you. . . ." It all came true.

# *The Past Has Not Been Lost in the Present*

On December 24, 1950, my father died. My family fell apart. There was nothing to live on. I turned to Evgeny Nikanorovich Pavlovsky, the director of the Zoological Institute, and asked him to hire me. I had worked zealously without pay on the Zoological Museum's collection of flies; now I would be happy to conduct tours through the museum—in other words, to have a position with a miserable salary usually held by unqualified people. A scientist with a scientific degree working as a museum guide would be a rather pitiable figure. "You are such a famous woman," Evgeny Nikanorovich told me, "and I'm a sick old man. I can't take you on." "How late that fame has come," I said. Perhaps Shtakelberg would have managed to give me a position. But he was seriously ill. When he learned of my father's death he suffered a heart attack. And when Father lay in an open casket in the Petrovsky Hall of the Academy of Sciences, there was a bunch of fiery cyclamen on his chest—the gift of one of his female admirers (my stepmother invariably called them celebrity worshipers). Father was buried in the Volkovo cemetery, in the literary section, not far from Turgenev's grave, right next to Miklukha-Maklay's.

I collapsed into insensibility. It wasn't consciousness that abandoned me, but the ability to feel anything. I didn't sleep, but I didn't suffer from that in the least. I could eat, but I didn't feel hunger, satiation, or the slightest desire to eat and sleep. I felt neither heat nor cold. I neglected my flies and botany.

The only thing that could return me to life was to enter into my father's life, his past life. But his archive was in my stepmother's disposition, which meant that it did not exist for me; instead she transferred it to the Academy of Sciences. All traces of the existence of his first family were excised. She wasn't a bad person. Jealousy and monstrous avarice, her vices, were the defective products of a mechanism which, if you deduct the mistakes in construction, was quite decent. It wasn't without reason that Sim spoke of her sympathetically. My regard for her was very different.

I'm going to tell you something now that will undoubtedly surprise you, dear reader. All my life I've done nothing but build idols for myself. One of my idols was my stepmother, Maria Mikhaylovna—Marmikha. My love for her had nothing to do with my estimation of her. Most likely it was an uncontrollable infatuation. I loved the sound of her voice, her looks, her smell—as though one had just opened a bottle of mineral water—her perfume. She was an educated woman and taught a course in the comparative anatomy of vertebrates at the same Herzen Pedagogical Institute with which my fate was also linked. She came from an ancient noble family. Her mother was born a Lapteva. Marmikha was educated at the Institute for Noble Girls at Odessa. To judge by his portrait, her father, a captain of the Black Sea fleet, was exceptionally handsome. The gentleness and femininity of her looks and her ability to comport herself enchanted me. Cleanliness and order were her natural elements.

I was seven when she appeared. I remember very well that I was happy for Father—his loneliness had come to an end. Her hatred of Sim and me could not moderate my warm feelings. She didn't even try to conceal her attitude, but, on the contrary, demonstrated it with maximum cynicism, even in the presence of outsiders, but always in Father's absence or to Father in our absence. I make that judgment on the basis of much oblique data. For instance, Father's assertion that I refused to study with the foreign language tutors whom he was ready to pay generously. Rumors about my laziness and the depravity in which I indulged, along with rumors about my refusal to receive any education whatsoever, occasionally reached me, sometimes without and sometimes with a reference to the source. This source was Marmikha.

My love for my stepmother, a maximally unreciprocated attachment, laid a very deep imprint on my whole life and opened the door wide for tragedies. In loving her, I became accustomed to loving in spite of a lack of reciprocity, and to this day a declaration of love for me produces a feeling of irritation in me. Sim used to say that I was a fool because I did not appreciate the people who tried to earn my love and were worthy of it. The heavenly angel Simochka was much more realistically inclined. His indifference to life made him dispassionate and wise. He didn't need the biblical commandment, "Make not unto thyself any idols." Making idols was not in his character, which was divorced from life.

My stepmother did not destroy all of the traces of my father's former love. As though she were performing an act of great kindness, she returned my baptismal certificate, my letters, and my letters of congratulation, which from early childhood on I had presented to Father for his birthday, all decorated and written out with great care.

I began to neglect my children and couldn't experience the slightest stirrings of maternal love in my heart, although earlier I had been happy to get up in the middle of the night to look after a child, rejoicing at the chance to fuss over the baby. But that single thing that could return me to life occurred. One day with Liza, on the way to the university Clinic—where we were registered as members of an academician's family, until they threw us out—we met Blyuma Abramovna Valskaya, one of Father's celebrity worshipers and the former head of the Geographical Society's archive. She is the author of many works on the history of geography, and had been fired from the Geographical Society. She said that she was teaching geography at a factory night school. "Go to the Geographical Society's Archive," she said, "and ask for Ignatov's archive. Your father wrote letters to him." I went.

What letters those were! They say that handwriting is the mirror of the soul. Father's small and extremely precise handwriting was that of a hardworking man to whom even the shadow of affectation was alien, who respected the people who would be reading what he had written. That trait—respect for the person, which was so clearly demonstrated in his handwriting, also determined both the style and content of his books and his manner of reading his talks. He did not rely on any oratorical devices and was extremely brief. His listeners were charmed. In his letters to Ignatov, Father described his travels around the Sea of Aral, his life in Kazalinsk, and his work as a fishing inspector on the Syr Darya river and the Sea of Aral. He and Ignatov traveled together in 1898. At the time they were co-authoring a monograph describing the group of western Siberian lakes. The letters contained the epic story of the creation of the first monograph on landscape study. For that book Father and Ignatov received the Geographical Society's small gold medal. The inscription on the medal read: "For useful work."

I never saw any of Father's other medals, but that one he gave to

Aunt Musinka in 1932 when once in a blue moon she had come to visit from Melitopol. That was a hungry year. Food was rationed. The starvation was artificially created by Stalin; grain was being exported. It was the tyrant's punishment for a country that had dared to resist his plan for total collectivization, his plan for reinstating slavery. Gold was pumped out of the population. Jewelry could be turned in at establishments opened especially for that purpose. In return, according to the weight of the items turned in, coupons were given out for special stores where items generally in short supply could be found in abundance. Musinka sold the medal—Father had given it to her for that reason—and shared the coupons with me. I bought a wool sweater and aspirin—you couldn't get them any other way.

At first I wanted to publish Father's letters. It turned out, however, that I didn't have the necessary educational background to do the annotating. And besides, the monstrous limitations that the censors would insist on frightened me. I decided to write a book and to describe not only my father's travels but also those of Ignatov. I had to start with the expedition they had taken together and then describe all the studies they had done separately. "Do you have a contract?" was the question that everyone asked me as soon as they found out I was writing a book. Everyone was amazed that I was writing without a contract. Only once did I ask Maria Mikhaylovna for help in regard to Father's archive. I asked her to give me Ignatov's letters. Maria Mikhaylovna said that they had been turned over to the Academy of Sciences archive. It wasn't without difficulty that I managed to obtain access to them.

Ignatov was two years older than Father. Both men had graduated from Moscow University at the same time. Ignatov was retained at the university for preparation for the rank of professor, as graduate study was called then. For Father the best place that could be found was as a fishery inspector on the Syr-Darya River and the Sea of Aral. In 1898, on assignment from the Geographical Society, they set out to study the lakes of western Siberia. The result of their research was the renaissance of geography as a science. Climatology, geomorphology, and hydrology became elements of landscape study. Finding themselves separated by thousands of miles, the two friends began to correspond with each other. Lev Semyonovich asked Ignatov to send him some books. Ignatov sent them. Expressions of thanks for the books and enthusiastic remarks about them are interspersed with exotic details.

Five cats had been named—four of them bore Kazakh names, and the fifth was called Vaska. "Ulkun has turned out to be an absolute swine, cat of a bitch, who ate my supper every night, and I've dismissed him and taken the black cat Karabulak into my confidence." I cited that sentence in my book. The editors of the Geographic Publishing House excised it. Militant defenders of Communist morality considered love of animals a sign of anti-Soviet inclinations. (Vegetarian cafeterias were abolished in the 1930s. Tolstoyanism was obviously seditious. Behind vegetarianism one sensed a morality that went outside the framework of the class consciousness of the proletariat.) A typist from the secretariat of the university Dean's Office typed the final abridged version of the book, where with a red pen I marked the passages that had been thrown out by the editor. She included the above-mentioned sentence—she liked cats a great deal. She made me a bet that they wouldn't throw it out—the winner would get a bouquet of lilacs. She lost.

After 1898, Berg and Ignatov continued making expeditions, but separately. In 1902, Ignatov died of tuberculosis, a horrible death, on the banks of Lake Shchuchye. Shchuchye is part of the same group of lakes as Borovoye. Forty years later, Lev Semyonovich found himself in Borovoye among the lakes that Ignatov had studied just before his death.

I didn't trust my geographical knowledge a great deal. The geologist Nalivkin, the hydrologist Lvovich, and the climatologist Khromov agreed to edit my book and give their evaluations of it. Their evaluations were positive and their editorial comments were taken into account.

The book was being considered at the Geographical Literature Publishing House. A few chapters of the book were published in the journal *Around the World*. At the journal's editorial office I was shown an evaluation stating that my father's conclusions, which I spoke of as definitive, had long since been repudiated and that the essays shouldn't be published. That evaluation was by a professional geographer, one of Berg's disciples. When I heard the reviewer's name, I was stupefied. It was Gekker—the titular editor of my book. I hadn't seen him once. I'd just spoken to him over the telephone and he declared his love for his teacher, my father, and told me how Lev Semyonovich had helped him in his time: He hired the poor student as a library assistant, paid him money, but never asked him to bring any books. I called Gekker and in a sweet voice asked him if he could advise me about the latest literature on periodic changes in climate, and whether he could give me the names of geographers who had repudiated Father's conclusions about paleoclimatology. "There are no such people," said the reviewer. I already knew that without him. "You don't know of a single one?" I persisted. No, he didn't. "But I know one," I said. "Who?" "You, my dear sir. Don't you think that in view of circumstances that have arisen with regard to your review in the journal *Around the World*, you ought to withdraw from your duties as the titular editor of the book?" He agreed. The Geographical Publishing House asked another one of Lev Semyonovich's students, Markov, the brightest luminary in the geographical firmament, to be the titular editor. He agreed readily.

I brought him the manuscript. He read some of it. At our next meeting it turned out that he had a request to make of me. A book was to be published, *The Pantheon of Russian and Soviet Geographers and Explorers*. He had been asked to write an article entitled "Lev Semyonovich Berg as a Geographer," but he had just written about Lev Semyonovich for an academy edition. Wouldn't I agree to write the article for *The Pantheon?*

He seemed to be doing me a great honor by implying that, to judge by the content of my book, my knowledge was sufficient for such a task. But his request was actually outrageous—a student trying to disassociate himself from the memory of his teacher. Not only that, his proposal meant that there wasn't anyone who could and would write such an article. Besides which, it sounded as though his proposal were a deal. He would be the titular editor for my book if I wrote the article for him. If I had been a geographer, that would have been one thing, but I was a geneticist, far removed from geography. I didn't feel it within my powers to write such an article. My book embraced but a small part of my father's boundless activity, a little slice of his life. Still, I agreed. Nothing in my life, either before or afterwards, cost me as much painful effort as did that article. The author's name that I put on the manuscript was K. K. Markov, but he only corrected it. The article was sent off under two names, as though I had provided the biographical data and he had written about the scientific aspects.

Now the matter was in his hands. I waited for what seemed to me a decent interval and then called him to find out how things were moving along with the editing of my book. I couldn't feel my legs beneath me when I heard his praise. He had done everything that he could. I should drop by to see his secretary to pick up the manuscript. He headed the Department of Geography at Moscow University. His secretary, a true princess, entrusted the manuscript to me, and not until the moment when I became acquainted with the package's contents did I feel my legs under me. With the manuscript, the package contained a letter to the editorial board, in which K. K. Markov declined to be the titular editor because of lack of time.

The editorial office asked Khromov, who agreed to serve as the editor. He had already edited the book earlier, before it was sent to the publisher.

Much time passed. Stalin himself was no longer at the helm, but his shadow remained. My relations with Geographical Publishing House entered into a bribe-extorting phase. The tactics were determined by the struggle against the Arakcheyevian regime raised to the lofty height of de-Stalinization.

Alexei Arakcheyev, Minister of War, Chief of Military Affairs, and Chief of the Department of Senate Councils, all in one person, was a mild prototype of Stalin. He lived at the beginning of the nineteenth century, during the reign of Alexander I. He even undertook the collectivization of agriculture. Georgy Vladimirovich Vernadsky, the son of Vladimir Ivanovich Vernadsky, writes in *A History of Russia* (New York, 1959):

> To make the army self-sufficient, Alexander I promoted "military settlements." General Arakcheyev had to establish them. "A military settlement" was established either by settling a military unit on a plot of land or by turning a community of state peasants into a military camp. Military drill was to be combined with productive work. The whole plan might be called an experiment in military communism. The advantages expected from the new system were that 1) the army would become self-sufficient, both economically and financially; 2) the soldiers would be provided with land and means of subsistence for their old age; 3) the majority of the population would be not liable either to pay taxes or to supply recruits for the army. Whatever might be the value of the military settlements in theory, in practice the system met with opposition from both soldiers and peasants. Each party regarded it at first as a new burden loaded on them by the administration. There were many cases of open revolt, which were ruthlessly quelled by Arakcheyev and his assistants. The system was continued, however, and at Alexander's death some 250,000 soldiers had been settled, or about one-third of the standing army [pp. 215–16].

The words *Arakcheyevian regime* are synonymous with bloody dictatorship. To everyone's amazement, Stalin himself used them in an article published in *Pravda* in 1950. Stalin had participated in a discussion of questions of linguistics. He hurled thunder and lightning at the heads of the unfortunate followers of the insane linguist Marr, whose mad ideas had been canonized. Stalin had apparently grown jealous of Marr. Stalin was the only one who could be permitted to be a source of truth. Stalin couched his super-decree in the form of answers to questions from a group of young people. Question: "Was *Pravda* right in opening a free discussion of linguistic problems?" The discussion, both before and after Stalin's presentation, was invariably called "free." The sound of the word sooner reminded one of the clanking of fetters than the tocsin of Liberty Bell from ancient Novgorod's democratic assembly. Answering the question, Stalin writes:

> It is universally recognized that no science can develop and flourish without a struggle of opinions, without free criticism. But this universally recognized rule has been ignored and trampled upon most unceremonially. A self-contained group of infallible leaders has developed which has begun to ride roughshod and behave in the most arbitrary manner after guaranteeing itself against any possible criticism. How could this have happened? This happened because the Arakcheyev-like regime established in linguistics cultivates irresponsibility and encourages such disorders. The discussion has proved extremely useful mainly because it has brought to light this Arakcheyev-like regime and smashed it to bits.*

Under the code name "battle against the Arakcheyevian regime," Stalin would declare war against scientific or scholarly authority in any field. Anyone who described scientific discoveries, who spoke about outstanding people in science and scholarship, and more importantly, who quoted them, rose in opposition to the great, infallible Stalin. And that opposition was not an error, but the subversion of the existing order, counterrevolution. And it had to be exposed. My book became a victim of precisely that madness. My description of the journeys of

* The translations of all articles published in *Pravda* on the controversy in linguistics during May 9–July 4, 1950 may be found in *Current Digest of the Soviet Press*, vol. II, nos. 18–22, 24–28 (1950).

two beginning geographers under unbelievably difficult circumstances that ended with the death of one of them now turned out to be the implanting of an Arakcheyevian regime.

According to the editor I attributed to Berg and Ignatov qualities they certainly did not have; I treated them as heroes that they were not; and I forced the reader to believe in their absolute infallibility, using citations from their works which pretended to depict unusual achievements. Next to every quotation, and I had a multitude of them, the editor wrote "overreliance on quotations" in the margin. He didn't use any descriptive words for this subversive activity; he wrote "quotationitis" (*tsytatnichestvo*), a term of abuse, a new word that was supposed to show the editor's vigilance.

The following joke was born in the post-Stalin era: "What is a telegraph pole? A well-edited Christmas tree." The editor at the Geographical Publishing House didn't remove decorations. He chopped off branches with an axe.

The book was cut to 132 pages. The contract was for 432. The parts relating to Ignatov were excised, on the grounds that "they were not of interest to the Soviet reader." The letter requested that I come to the publishing house to sign the edited text. I went. The rudeness of those who had formerly been extraordinarily polite and kind exceeded all expectations. "Your style is hopeless," they said. I offered to present them with evaluations written by members of the Union of Writers—Sobolev and Meilakh, who had read the book in manuscript and praised its translucent style. B. S. Meilakh had received a Stalin Prize for his research on Lenin's language.

"Your father was such a modest person," they said, "but you apparently didn't inherit that trait."

"Yes, he was modest, but he had the person who preceded him as an inspector of fisheries tried for bribe-taking, and when he lived in Kazalinsk he never left the house without a revolver. I write about that in my book." Those words of mine, disastrous for the fate of my book, testify to the fact that I hadn't guessed about the bribe-taking at the Geographical Publishing House. I refused to sign the edited text and left for Leningrad. I decided to ask for help from the Geographical Society.

Stanislav Vikentyevich Kalesnik, the vice-president of the Geographical Society, one of my father's students, a landscape specialist and glacierologist, said to me, "Ah so. And now let me tell you what the Geographical Publishing House did to me." The publishing house had suggested that they issue a collection of selected works of his. He refused, said that he was young, that when he died, then . . . , but they were insistent. He prepared two volumes, spending 2,000 rubles to have the manuscript typed. The publisher's reply read: "These words are outmoded. Write new ones." And that was the end of that.

A report on the activity of the Geographical Publishing House was made before a general meeting of the members of the Soviet Geographical Society by Budyko, the director of the publishing house. About 600 people filled the hall. It turned out that there was a great deal of dissatisfaction with the publishing house. Not a word was said about Kalesnik's selected works. Budyko complained that the responsibility for the text of a book was born exclusively by the editorial board. The author should also bear some responsibility. "The author is willing," I said. The hall roared with laughter. Right toward the end of the meeting someone asked why not a single book about Berg had been published. The director said that a book was being printed and named the author—me. A misunderstanding, he said, had held up the publishing, but the problem had now been eliminated.

The book *Through the Lakes of Siberia and Central Asia: The Travels of L.S. Berg (1898–1906) and P.G. Ignatov (1898–1902)*, came out in 1956. The publisher's plan listed two books about Berg: mine and a book by two geographers. Their names were Gekker and Murzayev. There is nothing mysterious that can't be made plain sooner or later. The riddle of Gekker's negative evaluation for the journal *Around the World* was solved. But Gekker did not become Lev Semyonovich's biographer. In 1976, on the occasion of the hundredth anniversary of my father's birth, Murzayev came out with a little book, *Life Is Action*, devoted to my father's activity. It was published in color, in a beautiful edition. All the maps illustrating Berg's expeditions were taken from my book. The source was not cited. And my book is not listed in the bibliography. I am an émigré; it's not forbidden to quote me. Why forbid that when it's enough just to recommend not to cite? I don't blame Murzayev. The author, no matter how much responsibility he bears for the book's text, still doesn't really bear any. The editorial board alone bears the responsibility. And it follows recommendations from above. What about conscience? you'll say. What on earth are you asking for? Conscience! After all, in addition to the editors who bear responsibility there is also the fearful censorship, and it bears responsibility. And there are so many people bearing responsibility that the great and infallible law of the dialetic comes into play—"quantity metamorphoses into quality" (excuse me for the jargon, but that's what we were taught at the university)—and not just into any old quality, but always "into its opposite," and the result is absolute irresponsibility before the truth. The more severe the punishment for disobedience, the easier it is for an author to make a deal with his conscience. Intractability is punished by death—the death of the book.

Murzayev's book is good. Unfortunately, he not only washed Father's linen but starched it for him as well. Imagine Krzhizhanovsky's cook as Gleb Maximilyanovich's biographer. Would she write that Krzhizhanovsky, at the age of sixty, while figure skating, did a split in the air in order to prove his virtuosity to the janitresses at the skating rink? She heard the story from Krzhizhanovsky himself. But she wouldn't have written it. Murzayev, had he been in her place, would have. He tells us that in his youth Lev Semyonovich wanted to become a singer and had already prepared a stage name: Zvyozdich-Struysky. The pseudonym exposed the point of the arabesque with extreme clarity—though Murzayev missed it. It's hard to imagine anything more banal than that combination. In English it would sound something like "Higglethorpe-Throckbottom." Father was only joking. But Murzayev cites the arabesque as a biographical fact. On the other hand, Murzayev apparently did not know that Father wrote marvelous surrealistic fairy tales.

Murzayev concludes his book with the words of A. I. Herzen: "The past has not been lost in the present or replaced by it: It has been fulfilled in it." It didn't take Murzayev so much as a minute of work to embellish his book with the citation from Herzen. I used those same words as the epigraph to the final chapter of my book. But unlike Murzayev, I conclude on a sad note.

Both on my book and on Murzayev's lies the mark of toadying. A Soviet biographer is obliged to burn incense to the status quo, the regime, and the authorities. The Great October Revolution brought happiness to everyone who wasn't destroyed by the secret police to which it gave birth. Biographies are written to glorify the wise leadership of the party. Awards, promotions, and elected posts are mentioned. Persecution, arrests, executions, and death by starvation in the camps—it's as if those things never existed. A half-truth is the worst, the most insidious sort of lie.

For the hundredth anniversary of the birth of Grigory Andreyevich Levitsky, a professor at Leningrad University who was destroyed by Prezent, A. A. Prokofyeva-Belgovskaya, one of Levitsky's most talented students, wrote an article.* The seventeen lines devoted to his biography, printed in brevier type, conclude with the words: "In 1945, in connection with the 220th anniversary of the founding of the academy, Levitsky was posthumously awarded the order of the Red Banner of Labor." The date and place of his death are not mentioned. Grigory Andreyevich was given the award by a bureaucratic mistake. His name turned up by accident in the lists of the associate and full members of the Academy of Sciences who were to be given awards. Had he been as well-known as Vavilov, he would not have received the award. In 1945, Vavilov did not receive the Order of Lenin that was due him by rank. Prokofyeva-Belgovskaya is dissembling. She knows that Levitsky was arrested and died in prison. His award was the result of shoddy work. In "those" times posthumous awards were not given. There's a joke that goes: "What change did Khrushchev make in the Constitution? Everyone has the right to posthumous rehabilitation."

The people at the Geographical Publishing House didn't forgive me for what had happened. You can't imagine anything more pathetic than the way my book was put together. It looked as though it had been published in the 1920s instead of the 1950s. And saddest of all—there was one misprint after another. The dates of Father's life and the name and subject indexes were thrown out. There isn't even a list of the illustrations that an artist did, using photographs that had never been published. And had I only realized that if I had given a bribe, the book would have looked wonderful.

The bribe! The mighty stimulus in the building of communism and the best means to raise quality in production. The great tradition of Russian life! The past has not been lost in the present. . . .

* "G. A. Levitsky and Soviet Research on Chromosome Morphology" (*Studies in the History of Biology*, vol. 7, Moscow: Nauka Publishing House, 1978).

# *Fish Swim Away from Death . . .*

If before Father's death, while exploring the mysteries of the formation of plant communities and raising my children, I had been frolicking at the edge of the abyss, then after his death I was gamboling at its bottom. I was without a job, without money, without a husband, and living in a communal apartment. A job was out of the question. If I had not been able to keep my place at the Fish Institute while Father was alive, then now I had absolutely no chance of finding work. I have already described how by protecting my rights as a person with a degree, the law doomed me to starvation. Family happiness came to an end. Being in the kitchen among my neighbors terrified me: I could only adore them at a distance.

Where had my husband disappeared to? Isn't it all the same? One of his many liaisons on the side turned out to be stronger than the others and he founded a new family. It was the post-war time. Men were hard to come by.

Maria Nikolayevna agreed to work without pay, but she was often ill, and when she wasn't ill she was off to see her son in Krasnoyarsk. I really don't know how we made it through the first half-year after Father's death, piling up debts. But then I received the inheritance. Sim did too—I persuaded him to take it. "Marmikha requests that you and I refuse our part of the inheritance," said Sim. "She showed me Father's will. He left her everything, so I agree." Maria Mikhaylovna made the same request of me, and she showed me a tiny little piece of paper on which in Father's hand, with his distinctive script, it stated that he willed everything to my stepmother and to her alone. "And do you know why Father willed everything to Marmikha?" I asked Sim. "Because she insisted on it," I went on, answering my rhetorical question. "And do you know why Father didn't have it attested? Because he knew of my situation and wanted me to receive part of the inheritance." Father knew that my family had fallen apart. So I didn't accede to my stepmother's request.

We hadn't even abandoned our house in Yazhelbitsy before the family crumbled. In the summer of 1951, I came into my rights of inheritance. The dacha, Stalin's gift to academicians—the gifts were handed out according to a list and Father's name was on the list—passed into my exclusive ownership. Father was posthumously awarded a Stalin Prize, first class, for his three-volume work on fish, and the prize money was included as part of the inheritance. Sim and my stepmother each received half of the prize money and I got the dacha to myself. And so the nightmarish disasters and wondrous joys of my years as a house-owner began.

To imagine Father as a house-owner is just as difficult as to imagine him as a singer coming out on stage in a starched shirt and using the stage name Zvezdich-Struysky. I once asked Father: "What would you like for your birthday? How about a beautiful cup?" "No," said Father, "I hate things. Give me some apples." When Father received the dacha as a gift, I had said to him, "You seem to be happy to have been given this thing." In reply Father said affirmatively, "It's air."

That air, that marvelous air, was located on the Karelian isthmus on the shore of the Gulf of Finland near Komarovo Station. The station had formerly been called Kelomyaki, but had now been renamed in honor of the botanist Vladimir Leontyevich Komarov. The Karelian isthmus, which had been the Nice of Finland, had become the property of the Soviet Union after the war. The granite foundations of buildings, stairs leading nowhere, lilac bushes next to the ruins, and pine trees in square formation alternating with plots of skillfully drained marshes—all of this were the traces of superb forest domestication.

In 1947, an academy village rose up near Komarovo Station. The nation's scientists and scholars, a tiny handful of elect, were provided with surroundings for work and leisure and I was provided with yet another opportunity to observe the relations among equals who had no wish to be equals. According to the Marxist doctrine as it is interpreted in the Soviet Union, the personality is formed by social conditions. Don't believe it. Social conditions are created by people in accordance with their human nature. The dachas were given to people as personal property: They could be inherited. Academicians were dying one after another. Pariahs began to appear in the academy village—inheritors. I was one of the first in that group. Weapons were sharpened by the Old Guard to be used against this strange category of people—those who had become property owners by virtue of inheritance.

Father's death postponed the fulfillment of the second half of the gypsy woman's prophecy. It wasn't in my husband's character to abandon me in my misfortune, right before the eyes of everyone. The inheritance untied his hands. He insisted that the money and the dacha be transferred to his name. Then his conscience would not have allowed him to leave his family. I understand that now, but back then I didn't understand it at all. I refused to transfer what I possessed to his name. He left his family for his new wife when nothing remained of the inheritance, but by then I was already branded as a rich heiress and the rupture did not stain my high-minded husband's good name.

Marusya, wonderful Marusya, whom I had hired to help us and to replace Maria Nikolayevna, left us, too. The scale of that misfortune was huge. Masha loved her so. A maid in a communal apartment is a touchy subject.

I had brought Aunty Sasha, honest and kind Aunty Sasha, from Yazhelbitsy. The drunken and thievish atmosphere of the communal apartment immediately dragged her down. Only born aristocrats could withstand the temptations. Truly royal blood had flowed both in Maria Nikolayevna's and Marusya's veins. But at the end of 1952, when I met Evgeny Lvovich Shvarts, Marusya was already gone.

Shvarts lived at Komarovo year-round. His little blue house was located between the railroad and the post office in the village of Komarovo, on the opposite side of the railroad from the academy village. The house didn't belong to him, but had been granted to him by the Union of Writers for use in his lifetime. A low fence surrounded the dacha. Very young mountain ashes were already bearing fruit. Little caps of snow crowned the scarlet clusters.

I became acquainted with him under the following circumstances. Liza and I were on our way to the post office to send off some registered letters. A man wearing a felt hat came out of the little blue house. He walked in front of us and the saleswomen from all the stalls greeted him. The line at the post office consisted of Liza and me and the man wearing the felt hat. We waited patiently while he filled out subscription forms for every magazine and newspaper on earth.

"I'm holding you up," said the stranger. "Mail your letter—I'll wait," he offered.

"Thank you, but we'll wait. I have a ton of letters."

"A ton," the stranger repeated.

Liza took control of the situation. "Can you add five and seven?" she asked the stranger.

"Yes, I can," he said. "Can you?"

"I'm asking you for a very good reason," said Liza. "There are seven leaves and five flowers on that poster." There was a poster adorning the post office. It was December. The poster read LONG LIVE THE FIRST OF MAY! and featured a flowering apple tree.

"All she does is count," I said to the stranger. "Today I heard the following conversation. My youngest one said to her, 'When there are a hundred empty mustard plaster packages, let's ask Mama for them.' And she answered, 'There aren't a hundred packages. There are five mustard plasters in each package. That means there are twenty packages.' "

"A hundred mustard plasters," said the stranger.

Just at the moment when the woman at the window finished writing out receipts for the stranger and I got the chance to mail my letters, Liza said, "Mama, let's go. I'm hot." I unbuttoned her coat collar.

"I'll take her out. We'll wait for you," the stranger said. Liza gave him her hand and they went out.

"Do you know who that is?" the woman at the window asked. "He's Evgeny Lvovich Shvarts."

I knew the name, of course—he was a children's writer and dramatist, an admirer of Andersen's, whose fairy tales he adapted for the stage. Through the post-office window I could see Shvarts buttoning Liza's collar, his hand trembling. When I went outside it turned out that their friendship was already on a firm foundation.

"Do you see that stall? A bear lives there. And up there—under the post-office roof—where the wires lead to the box? A bird lives in the box, and she talks to the bear over the telephone. We've become friends," Evgeny Lvovich said, turning to me. "I already know that you live in the pink dacha, number twenty-seven. Please come visit me."

"Mark well your words," I said. "Now I won't have a moment's rest: 'Mama, let's go see the new academician.' "

"Tell me," Shvarts asked ironically, "don't your children ever think in smaller categories?"

"You should have said 'other' instead of 'smaller,' " I corrected him ironically.

"No, really—you two come and visit," said Shvarts.

"I can't take one child visiting and leave another one overboard."

"Bring the other one, too."

"But you don't know how many children I have!"

"It doesn't make any difference."

Liza, Masha, and I soon visited Evgeny Lvovich for the first time. Liza and Masha were five and four. His housekeeper, helping Masha on with her coat as we were about to leave, asked what had happened to the young woman with whom she used to go for walks. "Marusya went back to her village and found another wife," said Masha. Masha had everything confused. It was her papa who had found another wife, and Marusya had gone back to her village and found a husband. Masha's funny words did not make Shvarts smile. It was obvious from the sad expression on his face that he attributed meaning not so much to the form as the content. It was clear to him that Masha's mistake was born of tragedy.

Now I'll prove to you that people aren't equal, even people in the same profession, even members of the same union—in this case, members of the Union of Writers. To do it I have to transfer you in time and space. The events occurred twelve years later and many thousands of miles away. In 1963, I was invited to Novosibirsk's Akademgorodok to create and head up a laboratory of population genetics. And until I was thrown out of there by the KGB, I lived there among the ranks of the privileged and was considered an interesting personality. I was favored with a visit by Granin, the author of the book *I Go into the Storm*, where he gave Trofim Denisovich Lysenko a bit of a kick, portraying him as the scientist Denisov. Granin was coming into power. He directed the Union of Writers' section of young poets. I told him about my acquaintance with Shvarts, and about how Shvarts, not paying any attention to his new young friend Liza's amazing abilities, repeated "One hundred mustard plasters." Granin also failed to pay any attention to the child prodigy's words. "Mustard plasters excite me," he started up animatedly. "On the inside of each mustard plaster it reads: 'Printed in one million copies.' "

You have to live under a totalitarian regime to understand this joke. Every printed line, though it were instructions explaining how to use a mustard plaster, or a program of a concert, is censored. The number of copies to be printed is decided by censorship. That number is indicated on the text. Restriction versus extension of numbers is among the most powerful means of ideological manipulation. It goes without saying that Shvarts reacted to my narration in a different way. Shvarts had been thinking about our pains in using plasters; Granin did not. We can only guess whether Granin's exclamation was a mockery about censorship or expressed a daydream of every author—a million copies of his book in print.

Starting to speak about Granin and the events of the Sixtieth, I can't stop. Permit me, please, to leave the unemployed geneticist with her two daughters in her dacha in Komarovo and continue.

No friendship arose between Granin and me. Soon after his visit I was supposed to go to Leningrad to defend my doctoral dissertation and Granin asked me to notify him ahead of time. "I'll give you a bouquet of roses," he said, "I'll give you a very large bouquet." However, the poet Brodsky occupied my thoughts much more than did my dissertation, which bored me.

The young poet Brodsky was not a dissident in the usual sense of the word. He did not fight for freedom. He was its embodiment. He lived according to his own laws, wrote poetry, and in literary circles was already considered a first-class poet and translator. Anna Akhmatova had written a poem about him, about the roses that he gave her. The journal *New World* had published that poem. At the time when Granin got to know me in Novosibirsk, the threat of inevitable arrest hung over Brodsky. In the Leningrad papers article after article appeared accusing him of what he was not guilty of. Here is a small example of the mendacious claptrap in the official press. In a poem that circulated in samizdat, the poet Bobyshev had denounced vice. The newspaper printed excerpts from the poem and claimed that it was a poem by Brodsky singing the delights of vice. The main accusation was of parasitism. The decree about punishing parasitism had just come into force. Brodsky's style of life did not really fit the preamble of the decree, but in a world of falsehood, asceticism did not save him from accusations of debauchery. I had every reason to be alarmed. I was on very friendly terms with Brodsky's parents.

I told Granin that I'd accept his gift if he interceded on behalf of the young poet. He told me that he had already tried to influence Brodsky, suggesting to him that he write poems about flowers, snow, and animals. Brodsky didn't object to the themes, but what came out from under his pen turned out to be absolutely unsuitable for print:

*FISH IN WINTER*

*Fish swim from death*
*following their eternal*
*fish route.*

*Fish don't shed tears;*
*they press their heads against*
*blocks of ice . . .*
*Fish are ever taciturn,*
*for they are mute.*
*Poems about fish, like fish,*
*Stick in your throat.**

And his wonderful poem "Snow Is Coming" is in a similar vein. But Granin promised to help. During Brodsky's trial the premiere of Granin's *I Go into the Storm* was playing at the theater. Granin neither went to the trial nor to the performance at the theater. No one knew where he was. In Granin's name Voyevodin Junior read a condemnation of the poet-freeloader by the members of the Union of Writers. The poet, as is well-known, was sentenced to five years of exile and hard labor. In announcing the sentence, the court based itself on the remarks from the Union of Writers. I said a couple of heated words to Voyevodin and another rogue, who would later be caught on some criminal matter. To that scoundrel I said: "What, you're happy? You've strangled a child with your own hands." He cried out: "Militia! They're threatening to strangle me!" Five uniformed militiamen surrounded me—five enormous things. They surrounded me and then retreated. I called Granin at home the day after the trial. He wasn't at home—he'd left town. I asked his wife to tell him that it was incumbent upon him to remove the black spot. . . .

I saw Granin once more after that, and again asked him to do something for Brodsky. He squirmed and asked why I myself didn't do what I was asking him to. I've been told that he none the less did intercede for Brodsky and that it is part thanks to him that Brodsky was returned from exile without completing his five-year term at timber rafting. Granin was not injured in the eyes of general opinion. Most felt he couldn't have done other than he did, since he was a candidate for a state prize at the time. But this shameful business didn't transpire during Stalin's time. Granin was not under a threat of death. On the other hand, not everyone is ready to risk losing a state prize.

Shvarts and I got along famously. I even nearly became a children's writer. The sharpest sensation of all the ones in my life was the feeling

* Translated by R. L. Berg.

of shame that I experienced when Shvarts listened to me read my stories and then said, "Literature such as that is punished by a term in a convict labor gang." I experienced the irresistible desire to disappear from the face of the earth; for an abyss to yawn beneath me and swallow me up. I understood the exact and literal meaning of the expression "to want to crawl into a hole."

Shvarts said, "Work. Come back in a month."

"You probably have a lot of people coming and reading all sort of junk to you. What do you tell them?" I asked.

"I tell them that getting published is so difficult that it's better not even to begin."

"And why aren't you telling me that?"

"Because you have something to say."

I went to see Shvarts again in a month and brought along ten little stories. As I was leaving, Shvarts said, "Convict labor groups are out of the question."

I put together a little book that was partially for children and partially for adults. An artist undertook to illustrate it. She spent a long time trying to find the appropriate style, and she never did find it. When she returned the manuscript to me much later, I no longer had the time or the desire to publish stories.

I often went to see Shvarts to talk. De-Stalinization was under way. Shvarts, a great authority on human souls, was mistaken about the scale of the freedom that had been granted, in exactly the same way that Krzhizhanovsky was. His plays, absolutely forbidden under Stalin, were now being produced by Akimov in the latter's theater. *An Ordinary Miracle. The Dragon.* A Moscow theater came on tour and performed *The Emperor's New Clothes*. Obtaining tickets required a great deal of luck. When the minister said to the naked emperor, whose only clothing was a ribbon passed between his legs, "Your Majesty, run while the people are still silent," the audience roared with pleasure. In Akimov's production at the Comedy Theater, the Dragon's three heads each spoke differently and were made up differently. One spoke with a Georgian accent, the second had a speech defect and a sweet voice, and the third barked. Stalin, Lenin, Hitler. Absolutely everything was mocked. Arbitrariness, seniority, inequality decreed from above, anti-Semitism, leaders' stupidity, the struggle for power, and the slavish instincts of subordinates.

Shvarts asked me to go to the performances and report audience reaction to him. I told him about the production details. In *An Ordinary*

*Miracle* the minister of enlightenment returns from exile. He has with him a little bundle—a few books, judging by the shape of the bundle. Sand is leaking out of his torn but elegant boots. He gathers up the sand, takes off his boot, and pours the sand back into it. He holds the boot up in the air and sand comes pouring out through the holes again. Shvarts laughed.

"Aren't you afraid that you and Akimov will be arrested and thrown into prison?" I asked him. "No," he said. "In the first place, there are too many people who would have to be sent to prison. And in the second place, I had German fascism in mind, not the Stalin era."

He really had written *The Dragon* during the war, as an attack on fascism. But totalitarian regimes resemble one another as closely as identical twins. The sting of satire that was directed against Hitler taunted the Soviet Union with equal if not greater right, and in spite of the author's intent, put the author under the tyrant's suspicion. It was just that Stalin was much more masterly than Khrushchev in the ways of ruling. In Stalin's time, Shvarts found himself among those in disfavor, and censorship forbade the staging of *The Dragon*. Now, in Khrushchev's time he had risen to the rank of one in half-disfavor. Editions of his plays were published, but only the privileged could obtain them. The books were printed in niggardly runs. A million people would have bought them.

The earth filled with rumors about the sufferings of victims of the Stalinist terror. Before Stalin's death, nothing was known about the fate of those absorbed by the GULAG. Those who came out were silent. Fear made them mute. In Khrushchev's era, tongues began to loosen. Shvarts told the story of a literary scholar who was a friend of his. In the train on the way to the camp he lost consciousness. The guards decided that he had died and turned him over to the railway hospital morgue at the nearest station. There he came to. The hospital director, on learning that the resurrected prisoner was a Doctor of Sciences, offered him a position as a physician at the hospital. The literary scholar was afraid to accept. He went off to catch up with the convoy that was taking him to the camp. There in the camp the common criminals organized a rebellion, smashed the cell doors, and won momentary freedom. They broke into a cell where there were many political prisoners and suggested that they come out. The members of the intelligentsia lodged themselves under the beds and refused to come out.

Shvarts also told about how people were released from the camps at

Khrushchev's orders. There was a commission at work. A prisoner was asked what he had been accused of.

"Of making preparations to murder Stalin." Under torture he had confessed that such-and-such a person on such-and-such a spot had given him a revolver.

"And how much of that was true?"

"Not a bit of it."

"You're free."

Then they were reviewing the case of an elderly, bearded Russian. "What were you accused of, Gramps?"

"Of saying that if I met Stalin, I'd strangle him with my own hands."

"And how much of that was true?"

"Every bit of it."

"Get going, Gramps. You're free."

Once I came to see Shvarts with a newspaper that had an article about the production of his plays. He had Yuri German, a quite well-known writer, there with him. German was telling about the adventures of a writer friend of his. The friend had begun to grow bald, and managed to obtain a hair tonic from abroad. His barber wanted to know how he had hair returning in a bald spot, and the writer told him the secret. That very same evening either militia or military officers appeared at his apartment. But instead of taking him to the Lubyanka Prison, they went to a mansion whose windows were completely curtained over. They led the unfortunate man into the bathroom and left him there alone. In a moment, a short, bald, ruddy man appeared in an unbuttoned shirt without a jacket, with general's stripes on his uniform trousers, and dead drunk. With a thick tongue he said that he had heard about the hair tonic. He wanted a bottle of it.

"And who are you?" the writer asked.

"I'm Stalin's son."

"Do you mean to tell me that your father can't order hair tonic from abroad for you?"

"And did you ever hear of my father ordering hair tonic from abroad for anyone?"

Next day the bottle of hair tonic passed into the possession of the offspring who was neglected by his papa. Stalin's son expressed his gratitude most lavishly. The writer received from him as a gift a Teutonic sword that had once embellished a museum, and Goebbels's short-haired purebred bitch.

I became friends with the literary scholar Boris Solomonovich Meilakh and his wife, Tamara Mikhaylovna, in 1939 in a vacation home in the Caucasus. His marvelous son Mishenka had not yet been born. (He was born in 1945.) I became acquainted with Mishenka in 1952. No one introduced us to each other. Boris Solomonovich, a professor at Leningrad University, a well-known specialist on Pushkin, a Stalin Prize laureate who was awarded the prize for his analysis of the language of Lenin's works, a member of the party, knew perfectly well how to combine the good tone of contemporaneity with decency. In that unfortunate year 1952, Meilakh's daughter, Mira, was trying to enter Leningrad University's Department of Philology. Despite her family's credentials and accomplishments, to the entrance committee Mira was not and could not be anything other than a Jew. Her failure of the entrance examination was fabricated—she was given a low mark for the composition, allegedly because it contained errors in spelling. After that operation had been performed on Mira's composition, Boris Solomonovich demanded to be shown the work. There were no mistakes in it.

But there was an examination in geography coming up, and Boris Solomonovich could not have prevented the fabricated failure at that examination. At my request, Professor S. P. Khromov, then the volunteer editor of my book, watched over Mira's examination and she passed.

Boris Solomonovich rented summer lodging in Komarovo, but not in the academic headquarters. His home was located in a village, on the other side of the railroad, far from the sandy shores of the Gulf of Finland. To help Mira prepare for her geography examination, I had gone there with some textbooks.

Evening was approaching. I asked a tiny little boy, "Do you know where Boris Solomonovich Meilakh lives?"

"It's very good that I'm the one you asked," said the boy. "That's my papa. Go around the glade, otherwise you'll get your feet wet."

That was how my eternal friendship with Mishenka Meilakh began. I don't remember much of what he said. People usually laughed when Mishenka spoke, formulating his own thoughts in such an unexpectedly intelligent way.

At age seven he went to Leon Abgarovich Orbeli's dacha to invite his dog to study at a school for guard dogs. Leon Abgarovich asked him why that was necessary. Mishenka said that most of all it was important for the dog itself, since any learning bears future fruit.

Usually I tried hard not to laugh at Mishenka's words that now I can scarcely remember any of his pearls of eloquence. We had conversations that ran something like this:

HE: That disgusting Mirka got tickets for the film *The Diary of Anne Frank* but didn't get me one.

I: Children under the age of sixteen aren't allowed to see it, and that's why Mirka didn't get you a ticket.

HE: Children ought to know the truth, and the whole truth. And it's better if they learn it from an artistic depiction than from hooligan friends. I'd manage it so that they let me in.

I: That film isn't the whole truth, but only half of it. The film will be the whole truth when they show us what went on in Stalin's concentration camps.

HE: That will never be. That's too bitter a truth.

He couldn't have been more than thirteen at the time, but he was skeptical about his wonderful parents. And once he came to see me and told me that he hated his parents. They were letting a mouse out of a mousetrap and watching a cat chase it. "I realize that that death is nothing in comparison to the enormous natural selection process that goes on in nature," he said. "But how can they make entertainment out of death?!" When he'd gotten a bit older, Mishenka became a vegetarian, to his parents' great distress.

He wanted to become a film director and was dreaming of entering the appropriate institute. His parents barely managed to talk him out of that. He began studying French, entered the university, and became a specialist in Old French literature and language. He did not limit himself to the medieval poetry of troubadours and the analysis of the Provençal language. He was interested in the poets Kharms and Vvedensky, victims of the Stalin terror, arrested and gone forever, completely exiled from works of literary criticism and history as though they had never existed. Kharms's and Vvedensky's books, with introductions and commentaries by Mishenka, were published abroad. In 1983, the University of Glasgow invited him to the United Kingdom to give a course on Kharms and Vvedensky, but he did not have the chance to go.

The desire to know had driven Mishenka all around the Soviet Union—from Bukhara to Vladivostok, from the Caucasus to the Kurile Islands. Tuberculosis would confine him to his bed for years, but as soon as there was an improvement, he would fly off to visit friends and learn more about the world. He wrote me very nice letters, the usual kind of letters that people write to their parents, or to their favorite grandmother. But once, at the dacha of the senior Meilakh, Boris Solomonovich read me a letter written to him by Mishenka from far away. And then I became jealous. It was a first-class work of art. The addressee clearly inspired the writer. The picturesque, flexible language gave evidence of a vast amount of artistic energy and of a tense emotionalism.

Mishenka borrowed books from me that were of the most specialized, professional sort. He needed to know everything. His worship of life was not limited to the defense of a mouse, care for the guard dog Malashka, or vegetarianism. He did not wait for requests for help. Seeing a person in trouble, he himself would seek a means to help him. When his French teacher, Elizaveta Solomonovna Golodets (she taught my children French, too), had nearly gone blind, Mishenka found a doctor and took her to see him. While still a little boy, he would arrive at Komarovo on his bicycle and, having received my permission, would head for the kitchen to announce that at the village store they were "giving out" chicken (in the new Soviet jargon "giving out" means "selling") and that he had taken a place in line for us. After I emigrated, he sent me books right up until his arrest.

Mishenka Meilakh was not a dissident. He signed no letters and participated in no protests against scandalous outrages. He maintained links with foreign colleagues, received books from abroad, and published abroad his works about Russian poets and the works of those poets. He allowed his friends to borrow books or simply gave them to them. He knew perfectly well that he was taking a risk. Soviet citizens are forbidden by law to read *samizdat* or *tamizdat* (books issuing from Russian-language presses abroad). But not only is he or she forbidden to read them, a Soviet person is not even supposed to dare to know of their existence.

In 1983, Mishenka was arrested. The investigation went on for a year. In 1984, he was sentenced to seven years of "corrective work" in a Perm camp, one of the most terrible camps in the GULAG archipelago, and five years of exile.

I know from reliable sources that Meilakh was offered an alternative to imprisonment: the public confession of his sins, including, of course, the confession of the alleged sins of others. In return he would be given his freedom and the opportunity to emigrate. At the trial, in response to the question whether he admitted himself to be guilty, Mishenka said, "No, I don't admit that." The moral test, a seductive alternative, did not lead him to capitulate to the forces of evil. He chose imprisonment and exile.

He spent only three and a half years behind bars. During those more than thirty years since a six-year-old tiny boy said to me, "It's very good that I'm the one you asked," Mishenka Meilakh became known and loved not only inside his motherland but far beyond its boundaries. The best representatives of brilliant Russian émigré writers, poets, literary scholars on both sides of the Atlantic protested against his imprisonment. This voice of public protest provided Meilakh with Western support at the moment when the Soviet government became sensitive to Western attacks on human-rights violations.

In February 1987, Meilakh was released.

But let us go back to the early 1960s. It was the height of the Khrushchev reforms. Genetics had been given the right to hold one nostril above water, under a hail of blows from loyal Lysenkoists. Another science, cybernetics, treated in Stalin's time as a lackey of imperialism, a tool for the enslavement of the working class by capitalists, had been admitted to the entrance hall of the Soviet Union.

Removing a modest part of the Iron Curtain of Stalin's era, Khrushchev lent an ear to the warnings of mathematicians that the failure to apply and develop cybernetics would harm the war potential of the country. Given its military applicability, cybernetics became a salutary safety valve for most branches of science suffocating in the atmosphere of Marxist-Leninist ideology and its integral partner, Lysenkoism.

Cybernetics is part of the theory of systems. Currents of information going back and forth to connect parts of a system are its main subject. The producer of any kind of production, including art work, and its consumer, are connected with each other through these currents. They are parts of a system whose name is society. Society provides channels to assure their contacts. Markets, banks, the system of education, publishing houses, theaters, art exhibitions are facilities to canalize the currents of information connecting them. The perception of art by its individual consumer—by you, by me—also has to do with these currents of information. Suffocating without the fresh air they felt cybernetics could give them, the intelligentsia rushed to apply cybernetic principles in technology, economics, and in most branches of biology. Time came to apply cybernetics to art.

Mishenka's father, Boris Solomonovich Meilakh, was the first to hoist a flag of cybernetics. The issue of freedom of creation, a freedom that he had absolutely exiled from his works when he analyzed Lenin's language, could now, it seemed to him at that time, be raised with impunity. Boris Solomonovich couldn't bring himself to use the word *cybernetics*. He found a euphemism. In 1963, he decided to organize a symposium on complex study of art production. Cybernetics was hidden under this word, *complex*. That would not harm anyone.

I was then heading the all-university seminar in cybernetics and was a member of Leningrad University's Cybernetics Council. Boris Solomonovich applied to the council to have the symposium organized under its aegis. I presented the plan to the council. The council rejected it unanimously. Cybernetics had only just then been allowed. Developing it as it applied to technology was fine, as it applied to biology—well, that was approved halfheartedly, but even that reeked of sedition. Everyone realized that defending genetics from positions based on cybernetics was easier than from the point of view of sociology, criminology, or even medicine. We have enough troubles with genetics, let alone a complex study of art. The council members were afraid to exceed the measure of freedom that had been granted to them. The most powerful, although hidden, reason for rejection was the battle for financial support of the newly formulated branches of science.

The symposium was ultimately held under the aegis of the Union of Writers, of which Boris Solomonovich had been a member since time immemorial, and took place at the most inappropriate time for discussing questions of art. Questions of ideology turned out to be at the center of Khrushchev's attention. The condemnation of abstractionists for depicting trees root side up had already rung forth. But it was just at this point that the difference between "then" and "now" was made clear. In Stalin's time, Falk, Neizvestny, and the young abstractionists simply hadn't existed. Under Khrushchev, the beautiful picture by Falk that Khrushchev had anathemized continued to adorn the exhibit at the Manege, as did Neizvestny's magnificent antimilitary bronze; in separate quarters, where only the elect were allowed, the young abstractionists were exhibited. It was there that Khrushchev anathemized them. Everyone had to know that Khrushchev had to have seen them to criticize them.

Meilakh's symposium fell in the interval between the rout of abstractionism in many of Khrushchev's speeches and the official dinner for writers organized in the tent at Khrushchev's dacha, where all the living forces in literature would be routed. Boris Solomonovich banished from the program everything that concerned painting—everything except for my talk.

I quickly wrote to Alexander Alexandrovich Malinovsky (I called him "Kot") in Odessa, telling him that if he would come to Leningrad, I would introduce him to Meilakh, and that there seemed to me a chance to include Alexander Alexandrovich's paper in the program. Exactly twenty years earlier, in 1943, Malinovsky had shown me a manuscript and asked me to read it and never breathe so much as a single word of its contents to a living person. The contents, quite inoffensive, were madly seditious for the times, and Kot was frightened half to death. Since he was the son of A. A. Bogdanov, he had a good reason to be afraid. Bogdanov, Lenin's companion in arms, an unacknowledged founder of cybernetics, misunderstood and condemned by Lenin, was perfidiously liquidated by Stalin.

Malinovsky's manuscript was devoted to a psychological analysis of art's effect and how one perceives art: why "a birch as green as an asp" is no good, while "grass as green as an emerald" is fine. Why, when telling a story about a queen, the author sometimes calls her by her name, and sometimes refers to her as "the queen." What's good about the sea "waving its white mane." The potent intellect of a philosopher and the talent of a true writer had created a work in which Marx, Engels, Lenin, and Stalin weren't mentioned, and free creative thought, independent of their statements, analyzed the intimate mechanisms of creativity. In February 1963, at the cybernetics symposium, Malinovsky's paper was read from the speaker's platform in the prerevolutionary mansion where the Union of Writers' administration and social center are housed.

Nina Alexandrovna Kryshova, a very well-known neuropathologist, spoke on inherited speech defects, basing her analysis on a study of defective speech in identical twins. You'd think that there couldn't be a more inoffensive topic. It was the height of sedition, however. It was precisely such papers that turned the palace on Voinov Street into a den of the enemy. The inheritance of speech defects points to the hereditary conditioning of speech and to the existence of genes that make a person a person. From there arises the question of the conditions for the accumulation of these genes in populations of ancestors and about their fate among populations of contemporary man. Racism? Nietzcheanism? Hitlerism? A doctrine of the primal and ineradicable biological inequality of people and the existence of a superman for whom everything is permissible? A perversion of truth? Yes, a perversion of truth, if one were to declare on the basis of scientific data, that Stalin, Hitler and Khrushchev were supermen for whom everything was permissible. But a sacred truth if one used scientific data to show that they weren't people, that in their case man had not yet left the animal stage, and that one needs to organize society in such a way as to erect insurmountable barriers against them on the path to ruling the society. Nina Alexandrovna went no further than the genetic conditioning of speech, but those who had ears heard.

My paper was the next to the last one. I spoke about the classification of the arts and about abstract art's place in it. I was interested in the hostile attitude that the majority of people have to abstract art. Why is music, abstract sounds, acceptable, while music lovers greet a nonrepresentational painting with bayonets? Speech accustoms our hearing to abstract sounds and in that way creates favorable conditions for the reception of music. Why hasn't writing played the same role? I analyzed hearing, sight, smell, touch, and the types of art intended for each of the sensory organs. I attempted to find an answer in the differences in fitness for a system of signals, providing information, of a sound, of a mark, of smell. I divided the arts into elementary ones and synthetic ones. All forms of art that rely on the word are synthetic. They are intended for reception in the imagination by all the sensory organs. The classification of the arts gives us the opportunity to predict new branches of art, just as new elements can be predicted with the help of the periodic chart of the elements. "To predict means to create," I said, "because the word is God and God is the word." People in the back rows applauded. Boris Solomonovich praised my paper—in the lobby, of course, not from the platform.

Mathematicians gave talks. Academician Kolmogorov analyzed art from the point of view of information theory. There were papers devoted to the mathematical analysis of verse forms. Poetry by Voloshin, the half-forbidden martyr to trampled human dignity, was heard from the podium. Innocent poems, of course, not the other ones.

Malicious tongues called the symposium a "Psychosium of Meilakholics." When I told Meilakh what people were calling the symposium, he was terribly distressed. " 'A Psychosium of Meilakholics'? No. 'Ave Maria' performed by angelic voices at a house of ill fame."

I was often invited to repeat my presentation. I was about to accept one of the invitations, from the young people's section of a house of scholars. But when they called to firm up the arrangements, I backed out. I was afraid for the organizers. It was not necessary for the publishers of the symposium proceedings to remove my paper. I did it myself, so as not to make trouble for Meilakh.

From 1963 to 1983, Meilakh organized and headed a Committee for Complex Studies of Art in the midst of the Academy of Sciences of the USSR. He chaired more than twenty annual symposia and was editor-in-chief of their proceedings. I know it from the obituary published by the *Literary Newspaper*. Boris Solomonovich died on June 4, 1987, soon after his son Mishenka returned home from prison. Reading his obituary I noticed that the year of his son's imprisonment, and the year he stopped heading his committee, coincided.

Immediately after my presentation at the Symposium on the Multi-systems Study of Art I received a note. In the most elegant French phrases the author asked permission to make my acquaintance. The note was written on a program. The author asked me to return the program to him. I wrote in Russian, "Come to my apartment tonight at eight o'clock" and gave the address. The contrast between the timidity of the elegantly expressed request and my offhanded accessibility amused me. Ivan Alexeyevich Likhachev introduced himself. He turned out to be a tall, lean old man. We walked to my place together and ended up having a dinner party. The children had grown up a bit, and they were making blintzes. Liza said to Ivan Alexeyevich: "How strange that you never visited us before. You are so right for our home."

Imagine that you're walking through post-blockade Leningrad to your home. You left your residence a long time ago and for many years have been dreaming of returning. And now you're standing in front of the building where you used to live. You can't enter it. A bomb has demolished the façade and you can see your room on the third floor. It's practically hanging over the orchestra seats of the street. It looks like a stage setting. The curtain has been raised. There are books, furniture. Would you have been happy that day? No. But Ivan Alexeyevich was happy, and not only because on that day he'd been released from prison where he spent almost ten years, and not only because he'd found his nieces, the daughters of his deceased brother, who had died during the blockade, and they had been glad to see him and declared their readiness to shelter him. He was happy because on that day, before sitting down at the dinner table, he washed his hands under the kitchen faucet.

Ivan Alexeyevich Likhachev, as the poet said, "trumpeted through those camps" for fourteen years. He knew eight European languages, did verse translations in both directions, and was a member of the Union of Writers—he claimed that he had strived greatly to become a member so that he would have a grand funeral. We had just gotten to know each other when I left for Novosibirsk. So in the summer of 1964, Ivan Alexeyevich came to visit me, and three of us set off on a wild camping trip to the Altai. (A trip such as that one is called "wild," as opposed to organized tourism.) The third person was Stanislav Ignatyevich Maletsky. It was he who dreamed up the trip. He'd grown up in an orphanage on the Altai. He was three years old when his parents died in besieged Leningrad. He had spent his whole life in a dormitory and had never sat down to dinner in familial surroundings. He received his education at the Institute of Agriculture in Novosibirsk. That should be corrected to "he received no education," since he was in school during the Lysenko period. He studied biology according to Lysenko and soil science according to Vilyams. He was a self-taught agronomist and geneticist. He learned English and mathematics completely on his own, and he was, despite his training, wise in the ways of genetics. He wound up in Akademgorodok by chance. When I arrived there in September 1963, and we met, he was a laboratory assistant. Leningrad, the city where he was born, had come to him in the person of me.

There was a storm raging at the Institute of Cytology and Genetics. The laboratory assistant, Maletsky, had hoisted the battle flag against the head of the laboratory, Miryuta. What a small world it is! In 1945, when I had been expelled from the doctoral program, I applied for the competition at Gorky, hoping to receive a position as an assistant professor in the department headed by Sergey Sergeyevich Chetverikov, the founder of population genetics. And then, suddenly, a handsome young man, Miryuta, a fellow worker of Chetverikov's, turned up at my room on Malaya Bronnaya Street. "You're trying to claim a position that I want to occupy, and I'm going to fight you. Why do you need it, anyway?" And then he heaped a tub of garbage on the heads of all those surrounding Chetverikov and on Sergey Sergeyevich himself. I fled from the field of battle without even entering into combat with the slanderer. Now it turned out that we headed laboratories at the same institute.

Maletsky accused Miryuta of ignorance. All the young staff members at the laboratory had resigned and asked to be transferred to other departments at the institute. A commission was at work. I was immediately included in its makeup.

Maletsky turned out to be right. When I first met Maletsky, he told me that he was leaving the institute. I dissuaded him. Miryuta brought legal action against Maletsky, accusing him of sabotage. Maletsky trembled. I reassured him—the court wouldn't prosecute him. That's just the way it turned out. A very few years passed and Maletsky defended his graduate dissertation and became the head of the laboratory at the academy institute, the same laboratory where six years earlier, by chance, he wound up a laboratory assistant. Now he's a Doctor of Sciences, a first-class scientist.

In 1964, he took it into his head to show Ivan Alexeyevich and me the beauties of the region where he had grown up. A camping trip in Russia is an insane undertaking. It takes great skill just to procure the most elementary things, like transportation and food. We had our fill of horrors. We were robbed. Our suitcase full of food, issues of the newspaper *l'Humanité*, paint, brushes, and notebooks, disappeared in a fierce thunderstorm. Nowhere on earth has there ever been such a tempest, lightning, and downpour as the storm that raged, crackled, and lashed away at us as we were spending the night on Lake Aya. Ivan Alexeyevich was bitten by a tick. We had to vaccinate him against encephalitis. Instead he came down with abdominal typhus. Stanislav Ignatyevich picked up Rocky Mountain spotted fever. I was the only one who didn't fall ill with disease. We came to know the most bitter fate of the people, went homeless and hungry ourselves, and enjoyed the beauties of nature. And how we enjoyed them! Maletsky had a special gift. He not only valued the beauties of nature. He evaluated them as well—on a scale of one to ten.

The Katun valley, light blue side scenes of mountains parting, vast distances, cliffs in the foreground. Flowers and feather grass. "How many?" "A three minus."

The Katun rapids—storming troubled waters, island spindles, overgrown with cedars, wooded mountains in the low distance. A slanting wide-open space framed by the ramrod-straight trunks of mighty pines in the foreground. "How many?" "A six plus."

We sailed around Lake Teletskoye. Ignatov had studied it. Not only in my book, but also in a special article published in the *Academy of Sciences Laboratory of Lake Study Works* I had described his expedition. The unexpectedness of the beauty tugged at one's heart. Olive-green water, like bottle glass, clear and transparent, played at the feet of the cliffs on the shoreline. Large multicolored pebbles could be seen on the lake's deep bottom. Shrouds of mist drifted over the surface, and their reflections drifted under them, refracted and agitated in the heavy water. The horizon was closed off by a mountain and the steep banks came up close to each other. The mountaintop was a plateau, not a peak. "How many?" "Ten," Stanislav Ignatyevich said with pride, as though he were the one who had created this beautiful world for the sole purpose of showing it to us.

The huge, deep river Chulishman feeds Lake Teletskoye. Another river, even more enormous, flows out of it—the Biya. Merging with the Katun, the Biya provides the source for the Ob, one of the longest rivers in the world.

We hitched a ride on a passing truck and headed toward the upper waters of the Chulishman. Two inebriated Altayans told us that the village Balykcha, where we were planning to spend a few days, was a fabulously wealthy cattle-breeding collective farm. We set up camp on the banks of the Chulishman on the outskirts of the village. We built a fire: We had decided to cook up groats. I set off to buy some milk.

"Would you sell me a liter of milk?" I asked a villager. No, she had six children and only one cow. Try that hut over there—they have two cows. "Would you sell me a liter of milk?" "Sure. Why not?" Another young woman came out of the house, with one child in her arms, leading a second by the hand, and a third one was hanging onto her skirt. In another moment yet another woman came out with a whole brood of children.

"Who are you?" the first woman, an elderly one, asked. "Where are you from? How many of you are there? Why have you come?" the young women joined her. A few more children came out of the house.

"How many children do you have?" I asked.

"Ten."

"You know what?" I said. "We have condensed milk. I won't buy any milk from you. You need it yourself." I left. We were cooking groats, using condensed milk. A little boy came running up from the direction of the village. He was carrying a glass jar full of milk. His grandmother was sending it to us as a gift.

The furious despair with which calves suck at their mothers after they've been milked disclosed the poverty of the residents of this "wealthy" region. The people who lived in the hut nearest our camping spot took us under their patronage. We were allowed to use their bathroom, better to say a toilet hut, that wooden construction fit to be locked from inside with a leather loop. The tiny little booth, where there was room for one person, had two holes in the floor over a pit. A leather loop, and not a metal hook. There was nowhere to buy a hook and nothing to buy it with. The two holes spoke of the severe Altai winter, its fierce cold. I leave the rest of the picture to the reader's imagination.

Laconicism is the soul of art. Those two holes in the lavatory's floor spoke about the miserable lives of the members of that fabulously wealthy collective farm no less eloquently than the eternally runny nose (judging by its redness) of our hosts' son. "He has tuberculosis," the woman told me. There was no doctor in the village. Honey and butter? There was no honey. Theirs was a cattle-breeding collective farm. They didn't have an apiary, and private apiaries were forbidden by law. And even when they hadn't been forbidden, the tax on each hive was so high that there wasn't any honey left for the beekeepers, so they gave up keeping apiaries.

In one of his lengthy presentations Khrushchev spoke of the disappearance of private apple orchards as a result of the state's taxation policy. He regretted the apple orchards. But he considered the total plunder of the collective farms legitimate and condemned the collective farm managers if they opposed him. "I asked the chairman of a collective farm what product of theirs was the most profitable," Khrushchev related, and *Pravda* printed it. "Magar." "What's magar?" "A kind of grass." "And why is it the most profitable?" "Because the state doesn't take any of it from us." Khrushchev cited the collective farm chairman's words as an example of flagrant irresponsibility. Khrushchev didn't know what magar was, but I did. Prisoners are fed a porridge made of magar. It's bitter and hardly suitable even for cattle-feed. That's why the state doesn't take it; but better magar than nothing, both in the camps and on the collective farm. That's the way things are. But I've digressed from our travels, rather, from our vacation camping trip.

First Stanislav Ignatyevich fell ill. When we were traveling on a boat back along Lake Teletskoye, both my companions were ill. We were lucky enough to be able to hitch a quick ride from Artybash to Turachak on a passing truck. Ivan Alexeyevich and Stanislav Ignatyevich bounced along on bare boards laid across the open back of the truck. It was especially hard on Ivan Alexeyevich. He was so thin that even without the bouncing and in a healthy state, it was painful for him to sit. Tiny little airplanes, much larger, however, than the U-2 with which my acquaintance with air travel began, flew from Turachak to Biysk. I found out at the airport ticket office that there wouldn't be a plane until the next day. I ordered tickets. We were the only people at the airport. I laid down the tent tarpaulin for one of them, and blankets for the other. They both lay in the shade under some fir trees. As the shade swung around I dragged each of them around to keep up with it. Toward evening Stanislav Ignatyevich recovered a bit and we went to the neighboring village to fetch a doctor for Ivan Alexeyevich. The doctor arrived at the airport in a car. He suggested that Ivan Alexeyevich be taken to the hospital, but we managed to avoid that. On the following day it turned out that there were no tickets. Unfortunately, I don't remember the details of the ruckus that I raised. But in the end we took off. By a miracle. At the very last moment a group of three passengers discovered that the flight wasn't the one they wanted.

In 1966, Ivan Alexeyevich and Stanislav Ignatyevich would accompany me on another expedition—this time to the Far East. We were going there to study plants. I'll tell you later about how I became a botanist while living at the dacha in Komarovo and occupied with the history of geographical discoveries. The mathematician Kalinin received funding from Leningrad University, flew to Novosibirsk and then flew to Kamchatka with me in order to participate in the gathering of statistical material. Lyusya Kolosova, a botanist and the secretary of the Biology Division of the Presidium of the Academy of Sciences, was preparing to do graduate work under me and was going with me. Two female students from Novosibirsk University received pathetic grants from the university, the delightful Klara with the turned-up nose, for whom the name Katya would have been appropriate, and Katya, with her dark hair and hazel eyes, who should have been called Klara. That's just what they ended up being called—Katya who's Klara, and Klara who's Katya, and they were marvelous. One couldn't imagine better members of an expedition. Or better people. But the splendor and pride of our expedition sparkled even on such a background.

The splendor and pride, Boris Genrikhovich Volodin, is a doctor, journalist, and writer. He is the author of a biography of Mendel—part of a series of biographies called "Lives of Remarkable People." The year prior to the Kamchatka expedition he adorned my fly expedition that followed the traditional itinerary: the Nikita Botanical Gardens, Erevan, Dilizhan. Now he had wangled an assignment to cover the Far East fishing industry for the *Literary Newspaper*. His sole reason for doing so was in order to become a member of my expedition. He was writing a book on Ivan Petrovich Pavlov for the same "Lives of Remarkable People" series, and said that he would be driven to drown his sorrows in drink, since the contract had been concluded, but the life of the great physiologist and the demands of censorship diverged in the most flagrant fashion. Internal censorship, the censor implanted by Stalin in the soul of everyone in Russia who writes for publication within the country, that drunken debaucher who squeezes out conscience, demanded lies from Volodin, but he could not and did not want to lie. He wanted to write about Pavlov as he had really been. His next hero was to be my father. When he asserted that he would write a book about Jesus Christ for the "Lives of Remarkable People" series, his wife and I interpreted his words as a joke. That joke was the key to his whole personality.

He had gone to prison when he was practically a boy. They had tried to force him to give false testimony, but he refused and was given a term of ten years. By a miracle, his father's friends managed to intercede for him and he didn't serve the entire sentence. While he was in exile, which had replaced prison, he graduated from a medical institute, again through a miracle. And by a miracle he returned to Moscow and became a journalist and writer. Without his beard he looked like Evgeny Lvovich Shvarts, and when he grew a beard, he started to resemble Solzhenitsyn. He was nineteen years old when the prison barber, shaving off all his hair, showed him the razor, on which there seemed to be nothing but foam, and said, "Well, young fellow, you're completely gray." That was news for the prisoner. There were no mirrors in the quarters where he lived.

In the Soviet Union, and everywhere in the world, I think, a woman explorer, the head of an expedition, arouses curiosity that is sometimes ironic. If she is accompanied only by women, as happened to me on my Tiraspol expedition of 1946, the presence or absence of a husband on the horizon worries the women with whom contact is established. In a village near Tiraspol the landlady asked me every day where Kirpichnikov was. He was supposed to arrive at any time but kept on not arriving. She called him "Kirpich" (brick), and every day began with the question "Where's Kirpich?" If men of various ages become part of the expedition, as happened that time, one of them, so far as the people around are concerned, definitely has to be the boss lady's husband. There were four men: three young ones—Maletsky, Kalinin, and Volodin, and one no longer young—Likhachev. My young years were far behind me, but no one considered Ivan Alexeyevich and me a married couple. "Your wife just left," Volodin was told at the cloakroom, and they meant me. He looked twenty years older than he was, and people took him for my husband. I felt badly for him, but he didn't suffer, or at least didn't show it. Women were very interested in him. At the hotel in Petropavlovsk on the Kamchatka Peninsula the janitresses, maids, and snack-bar waitresses surrounded me in his absence and asked for details about him. He had to wait a long time for me and was indignant. "I'll write about what a blabbermouth you are," he said. "Well, it's yet to be seen who will be writing about whom," I replied.

Having a biographer of Jesus Christ, if only a potential one, as a staff member of an expedition isn't always convenient. By the end of an expedition both your money and the government's have been exhausted, always unexpectedly and ahead of time. There's nothing with which to pay for the baggage. The last remaining rubles have been kept aside for food. If you use that money to pay for the luggage, you'll be threatened by the torments of hunger. You have to pack all the heaviest items in hand luggage, which won't be weighed, and carry it with a single finger, with the ease of a young hippo, according to my instructions, as though it weighed nothing. Volodin, seized by honesty, put his devastatingly heavy briefcase, loaded with bottles of Armenian Cognac and muscatelle from the Red Stone Valley, on the scales, condemning us to starvation. In Akademgorodok, he tipped the taxi driver who brought us from the airport to the institute. The money was paid out of my pocket, out of the last of the money that I had hidden away just for this situation. We were traveling from the south, where it was warm. We arrived in Siberia, where the temperature was something like 26°C. below zero. (It's impossible to make young people take winter clothes with them when you head south during the summer. They were all wearing light overcoats.) I had an agreement with the taxi driver—he would wait until we had unloaded and then take us all to our various addresses. He had been paid precisely for that—again, with my hard-earned money, because the state didn't pay for taxis. There were buses, they said, and please attach the bus tickets to your final report. The institute could have sent a car to the airport, but it's impossible to guess what time a plane will arrive, considering the meteorological conditions and the roguish practices of the Aeroflot flight controllers, who double up and triple up on scheduled flights when there are few passengers. The taxi driver understood the situation perfectly. I hadn't paid what was on the meter, but much, much more. I had foreseen everything—the temperature of −26°, the light coats, the absence of an institute car, and the threat of pneumonia (to them, not to me—I had my fur coat with me). My prudence only failed to extend to the altruism of a member of the Union of Writers who had reincarnated himself in Christ. The more I fought with Volodin, the more I loved him.

His book about Mendel was being published. And the new one—about Pavlov—was the reason that in any place, no matter where we arrived, Volodin found fellow toilers of the word and got drunk with them. But just as soon as his help was needed, or if he simply was supposed to be somewhere, he was right there, Johnny-on-the-spot.

Perhaps it would be better not to write about how I became disgusted with Ivan Alexeyevich Likhachev, the best of the best of all the people I had ever met? Perhaps it would be better not to write?

Ivan Alexeyevich is a Don Quixote of a special type—a higher one, it seems to me. He served people, not an idea. And he possessed a special knack for the business of serving people. He devoted his life to other people, but without the gloomy consciousness of a duty fulfilled, not because of high-minded considerations, and without elevating himself. In giving help to people there is a great moral profit, pride, and self-elevation. Ivan Alexeyevich's spiritual generosity knew no limits. He didn't give alms. To a legless cripple he gave shelter in his own impoverished quarters, offering the cripple his own bed. The cripple ought not to be deprived of any of life's joys—along with the young people who crowded around him, Ivan Alexeyevich carried the cripple to the top of the tower on St. Isaac's Cathedral. He would go to the hospital to serve as an interpreter for people who had fallen into misfortune in a country where they spoke an alien language. He adopted and raised a boy who had lost a leg in early childhood. If some unfortunate one didn't feel himself to be a person, he soon began to in the presence of Ivan Alexeyevich. And not just a person, but an inspirer of spiritual comfort, subtle joking, and sparkling witticisms. Ivan Alexeyevich was too gaily ironic for a Don Quixote. Nothing could spoil his nature. But a vice appeared in him. To my misfortune, the vice disclosed itself during our Kamchatka expedition, in this way:

There were eight of us—Katya, Klara, Lyusya, Kalinin, Maletsky, Likhachev, Volodin, and I—but then my expedition began to disperse. Only five of us flew to Vladivostok. Katya and Klara took the train because the university would only pay for the least expensive means of transportation. Volodin flew from Moscow to Petropavlovsk in Kamchatka in order to join us there. Katya and Klara were sailing from Vladivostok on the *Ilyich*. We sent Volodin a telegram in Petropavlovsk: "Meet girls, tall, one dark, the other medium blond." The girls were supposed to be able to recognize Volodin by his *Literary Newspaper* badge. But Volodin had already given the badge away to some collector, and there were more tall dishwater blondes and brunettes on the *Ilyich* than you could count. Nonetheless, the little group met up with each other.

My own group flew from Sakhalin to Petropavlovsk via Khabarovsk. Oh, do I remember that repulsive Khabarovsk airport, which I visited on more than one occasion! But there is something pleasant to recall too. The swallow population at the Khabarovsk Airport had solved the problem of mastering space in an ideal fashion. There were fifty-four nests under the roof of the building's façade, and they were located at absolutely equal distances from each other. How was that process accomplished in time? A mystery of mysteries.

I came down with something like abdominal typhus, which I caught in a Khabarovsk café. Very ill, I watched out of the airplane window as we approached Petropavlovsk in broad daylight. Indescribable beauty! The enormous size of the huge mountains that rise over vast planes of blue water conceals the distance. Most amazing of all are the glaciers that look like antlers. I had seen lots of mountains, but I'd never seen glaciers that looked like these.

I stayed in bed in a hotel room at Petropavlovsk and all the others set off for Paratunka to the hot springs. Out of the seven, five returned. Kalinin and Lyusya had disappeared. A group of geologists took them to Avacha and promised to get them back to Petropavlovsk in two or three days, but insurmountable obstacles arose on the path when it was time to return, and Kalinin and Lyusya dropped out of the expedition makeup. Maletsky, with my blessing, left for Novosibirsk to work on his own things, which were much more important than my expedition's tasks. The idea of going to Bering Island was Volodin's. Four of us sailed. Volodin, Ivan Alexeyevich, Nina Alexeyevna Efremova, and I. Nina Alexeyevna is a botanist, a staff member at a wild-life sanctuary, a local resident, and well acquainted with the local flora. We left Katya and Klara in Petropavlovsk.

The chromosome sets of the many species of plants that grow in the Far East, including the southernmost tundra in the world, the tundra on Bering Island, are well-known. Certain species are distinguished from their closest relatives by the numbers of sets of chromosomes. One of the related species has, as is usual for every species, two sets of chromosomes. But another has four sets instead of two. The latter species came about as a result of mutation. The doubling of the number of chromosomal sets has great significance for the formation of species and the selection of cultured plants. The flora of the Far East is rich in such related plants with unequal numbers of chromosome sets. The goal of my expedition was to see how the appearance of a plant changes when its chromosomes double. Most of all I was interested in whether species of the same genus but with different numbers of chromosomes differed in the proportions of their parts. I needed to find appropriate pairs of species—and in that regard Nina Alexeyevna was absolutely essential—and then measure the various parts of the plants and calculate the correlations between the sizes. We measured stems, leaves, petals, stamens, and pistils thoroughly. We devoted special attention to measuring pollen grains. For that purpose we carried microscopes with us.

While we were sailing I gradually recovered from my illness. It was cold, overcast, and there was a gale-force wind. The sea was waving its white mane and there was nothing good in that. Ivan Alexeyevich joked and kept trying to attract my attention to objects that were changing place. In the dining room the soup shuddered in the bowls. "Look, look!" said Ivan Alexeyevich. "It's a seiche," I said. *Seiche* is the periodical oscillation of the surface of a body of water. The entire mass of water of a lake is drawn into a single oscillating motion. In the last analysis, it is not often that one has the chance to see the soup in one's bowl agitated by the Pacific Ocean. It was a most interesting spectacle for such born wanderers as Volodin, Ivan Alexeyevich, and me. It hadn't occurred to me that that was exactly what was going on.

Taught by the bitter experience of my trips through the Caucusus, when not a single expedition to the Nikita Gardens had transpired without terrible attacks of seasickness, I knew that observing objects rocking before your eyes, when both the objects and you yourself rock for hours, ends in vomiting. In addition, the flight from Kishinev to Uman on the memorable U-2 had also taught me a few unforgettable things. So fleetingly I asked myself if, just possibly, Ivan Alexeyevich could have been trying to attract my attention to the sea's waving mane and to the seiches with the goal of causing vomiting rather than out of love for unusual sights? Soon after, I saw with my own eyes that he did act upon the urging of a strange vice.

I was lying in my cabin on the upper berth, and opposite me on the bottom berth lay a girl who must have been about ten. I was very ill, and the little girl was suffering terribly from seasickness. Ivan Alexeyevich came in. "Look how the curtain is rocking," he said to the girl in a sweet voice.

"Ivan Alexeyevich, what are you doing?! Get out of here right now!" I yelled.

He left. In response to my bitter "How could you?" he didn't even attempt to justify himself. He just said, "I didn't notice you."

I had not heard of such a strange perversion before that, nor have I heard of it since, and Volodin, a doctor and former camp prisoner, refused to believe me when I told him about that bizarre pathology. Ivan Alexeyevich was caught red-handed, and in such a vile fashion that I am not surprised at Volodin's disbelief. I wouldn't have believed it either if I had not been an accidental witness.

All my life I had dreamed of seeing the tundra. And now there it was before me. Tundra in the middle of the ocean. All covered with flowers. Violets, irises, and buttercups. There were rhododendrons and saxifrages growing on a terrace cliff. Snowy mountain caps. Multitudes of animals—in the water, on land, and in the air. Gulls and cormorants. "See—there's an owl," a frontier guard told me, pointing to a little white post on a knoll. "You're joking," I said. But the little post suddenly unfurled its wings and flew off. Large fish in wedding garb swam through the water, looking like a red apparition. Arctic foxes were inspecting the line of the surf. We gathered orchids at a stream: two types, very close relatives of the butterfly orchid. Marsh marigolds and violets bloomed among mounds of tundra, close to one another. The contrast between their orange-yellow and violet blossoms revealed clever strategy in their contiguity.

Unfortunately, Nina Alexeyevna's love strategy didn't bring her to her goal. She had fallen in love with Ivan Alexeyevich and alternately waited on him and made him wait on himself, did all the dirty work in the laboratory for him or proposed that he be the one to clean up. Ivan Alexeyevich submissively did everything that was demanded of him as though that was the way things had to be, although I assured him that we were spared washing the floors. I hired an Aleutian girl to do that. Failing, in spite of it all, to achieve her purposes, Nina Alexeyevna left for Copper Island, while Ivan Alexeyevich and I headed off to the seal beds. Volodin had headed there earlier.

I recall with great pleasure the stories told by Pyotr Georgiyevich Nikulin, the head of the seal laboratory. In the bloody year 1937, he worked in a research institute on Chukotka. When the NKVD agents came to arrest him and his comrades, the scientist hunters decided that a counterrevolution had occurred and, with arms in hand, they began defending the Soviet regime. The NKVD agents retreated without accepting the battle and no attempts at arrest were renewed.

Good company for Volodin was soon found, and the two of us, Ivan Alexeyevich and I, worked on measuring. On account of our stay the island's electro-generating station worked both day and night—Volodin arranged for that. At night, shortly before departure, we were measuring bluebell leaves, stems, and blossoms. Suddenly the door opened and, without knocking, the chairman of the state fox fur farm entered. (Besides a seal laboratory, there was a fox fur farm on the island. Its director gave us our apartment.) "Why are you here?" he asked.

"And where do you think we ought to be?" I responded, not realizing with whom I was speaking.

"You haven't been listening to the radio. In twenty-five minutes the settlement of Nikolskoye will no longer be. A tidal wave is on the way—nine meters high."

We packed up the expedition equipment, the binocular microscopes first of all, the plant guidebooks, expedition journals, and a few of our own things and went up the hill, where Volodin was drowning himself with vodka over the impossibility of writing the truth about Pavlov. A cat in the village cried in a strange voice. The alarm turned out to be a false one, however.

We made our way back to Petropavlovsk by plane. The weather was clear. The volcanoes of Kamchatka were smoking. Ivan Alexeyevich flew off to Frunze. I joined up with Katya and Klara, and we worked like slaves. And my stomach stopped bothering me. Katya and Klara found out from local people that the large blue fruit of Far Eastern honeysuckle is medicinal, and they cured me with it. The four of us, Volodin, Katya, Klara, and I, spent about ten days at the hydrobiological station on Lake Dalneye. Two members of the intelligentsia, the ichthyologist Faina Vladimirovna Krogius, and her husband, the hydrobiologist Evgeny Mikhaylovich Krokhin, had founded the station thirty years earlier, and Lake Dalneye, with its spawning grounds of Kamchatka salmon, has become one of the most studied reservoirs in the world. I remember the draught dogs, rings of clouds rising from the volcano tops into the sky, two Scottish shelties running through the field while Volodin and Faina Vladimirovna's nephew played tennis, flowering thickets of rosebay and the most delicate hues of its mighty inflorescences.

I have an irresistible desire to describe the humorous details of our stay at Lake Dalneye and our departure. Volodin and I quarreled constantly, and very funny situations arose in connection with our debates. But you can't tell everything. Nonetheless, I'll tell about one instance.

Faina Vladimirovna and Evgeny Mikhaylovich offered us hospitality in unbelievably difficult conditions. During the summer their station was practically cut off from the world. There were swamps all around.

We would make our way to the station getting stuck in a quagmire up to our ankles and sometimes to our knees. You could get through on foot, but it was impossible to carry goods. In the wintertime, food was delivered on dog sled. In the summer, a motorboat very infrequently brought food from the other side of the lake, which one could reach by car with difficulty. Faina Vladimirovna offered to have us all stay at her apartment. She would cook for us herself. She was ready to make any sacrifices for the sake of her deceased teacher's daughter. She looked at me and cried, mourning my father's death. I had no intention of burdening her and did not allow myself to be persuaded, but Volodin did. Katya, Klara, and I settled in the library and cooked on a hot plate. Each day we were given a salmon each from the station. And Volodin lived at Faina Vladimirovna and Evgeny Mikhaylovich's.

Ten days passed. Work was in full swing. I remember very well that we only needed to get a few dozen samples of monkshood up to the necessary quantity—two or three days—and we would be able to return to Petropavlovsk. And just at that point, a motor boat came from the other shore and Volodin made arrangements for us all to leave that very day. He considered it simply indecent for him to burden his hosts with his presence any longer. Volodin needed to leave no matter what, and no matter what, I needed to stay and finish measuring what was left to measure. With his love of luxury, and most likely quite by accident, he had put himself in a hopeless situation. I had suggested many times that he move into the library. A folding bed could have been found. But he delayed. Now he had been punished. His despair, when I announced that I would not leave, tickled me. I understood the noble impulses that were feeding his despair. Having acted a role for the sake of appearances, I decided it was best to leave. Besides, I was not looking forward to tramping for hours through swamps on the way back. I'm sitting here writing, and I can see the sundews from the swamp right before my eyes. Sundews also grow near Leningrad. They are predatory plants that feed on insects. The Leningrad sundews have a small and round capturing apparatus, but the Kamchatka ones have a much larger and oval one.

In January 1974, the members of the Union of Writers held a meeting in the palace on Voinov Street. The gathering was dedicated to the memory of Ivan Alexeyevich Likhachev. He had passed away not long before that. He'd gone into a drugstore to buy some medicine, fell down, and died. He was buried with all the honors appropriate for a member of the Union of Writers. The coffin was displayed in one of the palace halls. The honors, however, were cut down to a minimum. Instead of an orchestra, there was a record player. An extremely short time was allowed for the mourning assembly, and only a small number of buses were available to take mourners to the cemetery. And there were masses of mourners and flowers. Ivan Alexeyevich was buried in Komarovo, not far from Anna Akhmatova's grave.

The same month, January 1974, the Union of Writers had expected to have an assembly devoted to the seventieth anniversary of Ivan Alexeyevich's birth. Instead, the assembly became a memorial meeting. My recollections of him were included in the program. People spoke a great deal and well. Ivan Alexeyevich had helped some eccentric select a name—it had to be Indian—for his pet snake, and their correspondence was carried on via telegrams. The next to the last person to speak—my presentation was to be the last one—was a writer and doctor who had been a great friend of Ivan Alexeyevich's. This friend had been a doctor in the camp where they were both serving their terms. An order came to move Likhachev out of the camp. Both of them realized that things would be significantly worse in another camp. The doctor delivered his opinion that Likhachev was mortally ill, that he needed to stay in bed, and that he wouldn't survive being transported. That clever stratagem worked for a while. But then an order came again: Move the prisoner—dead or alive. And so Ivan Alexeyevich was taken away, but he didn't know where he was going. In letters to his friend, Likhachev described his journey in a regular train car, without compartments, where his cell was improvised—one of the walls was a sheet that curtained off the aisle. He was accompanied by two guards, to whom Ivan Alexeyevich gave the food that was for him, and for whom he ran, teapot in hand, to fetch boiling water at the station stops. Ivan Alexeyevich was brought to a prison in a large city and told that he was free and could go wherever he wanted. He went out into the street without guards and found himself for the first time in many years in his native city, Leningrad. He walked to his home, saw the bombed-out building and his room suspended over the canyon of the street. It was on that day that he washed his hands under the kitchen faucet before sitting down at his niece's table. They had been tiny little tots when he went to prison, and now it was as though he were seeing them for the first time. "You look so much like Papa," the two nieces took turns exclaiming.

The very language of those letters seemed to grin. There was irony, directed first of all at himself. The words of the prisoner fluttered unfettered over the perishable nature of existence, over the contrast between an insane reality and the joy of being. I sneaked up quietly to the meeting's presidium and requested that the time allotted for the presentation by the addressee of Ivan Alexeyevich's marvelous letters be extended and that my speech be excluded from the program. The chairman refused.

I told about our pleasure trip through the Altai region. The train station at Biysk. We needed to get to the airport in order to fly to Turachak. Maletsky went off to look for a taxi. We waited for him in the park in front of the train terminal. Night was coming on. Ivan Alexeyevich said that it was better to spend the night at the train station than at the airport. It's never boring with the people, he said. A fellow walked up to us—to judge by his clothing, a construction worker or a painter, all covered with whiting. "Give me what I'm entitled to by my age," he said to me. "Sir," I replied, "I'm ready to meet any of your wishes more than halfway, but it remains a mystery to me what precisely your age makes you deserving of." "A match," he said. It was obvious from the expression on Ivan Alexeyevich's face that life was wonderful.

We drove up to the airport in a taxi late at night. There was no hotel. Pitch darkness. We spread out the tent tarpaulin and settled down to sleep. When we awoke, we discovered that it was only by chance that we had avoided death. We were sleeping in a potato field at the very edge of a one-hundred-meter drop off. The huge Biya River flowed on magnificently below. "What do you see before you? You see before you an abyss," said Ivan Alexeyevich. He was quoting an English textbook of Russian. One pearl after another.

Once an elderly man came up to us as we waited for a boat on the shores of the Chulishman and tried to sell us Maral antlers,* which are considered of medicinal value. He called them "moral horns." "Horns are always amoral," said Ivan Alexeyevich.

On one of my expeditions, I was in Frunze with Yuri Nikolayevich Ivanov and Tanya Bochkareva, a shy little lab assistant from Novosibirsk. My co-worker Golubovsky invited Tanya to come to work in our laboratory after he'd noticed with what deftness she washed dishes in the cafeteria. The little dishwasher didn't let us down. She coped marvelously with the duties of a worker, preparation assistant, and laboratory assistant. I don't remember how it happened that the two of us, Tanya and I, had Ivan Alexeyevich come to the laboratory. We removed the jar of machine oil—a fly morgue—so as not to upset Ivan Alexeyevich

* Maral is a deer inhabiting the Altai Mountains.

by the sight of it. We decided to show him the sleeping flies and return them to their vials, even if we no longer needed them. I pride myself greatly on my ability to treat flies gently. "What are you doing, grabbing such tiny little things with your enormous pincers [this is what he called our tiny tweezers]—stop it!" Never had anyone interceded on behalf of flies with such passionate emotion as Ivan Alexeyevich.

He had just then returned from Paris, where he had visited his cousin. He told Tanya and me about how his relatives had asked Rothschild to invite them to their place, because of Ivan Alexeyevich's visit. The snobbish reply was conveyed through a servant: "Ivan Alexeyevich is most welcome. His relatives aren't." Rothschild's wife was Russian. Rothschild was a poet and translator. The gray-haired lady received her guest wearing knee-length white lace pants. Her hair was separated into locks, and on each one there was a clip topped with a black velvet ribbon. These ribbons kept falling onto the floor. Ivan Alexeyevich depicted the lackey picking them up and gallantly offering them to her. There was another guest as well—an English poet and translator. Rothschild decided to acquaint his colleagues with his work. The millionaire was in the process of translating John Donne's poem addressed to an inaccessible beauty. Time in its passing chewed her charms with a toothless maw according to Rothschild's translation. Ivan Alexeyevich had translated just that poem into Russian. In it there was no time with its toothless maw. Time for John Donne is a burden, an excessive weight that shatters beauty. Ivan Alexeyevich pointed out the error. The invitation to visit the Rothschilds again, which had been made several times during dinner, was not renewed when farewells were said.

Every time that I was in Frunze, our expedition laboratory found a haven at the Medical Institute's Department of Biology. In all the previous years, Ida Adolfovna Gontar, an entomologist and the head of the Biology Department, invited all the members of the expedition to her apartment and arranged a marvelous dinner for us. That year Ida Adolfovna's generous invitation included Ivan Alexeyevich. The Gontars knew how to receive guests as well as the Rothschilds. The vases full of fruit, the antique glass and crystal wineglasses, and the various types of Cognac and wine seemed to have materialized from the canvases of Dutch painters. Ivan Alexeyevich traded impressions with the host about brands of French wines and then turned all his attention to the hostess, dressed in lace. "Do you like slang?" Ivan Alexeyevich asked cordially. "How do you like the expression 'Don't roll a barrel of beer at me,' or 'Shut your trap or you'll freeze your intestines'?" And on and on in the same vein and only in that vein. The next year there was no banquet in honor of my expedition. Tanya was won over completely. By John Donne, of course, and not by the barrel of beer or the trap, as I understand.

And with that I concluded my presentation at the memorial gathering.

# *In the Beast's Lair*

Isn't it time for me to return to the narrative long since interrupted? An abandoned wife was spending the winter with her two small daughters, Liza and Masha, at the dacha inherited from her father, and she was writing his biography. Stalin had died and it was no longer he, but his specter, that stood at the director's podium.

Vasily Sergeyevich Fedorov, a lecturer at the Genetics Department of A. A. Zhdanov Leningrad State University, was afraid, frightened to death, and continued to be afraid even after the times had changed. Comparing the old times and the new, post-Stalin ones, Genochka Shmakov, the poet, critic-scholar, and translator, said: "It used to be that the fear of death stood at the director's podium. Now the fear of receiving pressed caviar instead of soft is pounding away at the typewriter."

The office of Kirill Mikhaylovich Zavadsky, chairman of the Department of Darwinism at the Leningrad State University, is separated from the main corridor by five doors. Vasily Sergeyevich made sure that they were all locked, that there was no one behind any of them, no one to overhear, and that we were alone in the departmental office. Then he said, "You can't imagine what they are doing at the Genetics Department." The Genetics Department at that time was chaired by N. V. Turbin. "They are impairing heredity by means of suffering. Suffering results from incomplete pollination. The incomplete pollination is achieved with the help of a hair from a certain dog's tail—we have a dwarf Pomeranian. You have to pull out a hair, collect pollen with it from the anthers, and apply it to the stigma. The seeds resulting from this procedure will produce plants with impaired heredity. There is huge variation, and all that is needed is to select the forms needed for practical applications. I suggested that they conduct a quantitative measurement of suffering—count the number of pollen grains carried to the plant stigma, but they rejected that as 'Mendelism' and continued to use the hair. Just don't tell anyone I told you this."

This was in 1954, when under the influence of fear I was hired at half-salary, as a lecturer for this same Department of Darwinism at A. A. Zhdanov Leningrad State University without recognition of my degree or my previous years of service. The honorable invitation to teach at one of the most prestigious universities of the country interrupted my stay at my father's dacha and the work on his biography. Vasily Sergeyevich came to make my acquaintance.

In the past, Vasily Sergeyevich had been a co-worker of Nicolay Ivanovich Vavilov's, and had studied the genetics of peas. He was a simple, blue-eyed Russian, a party member, and of proletarian extraction. Decency destroyed him. When Vavilov was arrested in 1940, Vasily Sergeyevich refused to besmirch his mentor. He kept his job until 1948. Then he was thrown out of it. N. V. Turbin, who had taken over the Department of Genetics at Leningrad State University in 1948, met Vasily Sergeyevich on the street when the latter had reached an extreme state of impoverishment, and gave him a job. The conditions dictated were harsh: He would have to preach Lysenko's teachings. Vasily Sergeyevich preached. While I was teaching the introduction to evolutionary theory for Leningrad State University's Department of Philosophy, Vasily Sergeyevich was teaching genetics for the same department. Students would show me their notes from his lectures: "Heredity is the property of a living substance to demand definite environmental conditions and to assimilate them. It changes if what must be assimilated changes. If you operate with cold—it will be assimilated, and the offspring will be cold resistant—without any participation of natural selection." He repeated word for word the illiterate Lysenko nonsense. In the Genetics Department, Vasily Sergeyevich taught students real genetics, and they were grateful and liked him for it. But he did not argue with the local powers that be, and they used him as their puppet.

I said that I was given a job under the influence of fear and I, an inveterate anti-Lysenkoist, found myself right in the beast's lair—the Department of Darwinism. That department had been previously headed by the great, invincible I. I. Prezent. But Prezent, the brightest star in the skies of Lysenkoism, its ideologist, had been hated by the whole Department of Biology and Soil Science. Apparently he had severely mistreated them, and they succeeded in having him removed.

"Well-known people in science, your humble servant included, had tried to do that," I was told by Valentin Alexandrovich Dogel, a scholar among scholars and head of the Department of Invertebrate Zoology at Leningrad State University, "and they couldn't do a thing, but three unknown assistant professors found a way." One of those three turned out to be Leonty Evmenovich Khodkov. He was the one who, in response to Prezent's question to me at a lecture, "What is the Marxist-Leninist theory of cognition?," had yelled about the workers who built the Volkhov Hydro-Electric Station through sheer enthusiasm, without mathematical formulas. Now Khodkov was using his knowledge and party membership card to strengthen Leningrad State University's Department of Darwinism. The Lord works in mysterious ways!

Prezent continued to reign in Moscow under the aegis of Lysenko. In Leningrad he was replaced by Zavadsky—a man whose convincing tone and irreproachably masculine voice could convey a threat to anyone who still did not understand before whom he or she should be groveling at a given moment. Zavadsky, better than anyone, could sense which way the wind was blowing. He fell under suspicion as a wrecker, escaped by betraying others, and without embarrassment, boasted about it to me and named his victims. Zavadsky was also hated by the entire department. But it was a different type of hate than the hostility toward Prezent. Zavadsky was quite intelligent in comparison with the people who took the places of the purged professors from the old St. Petersburg University. And those new people hated Zavadsky just for his superiority.

To attain the position of head of the Department of Darwinism he had employed very cunning means—a skillful combination of servility and the appearance of scientific polemic. He praised Lysenko and Lepeshinskaya, but on one point—that of the nature and origin of species—he differed from them. This made him the scientific adversary of Prezent and Lysenko. Even so, he had given Lysenko and Prezent such unequivocal proof of his loyalty that now, after Stalin's death and with Khrushchev's coming to power, it was his turn to tremble for his skin.

But as it turned out, he was in absolutely no danger. Lysenko temporarily fell into disfavor. Now the Lysenkoists shook with fear. They were afraid that their all-union alliance would be liquidated. Dismissals, imprisonments, public defamation—everything that happened to each person and category that fell into disgrace in Stalin's time—seemed inevitable. And the rise of those who were previously defamed seemed just as inevitable according to the laws of the transfer of power.

The laws of the transfer of power from one ruler to another one are ingeniously portrayed by the Russian poet and historian A. S. Pushkin in the advice which Boris Godunov gives to his son and heir. Here is my brief translation from Russian into English:

> *With fear they trembled at my frown,*
> *To come to action Treason did not dare.*
> *I had to introduce disgrace and executions.*
> *You may abolish them, to win the people's love.*

Lysenkoists were sure that their turn had come to be the victims of the new tsar's "disgrace and executions," and that the Mendelist-Morganists-Weismannists—those fly-loving, people-hating, corrupt mercenaries of Wall Street—would now be given the chance to bless the inaugural throne of Khrushchev.

Zavadsky prepared me in case of a purge. A secure future was assured. If they should start to come down on the Lysenkoists the faculty had an iron-clad defense: "Please, what kind of Lysenkoists can we be? Look who we have teaching—the geneticist Raissa Berg." And if things remained as they were and the department were reprimanded for allowing an obscurantist into the classroom with the students, then, "What do you mean? She is here on an assistant's half-salary and teaches biometrics. We took pity on this woman and her children." And everything would be fine.

He told me: "I'll grab you by the throat and throw you out"—that's exactly what he said—"the minute that the situation changes. Agreed?" And I answered, "Agreed. The missing half of my salary will be payment to you for my freedom. I will teach my own material, and if the situation changes and you are ordered to use dialectical materialism to discredit the chromosomal theory of heredity, you will do so without me. I'll find something else to do."

His cynicism was boundless. He would sit me down in his office—the same one where Vasily Sergeyevich and I had talked when everyone else was gone—and say, "Now I'm going to read a bit to you, and you will tell me which scoundrel wrote it." Of course I don't remember word for word, but what he read went something like: "Only Lepeshinskaya led Soviet science out into the front ranks of world science. It was she who substantiated the possibility of one species arising from the living substance of another species by the means of a change in the particles of the living substance, as a result of the assimilation of external conditions." And so on in that vein.

I told him: "Your question is unfair. What you just read to me is not science, but the profession of a doctrine under the fear of death. Those are the ideas of Lysenko, and any hanger-on could have written that—Glushchenko, Nuzhdin, Stoletov, Kushner, Prezent, Dvoryankin, Boshyan, Dmitriev, Babadzhanyan, and so on ad infinitum. It is impossible to tell them apart."

"I wrote that," he said. "It was a symbolic protest."

"But where is the protest?"

He maintained a majestic silence. He feared that he would not be re-elected to his chair, and complained to me that the faculty dean, Georgy Alexandrovich Novikov—George was his nickname used by his subordinates—had come to him and suggested that he decline to be re-elected, since he wouldn't be re-elected anyway. I told Zavadsky: "He's lying, pulling your leg. You'll be re-elected for certain, because the moment you're blackballed, they'll get Prezent like fire. So they'll re-elect you." And they did.

The graduation theses that the department's faculty members directed were a monstrous profanation of science. The only subjects allowed were those of indubitable importance for agriculture—clover and rabbits. Goosefoot, too. Whether the experimental subject was appropriate for solving the problem involved was of no interest to anyone. One unfortunate female student was studying vegetative hybridization in black and white rabbits. It was presumed that if a white female rabbit were injected with macerated tissues of a black male rabbit, vegetative hybridization would result and the white female would bear black offspring from a white mate. And the baby rabbits really did come out a little dark. They were of the breed that grows darker when it's cold, and the heating system had broken down in the rabbit hutch.

Another student who wrote her thesis on the influence of altered metabolism on heredity defended it with tears in her eyes. Clover produces two types of seeds. One sort germinates the following year, the other kind in two years. It is a great device, and each plant ensures itself in that way against the happenstances of existence. Segregation into types is placed under strict genetic control: It doesn't depend on a given plant's environment and is not adapted to it. Seed production in clover has been complicated by that clever ruse on the part of nature. Our esteemed leaders wanted to fix it so that only one type of seed was produced. Instead of trying to create the desired mutation by means of irradiation or a chemical agent, L. E. Khodkov, the thesis director, proposed collection seeds from plants whose leaves were systematically removed. Suffering, it was thought, would alter heredity in the desired direction. But the results, of course, were negative.

Leningrad University became the advanced post of anti-Lysenkoism partly because there were people capable of reviving genetics left there, and also because scientists in contiguous areas found in themselves the courage to defend genetics. They weren't risking losing their jobs, but Polyansky, Nasonov, Takhtadzhan, Terentyev, Danilevsky, Malchevsky, and Shvanvich were still risking a great deal. None of this would have worked had it not been for the fear of getting Prezent back. How appropriate Zavadsky was in that case!

The time between Stalin's death in 1953 and Khrushchev's fall in 1964 was the period of hope and disappointment. Daring words were sometimes sounded even in lecture halls. But as soon as a document had to be produced, even the most courageous tried to foresee the future narrowing of the limits of what was authorized. Most bold were the speeches at funeral meetings where minutes were not taken and no resolutions were approved. Liberty doomed to extinction, and burial.

Veselov's biology textbook for secondary schools was discussed at a meeting of the Academy of Pedagogical Sciences, in 1959, I believe. Genetics and the theory of evolution were given the Lysenko treatment. There wasn't a word about Mendel in the book. Schoolteachers thought that the new edition was a little better than the old one. It didn't contain the most shameful chapter of Lysenkoism, the origin of the species through assimilation of the unusual environment by an individual, which starts to produce seeds and vegetative parts of another species. There were, in this next textbook, more illustrations and more of them in color. Then came time for the university professors to speak, including Yu. I. Polyansky. He said that a biology textbook without Mendel's laws was the same thing as an arithmetic text without multiplication tables, and that the treatment of genetics and the theory of evolution was all wet, and then he proposed a resolution: to approve the new edition. When I protested and cited his remarks, everyone was very offended that I didn't respect the majority opinion. I could only agree with that. If the opinion is based on ignorance, then so much the worse for the majority if they hold to it.

Khrushchev fell under Lysenko's influence, and the situation was growing worse and worse. Censorship was tightening the screws. Articles attacking geneticists and biologists who supported genetics appeared in the central press. The editorial board of the *Journal of Botany* was bombarded and an article in *Pravda* inveighed against the *Journal*'s mistaken line. Two outstanding scientists, Zhores Alexandrovich Medvedev and my former husband Valentin Sergeyevich Kirpichnikov, dared to speak out against Lysenko, and in the confusion of the Khrushchev era the journal *Aurora* published the article. Both scientists and *Aurora* got into a lot of hot water.

Thunder was crashing down from on high. Lysenko received a rostrum on the pages of *Pravda*. The sad, sad epic of the bull calf of the high-fat milk Jersey breed that was to be crossed with low-fat, but productive milk cows in order to derive in the very first generation a new Soviet breed of high-fat milk cattle that would give extraordinary quantities of milk, was played out precisely at that time. Before the hybrid cows started to refute Lysenko's expectations and the milk yield started to drop, Lysenko was in a hurry to add the laurels of a livestock expert and soil scientist to his agronomist laurels. One proposal followed another. His *Pravda* article in the spring of 1963 took up four entire pages of the newspaper, which had an insert for this occasion.

I was still working at the university. Nikolay Lvovich Gerbilsky, a talented bioendocrinologist, a subtle researcher, a marvelous pedagogue, and the chairman of the Department of Ichthyology at Leningrad State University, made a request of our department, where I had been the only geneticist for nine years. He asked us to organize a seminar where he could make a presentation. We called a seminar together, and he spoke on the significance of Lysenko's works for science and practice. He said that we needed to reform as quickly as possible, since a new rout of genetics was inevitable. We needed to follow Lysenko's ideas with the greatest attention and try to understand him. With great smugness, this elderly man, a member of the intelligentsia, very slender, with a good profile and a beautiful smile, said that he had occasion to chat with Lysenko at a conference. They found themselves at the same table in the snack bar. And Lysenko turned to him and addressed him using the informal form of the pronoun "you." He said something about Gerbilsky's presentation, and Gerbilsky talked about ways of getting mature span and milt from fish heading for spawning grounds but unable to reach the grounds because of hydro-electric dams. And then Lysenko suddenly asked, "Do you know how to raise a puppy?" (I should point out that Gerbilsky is a specialist in raising puppies. He's an enthusiastic amateur dog breeder, and his home is full of dogs. I once met him by chance in a line at the West German book exhibit at the Russian Museum, and we talked for four hours about dogs. It was so interesting that I didn't even notice how long I'd been standing.)

"Yes, I can," answered Gerbilsky.

"Well, how do you begin?" asked Lysenko.

Gerbilsky said, "I go to the drugstore, buy a nipple, and make a milk formula."

"Right," said Lysenko. "And would you buy a toothbrush?"

"Why?"

"You have to brush a puppy on the underside of its tail, otherwise there won't be any defactation."

Gerbilsky repeated the word just like that—"defactation," instead of defecation, because that was the way Lysenko pronounced it. Those were the last words of Gerbilsky's report.

The leader had to be idolized in spite of his illiteracy. Gerbilsky triumphantly awaited our reaction. I said that this was shameful, that a great edifice of genetics had been erected in the Soviet Union and that it had been erected by great people, and that now it had been destroyed and on its rubble resounded the voice of an unbridled ignoramus and arsonist. Gerbilsky was not at all expecting that and he suddenly seemed to wilt, his vigor spent. He was very ill by then and apparently it had taken every bit of his waning energies to swagger as he had, and now he suddenly began to look very sick and only said that he considered my statements a personal attack.

He was part of a current heading in the direction of a new disaster. But there were crosscurrents as well.

The history of genetics is a chain of triumphs. The end of the 1940s and the beginning of the 1950s was the era when the physicochemical bases of heredity were discovered. Parents transmit to their children a certain substance, a cell, or a part of one, and an organism develops from that substance. What was this: a ready-made person who only needed to grow, or something not at all resembling a person, but capable of developing and becoming a person? It turned out that it was neither the former nor the latter. Along with cytology—cell science—genetics showed that parents transmit to their children a coded record of their traits, a cipher of hereditary information. The development of the embryo is the translation of that record from the language of genes into the language of traits. That was established by geneticists even before their science entered its physicochemical and molecular phase of development, when it was still in its prechemical phase, I would say. The gene is the basic unit of hereditary information. It changes, mutates, and new characteristics show up in the offspring. Genes received their names according to the traits whose development they regulated. The sequence of their arrangement vis-à-vis each other had become known. But no one knew *what* they were. The situation with vitamins was similar. Diseases caused by vitamin deficiencies indicated that they existed. It was known where one vitamin or another could be found in abundance. Finally, their chemical composition was recognized.

The year 1948, a fatal one for Soviet genetics, inaugurated a brilliant era in the understanding of genes. The physicochemical nature of genes was studied step by step, the intimate mechanisms of linkage between genes and traits were uncovered. Both languages—the one in which the recording of hereditary imperatives was made, and the other one, the language of developing traits—were decoded, and not just in principle, which in and of itself would have been a great discovery, but in detail. And at the same time that great things were being accomplished in the West, V. P. Efroimson was displaying insane boldness by trying to prove that the accuracy of distribution of traits in the progeny of differing partners, the very basis of gene theory, was not an invention on the part of the Mendelists and Morganists, but a fact, and that the numerical equality of males and females necessarily follows from it.

In the light of the ever-larger achievements of science, Lysenkoite witchcraft looked quite indecent and it was becoming more and more difficult to pass off Lysenkoism as the most advanced science in the world. Information was trickling through, although in the libraries on the display shelves, one found excised photocopies of journals, not the originals. It was as though the offending articles didn't even exist. The earth was filled with rumors. Those who had ears heard. The very background of science changed, I would say. Thanks to the works, talents, and courage of Kolmogorov, A. A. Lyapunov, and Sobolev, another seditious science, cybernetics, that lackey of imperialism, began to win the rights of citizenship and send out offshoots in the "enslaved furrows"* of the Soviet field. And genetics entered into contact with it.

* "Enslaved furrows" is an expression used by Pushkin in one of his poems: "Into enslaved furrows my animating seeds were thrown in vain."

Cybernetics, biochemistry, biophysics, mathematical methods in biology, systems theory—all of that was now possible, and the languages of these sciences became the Aesopian language of genetics. A "unit of hereditary information" sounded less anti-Lysenkoist than "a gene."

By command from Alexander Danilovich Alexandrov, the long-suffering chancellor of Leningrad State University, an all-university cybernetics seminar was organized. By a whim of fate and the play of chance, I became the seminar's chairman, though neither by rank nor area of specialization could I lay claim to such a high post. The first of a series of presentations on the topic "Cybernetics and Genetics" was set for December 24, 1958. The code of hereditary information, its physicochemical and biological nature, the structure of self-reproducing systems, and cybernetics laws that govern the mechanism of evolution were to be the contents of my papers. In addition, there was supposed to be the formulation of the law of nonhereditary nature of traits acquired during individual development. The Penroses, a geneticist and a mathematician, had just created a very interesting machine. The machine arranged parts in a linear order, copied its own structure, and tore its copy sequence away from itself, to begin self-copying again, as long as there were parts. The machine functioned exactly like a chromosome. Darwin's theory had just been translated into the language of cybernetics by my teacher and friend I. I. Schmalhausen. According to Schmalhausen, the feedback mechanism from the trait to the gene that forces the gene to improve itself, creating those clever genes we possess—genes knowing what has to be done in life and development—is natural selection, operating in a population with its alternating generations. The signal that something is wrong in gene action, the feedback signal, is taken into account by the offspring population, whose genes are improved because the possessors of stupid genes died or did not succeed in producing offspring.

On the eve of the talk, Pavel Viktorovich Terentyev called me and said: "Don't deliver the paper. An article attacking the editorial board of the *Journal of Botany* for publishing articles in true genetics has been printed in *Pravda*. If you give your paper, the seminar will be prohibited, and there will be trouble for the chancellor. Say that you're ill." But I gave my talk. I called the chancellor beforehand, and it turned out that he wasn't afraid.

The seminar wasn't prohibited. And I invited N. V. Timofeev-Resovsky, Efroimson, with his theories of immunity and cancerogenesis, and the great thinker Malinovsky, the son of A. A. Bogdanov (the same

Bogdanov with whom Lenin polemicizes in his book *Materialism and Empiriocriticism*).

A. A. Bogdanov was the founder of a new branch of science—tectology. Tectology, or as Bogdanov called it, general organizational science, is the theory of systems. Cybernetics—how systems are managed by feedback links—was part of its makeup. Malinovsky expounded the history of cybernetics, including his father's ideas. This all was heard at a seminar which had become a citywide affair and was printed up in the university newspaper. There was even an article about the seminar in *Izvestiya*, by O. Pisarzhevsky. My paper on living systems attracted attention, and I repeated it at a scientific session at Leningrad University. In particular, speaking about metabolism, I departed decisively from Engels's formulation, which introduced metabolism into the definition of life. The exchange of matter is not the basis of life; rather, it is the reason for death. The basic structure of life—the chromosome—does not have an exchange of matter. No flow of matter passes through it. It builds copies of itself and tears them away from itself. It is immortal. The cell, too, is immortal, as the result of divisibility. The cell is the only system that combines an exchange of matter and immortality. A multicellular organism is a system that, like the cell, includes metabolic devices—lungs, kidneys, intestines—in its makeup. They wear out. And the more complex the system has grown and the more individualized, the more divisibility is forfeited. Death and the immortal soul came out into the arena of life arm in arm. Death, payment for immortality of the soul, is the child of the exchange of matter. All that is primitive is immortal—the cell, the chromosome. Only the multicellular organism is mortal. At the foundation of life there is only the accumulation of matter and energy, but there's no exchange of matter.

I received a note during the presentation. On the outside it read "Not to be answered." The text read: "It's because of people such as you that genetics perished. A. S. Danilevsky." That meant: You and your kind exceeded the measures of the allowable, and the rout of genetics was retribution for violating the limits of freedom outlined on high. A. S. Danilevsky—Alexander Sergeyevich Danilevsky—is A. S. Pushkin's linear great-great grandson and a descendant of Gogol and Danilevsky.

He was moved by the best feelings. He wanted to show me what he thought was the correct strategy for maneuvering between the truth and its violation upon order from above. He warned me, wanted to defend both genetics and me from a blow. But I had no intention of maneuvering and never considered maneuvering a way to victory. I didn't answer his note. After my presentation, Danilevsky came up to me and repeated his reproach. "No," I said, "it was because of rabble like you that genetics perished. Because of the treachery of scholars and scientists. Physiologists praised Lysenko as a geneticist and an evolutionist. The geneticists hailed him as a physiologist. The evolutionists praised him as a geneticist and a physiologist. He received the highest attestation from biologists in all branches. Each one followed the principle of dogs who chase a thief to the neighbor's gate. And they got what they deserved." He looked at me sadly. His face was so refined, very much like Pushkin's pen-and-ink self-portrait in profile. He didn't say anything and walked away. I felt sorry for him and I was sorry that I'd been excessively rude to him. He was an outstanding entomologist. And he had never participated in any vileness. He'd kept silent. He wasn't angry with me. He forgave me—he's an aristocrat. When I saw him for the last time he said that he was disillusioned about people. That's not a very profound assertion for a person with his genealogy and at his age—he was about sixty.

In 1960, I was invited to give a presentation for the Dialectical Materialism Department of the School of Philosophy at Leningrad State University. Nothing notable took place there except for my acquaintance with a woman philosopher whom I'll call Maria Ivanovna. She invited me to lecture on "Genetics and Cybernetics" for the USSR Academy of Sciences Institute of History of Natural Science and Engineering, along with a philosopher whose specialty was the history of chemistry. He was to speak on "Problems in the Formation of Models."

Maria Ivanovna met me when I arrived at the academy at the appointed time. Her look was so frightened that I could hardly recognize her. She informed me that the District Committee of the party had sent a representative to the seminar. "What do we do now?" "Well, I'll try to use lots of scientific language, and I won't mention Mendel or Morgan—maybe he won't catch on." We agreed. But I couldn't avoid mentioning Schmalhausen. His was a new word in science—I wasn't the one who had thought it up. And the representative caught on. The director of the institute poured more oil on the fire, saying, "The geneticists didn't give up. They kept on fighting. Scientific progress keeps giving more proof of their rightness, and from a wider perspective it is now clearer just how right they were." After the lecture Maria Ivanovna came up to me and said, "It's a disaster. He understood. He wants to know where such a person came from, and why she's been honored by being allowed to preach revanchism at the academy. He called us reactionaries. He wants to speak to you."

"Is he old or young?" I asked.

"Young."

"Well, I am a lady of a respectable age," I said, "so if he wants to have a talk, let him come to me."

They led him in. Although young, his face was haggard, pale and wrinkled—the portrait of Dorian Gray after his crimes had left their imprint on it. He started talking. Yet I could not catch what he was saying, and not because he was speaking softly or indistinctly—I don't know how he was speaking because I couldn't hear the sound of his voice at all. I looked at his face, which expressed such an extreme degree of spite and expressed it so ridiculously that I had to use all my strength to keep from bursting out laughing. That's why I didn't hear anything. The other lecturer, a scholarly, decent, and very handsome young man, approached us, and the representative turned his spiteful gaze away from me. At that moment his voice cut through. "You said that it is possible to form a model for anything," he was saying. "You are mistaken. You can't model infinity." The handsome young philosopher replied that he had had in mind finite objects.

Here I jumped in: "What are you saying? That's totally wrong. When my daughter was in the third grade, fractions were being taught. One assignment was to draw squares and circles, divide them in halves and quarters, and color them in. And she colored and said, 'You can divide a square in half with four different lines, a pentagon with five, and hexagon with six, but a circle can be divided by an infinite number of lines. That's because there are no angles in a circle, or you could say it has an infinite number of angles. So zero and infinity are in some ways the same thing.' And there's a model of infinity for you. And the theory of limits? Not to mention everything else."

"Your daughter will be a mathematician," said the representative. "Or a logician," said the handsome philosopher. And the party representative turned to me, and there was not a trace of spite on his face. He asked, and this I heard perfectly well, "Tell me, are you the daughter of academician Aksel Ivanovich Berg?"

"No," I said. "I'm the daughter of another academician."

"Then you're Lev Semyonovich Berg's daughter? I was a member of the Geographical Society at the Palace of Pioneers, and your father was honorary president."

So how did you fall to such a level of subservience? I wanted to ask. But I held my tongue, and everything worked out all right.

What did I do at the Department of Darwinism? I gave lectures for the Biology, Geography, and Philosophy departments. My courses were called "The Genetic Bases of Evolution," "Evolutionary Genetics," and "Darwinism." In certain years, the title "Darwinism" was replaced by "Introduction to the Theory of Evolution."

At the Philosophy Department I taught a huge course on Darwinism and Genetics with laboratory work using the fruit fly to study genetics, and I took great pleasure in sowing seeds of doubt in the souls of future workers on the ideological front. Had the spiteful philosopher been a bit younger, he would have turned up as one of my students. My relations with philosophers were excellent. The seeds that I sowed were germinating. Some of the students abandoned philosophy and became geneticists.

The one and only time in my life that I felt an insurmountable desire to go to an open party meeting had been in March 1956. The meeting was originally to be held in the assembly hall. Then it was announced that the meeting was moved to the History Department's large auditorium. The assembly hall in the university's main building was not suitable for audiences in the thousands. Its mezzanine was in danger of collapsing. Prince Menshikov and his treasury pilfering were to blame for that, or perhaps two-and-a-half centuries had taken their toll. I didn't even try to sort that out, but just went to the History Department. The auditorium was filled to overflowing. An entire row was occupied by geneticists: M. S. Navashin (a beautiful silvery-gray head, a noble form that did not correspond to the cowardly content), Olenov, Chuksanova, Feodorov, and other of Turbin's guard officers. Turbin himself was no longer there; he was in his native White Russia, enjoying the blessings which his rank of a full member of the White Russia Academy of Sciences entitled him. The geneticists saw that I was looking for a place. The auditorium's beautiful benches gave them the chance to crowd together a bit, but the geneticists didn't make a move. "Come sit with us," my philosophers called to me. The hall was getting more and more full. The doors were closing and opening wide under the pressure of the crowd. There was agitation outside the doors: "Let us in!" A fire hose was dragged in, not to aim at and frighten away the people trying to gain admittance, as it first seemed, but to tie shut the doors, which no longer latched shut. The stream of people was cut off, but there were too many people in the auditorium, that is, from the point of view of those who had to lead the parade. They made a wise decision. A one-armed Savonarola announced: "All those who aren't party members are to leave." I was about to go, but the philosophers held me back. Time passed, and no one moved. Savonarola repeated his demand, to no avail. The bosses' complete organizational helplessness was obvious. "They don't know how to deal with the population," I told the philosophers. "I'd put things in order in ten minutes." "How?" "The person farthest to the left in each row should be ordered to compile a list of all the people sitting in the row, and a punishment should be announced for disobedience." But a more humane strategy worked. A thin formation of nonparty people was making its way up the aisle. A beautiful woman of very refined looks was walking on crutches. People guessed that these were obedient teachers from the foreign language departments—people who had survived the arrests. They didn't reach the exit. Savonarola's heart apparently was moved. He said quietly, "Go on back to your places." And they did.

Then the agenda was announced: the text of Khrushchev's secret speech at the Twentieth Party Congress, where Khrushchev heaped all the blame for the misdeeds of the Socialist regime on Stalin. Savonarola read the speech. When it came to the part about Stalin's role in Kirov's murder and about torture, the hall groaned. Were these groans expressing grief, pain, compassion for all the suffering of the tortured? Or were they an indication of disbelief, a protest against the false accusations smearing their idol, Stalin? An analysis of the transcript of the speech gives the answer.

Khrushchev's de-Stalinization speech was never published in the USSR.* But here before my eyes I have the full text of it, *Khrushchev's De-Stalinization Speech, February 24–25, 1956*, published by the U.S. Congress† and reproduced by Basyl Dmytryshyn in his book *USSR: A Concise History*.‡ In it, the Kremlin audience's reactions to Khrushchev's presentation—tumultuous, prolonged, or loud applause, laughter, murmuring—are indicated. The audience I was in at Leningrad State Uni-

* It is possible that in the recent era of *glasnost* (openness) and *perestroika* (rebuilding), heralded by Gorbachev and marked by a new peak in condemnations of Stalin, this epochal document will finally be made public. To ascribe all the failures of Soviet life not to the false idea implanted by the ruling clique through terror but to the imperfection of Stalin's character was Khrushchev's strategy in his struggle for power. The circumstances of Gorbachev's era are just as unsettled as they were after Stalin's death.

† *Congressional Record*, 84th Congress, 2nd Session, vol. 102, part 7, pp. 9389–9402.

‡ Second edition. (New York: Charles Scribners Sons, 1971).

versity reacted identically. References to the torture of innocent people were met with indignation, but no indignation was expressed as Khrushchev passed along Stalin's telegram to the Party committees sanctioning torture and oppression as "both justifiable and appropriate." People in both audiences had no reason to be surprised.

At the Department of Darwinism the part-time assistant lecturer, with her salary of 87 rubles 50 kopecks, had other duties besides teaching courses. I was supposed to engage in research and deliver my scientific production in the form of articles. The university was preparing a collection of works on the topic of species and I was to write an article on types of polymorphism. The collection was to come out in May: I had arranged my time so as to finish the article by that deadline. It turned out, however, that the deadline was March 1. Anticipating a repression, and knowing that I was a thorn in the side of Zavadsky, George, and many others, I wrote an official note to the Dean—George—explaining the reason for my delay in fulfilling the plan. I listed all the conferences and conventions in which I had been a participant and attached a list of publications. There were many articles, but they were not on the topic called for by the plan.

In May, I received a summons to go see the Dean of Sciences of the university, not just of the Biology Faculty, Kondratyev. Not long before that, I found out that a document had arrived from the Academy of Sciences Council on Genetics. The council proposed the organization of a radiation genetics laboratory, and my name was mentioned as the potential head of the laboratory. I knew Kondratyev's book, *Solar Radiation*, very well, having gone through it when, exiled from genetics, I was studying the ozone screen in connection with the origin of life on earth. I foresaw a pleasant conversation and an offer to head the laboratory. It turned out, however, that I was profoundly mistaken. George had recommended that I be discharged and Kondratyev had taken upon himself the task of telling me about this. A production plan was a production plan. The article for which I was being discharged for not having written was ready. I had the manuscript in my briefcase, but I wasn't about to show it to Kondratyev.

"Did you see my note of explanation?" I asked Kondratyev. He had seen it, but it was only proof that I had failed to fulfill the plan. "Tell me," I asked him, "do you realize to whom you are speaking?" I meant my position, not my name or profession, and of course not my merits. He understood. "You're an assistant professor in the Department of

Darwinism." "You've been misled," I said. "I'm a half-time teaching assistant, with my degree and experience waived. I receive eighty-seven rubles, fifty kopecks a month, nominally. I was certain that you had called me in here to propose that I head up a laboratory." That was the end of our conversation.

I went to the departmental office, gave the typist the first two pages of the manuscript and the tables, showed the manuscript to Zavadsky, and went home. I still had to compile the bibliography. When I got home, however, the manuscript wasn't in my briefcase. I'd left it at the department. I called the person on duty that day—it was the graduate student Bolotova-Khakhina. The telephone wasn't working, so I went to the department first thing the next morning. The manuscript was nowhere to be found. Bolotova-Khakhina said to me, in the presence of the typist, "Maybe the manuscript never existed?" "This is a question of honor," I told her, "only yours, not mine."

The editorial committee was to meet in three days. I returned home, sat down at my desk, and spent about fourteen hours reproducing the entire text from memory. I even added a bit to it. And the typist didn't let me down. Fifteen minutes before the meeting I was pasting the tables into the article that was hot off the typewriter, and making final corrections in the typed text. At the last moment everything was ready. I flew up the stairs. I made it in time.

I had left my Leningrad-brand fountain pen on the table where I was gluing and correcting the manuscript. I returned there five minutes later. The pen was gone. The staging was extremely obvious. I would ask Antonina Pavlovna where my pen was. Antonina Pavlovna was Sofochka's analogue, but she was completely illiterate, although on the other hand, as she proudly announced to every newcomer, her brother was an investigator at the Big House.* So, in response to my query Antonina Pavlovna would say, "You lose everything yourself and then blame other people." I had never accused anyone of anything; the words about honor had been my sole reproach. But I had no intention of being an actor in her play. Having determined that the pen was missing, I didn't make a sound.

A lot of time passed—a year or a year and a half, I don't remember. I walked into the room where the final scene of her play had failed to come off. Antonina Pavlovna was sitting, lost in thought, over a pulled-

* The Big House is the Leningrad analogue for Moscow's torture chamber, the Lubyanka.

out desk drawer and holding my fountain pen in her hand. She looked absently at the pen and then transferred her wandering glance to me. I looked absently at her and pointed at the spot on the table where the lost pen had once lain, and said in a low voice, slow, and with no modulations, as if remembering a forgotten dream: "It was right here that I left a pen just like that one on May fifteenth at 12:59 P.M." Without the least bit of animation, she handed the pen to me and said scarcely audibly, "I'm not the one who took it."

No, she wasn't the one who took the pen. Someone else took it and told her to hide it, which she did. I know who took the pen and who destroyed my manuscript. It was Khakhina. Circumstantial evidence of her involvement did not keep me waiting, but enough of that. I remember to this day and I'll never forget the numbers of the fly strains that were destroyed, experiments that came to nought because of fly food that had been spoiled, falsified records of courses taught, false denunciations. All with Zavadsky's kind blessings. He didn't care if the department's prestige suffered, as long as nastiness was done to me.

Ultimately, Kondratyev did not sign the order for my dismissal, as George had insisted. Zavadsky didn't want me to be discharged. I was extraordinarily useful to him. He had no objections to having me as a slave. With his typical cynicism he told me that after my conversation with Kondratyev, the latter organized a committee to examine the department's work and headed it himself. Zavadsky began complaining to him that I wasn't fulfilling the plan. He wasn't the least bit embarrassed to tell me about his lie. Kondratyev said, "You say she behaves like a free artist? Let her remain a free artist and do whatever she wants."

The department's fate was hanging on a thread. Not a single student wanted to work under the guidance of Zavadsky, Khodkov, or Gorobets. It was only at my insistent request that the chancellor agreed not to close down the department. When I went to work in Novosibirsk, the department dissolved.

# *Correlation Pleiades, or the Model of a Husband*

While the Department of Darwinism did exist, I was busy with a host of other things besides teaching courses and publishing articles required by the plan (called "delivering scientific production"). I was continuing research that I had begun in "exile" (that is how I refer to that unfortunate period when no director of any establishment or institution had the right to offer me a job). Doomed to unemployment for life, I had discovered *correlation pleiades*. The term was coined by that great wit Pavel Viktorovich Terentyev. When the exiled Terentyev was out of a job and without laboratory equipment, he measured frogs. He found that the size of certain parts of the organism did not depend on the size of other parts or on the magnitude of the organism as a whole. He grouped the indexes of dimensions of various parts in such a way that the indexes that fluctuated independently of each other fell into different groups, while all the dependent ones were put in a single group. That was how correlation pleiades came about. The idea and the term date from the end of the 1920s, the same period when Schmalhausen, simultaneously with Huxley but independently of him, was working out his theory of heteronomous growth. Heteronomous growth is the independence of the rate and duration of growth of certain parts from the rate and duration of growth of other parts of an organism. In measuring the dimensions of various organs in chicken embryos, Schmalhausen did not even suspect that he had discovered the same phenomenon that the exiled Terentyev called "correlation pleiades." And Terentyev did not see the link between correlation pleiades and

Schmalhausen's theory. It is most likely that at that time he did not even know of its existence.

Correlation pleiades are one of the manifestations of heteronomous growth. Schmalhausen published his works in Ukrainian (he began his career in Kiev), in Russian, and in very respected German journals. Fame and honors came his way. The exiled Terentyev submitted his article about correlation pleiades to Russian and German journals. No one understood him and no one wanted to publish him. Finally, the Russian scientist's article was printed in German in the British journal *Biometrica.* That happened in 1931, exactly a quarter of century before my article was published in *The Journal of Botany.* Terentyev's article didn't attract anyone's attention, including mine. The fate of ideas resembles that of their creators. By the time that I turned up in the beast's lair—excuse me, the Department of Darwinism at the A. A. Zhdanov Leningrad State University—in 1954, a year after Stalin's death, Pavel Viktorovich Terentyev was the head of the Department of Vertebrate Zoology there, and his correlation pleiades were at the center of attention for biologists of the most diverse specializations. I was responsible at least in part for the latter development. I had discovered correlation pleiades in plants and begun to publish articles without suspecting about the hundreds of frogs that Terentyev had measured and remeasured, lengthwise and across, inside and outside. I soon had the good fortune to make a major contribution to the history of sciences. I forced Terentyev to remember his long-forgotten work.

The idea of organizing a seminar in mathematical biology belonged to Terentyev. He became its chair, and he invited me to help him in the capacity of secretary. The mathematician A. A. Lyapunov gave a presentation at the first session. He spoke of how the process of inheritance looks from the point of view of a mathematician, cyberneticist, and engineer. It was a great pleasure to listen to him. At the second session, A. A. Lyubishchev was presenting a paper. He talked about the selection of traits by which it was easiest to distinguish species. It was necessary that the traits not be linked with each other by correlation dependence.

The discussion of Lyubishchev's paper became the starry hour for the theory of correlation pleiades. I spoke about correlation pleiades, rather, about plant characteristics that fluctuated independently of each other, since Pavel Viktorovich's astrally beautiful term had not yet ever been pronounced within the range of my hearing. Pavel Viktorovich spoke about frogs. We joined up. Our seminar's second session became its last. In its place, an annual conference on the biological application of mathematical methods was started. The first conference was devoted to correlation pleiades. Terentyev, Lyubishchev, and I gave papers.

I set myself the task of trying to understand the origin of independence in the process of evolution. Independence as adaptation. An absurd phrase? No. Independence from certain components of the environment ensures adaptation to other components of the environment. In some cases life and death depend on the strictness of the standards. I examined Terentyev's correlation pleiades in the light of Schmalhausen's theory of stabilizing selection. And I returned the theory itself to its sources, to the principle of heteronomous growth. Schmalhausen himself had never emphasized that link to his cardinal ideas. It was my job as a historian of science to uncover it.

The evolution of ontogenesis—the increase during the process of evolution in the degree of independence of certain parts of the organism in regard to other parts of the same organism! Filatov had discovered this law of progressive evolution. Schmalhausen provided the general principles for understanding independence and strict genetic conditioning with his theory of stabilizing selection. I applied his train of thought in order to show how Terentyev's discovery, the most surprising of phenomena, the independence of the size of certain parts of an organism from the size of other parts, is generated in evolution, i.e., in the process of generations replacing one another. The parts all grow together, they make use of nourishment together, but there suddenly comes a moment when the growth of one organ comes to a halt, while the other organs continue growing. Nourishment reaches the organ, and the organism is young and growing, but something independent of the environment and of the nourishment tells one of its parts "Stop." That something could only be a gene. A shut-off valve is installed inside each cell. All the representatives of a species possess it. Didn't that mean then that the genes responsible for the independence of certain parts in relation to others give their owners better chances in the struggle for life, that they increase the probability of leaving progeny? I had solved the whole riddle of why independence was necessary and how it was generated in the process of selection for the most stable states.

I've spent my whole life going from one surprise to another. After the expedition to Dilizhan and Erevan in 1940, I was ill for a long time, and the spring found me in the Karadag on the coast of the Black Sea. The poppies were blooming. If there was a rain, the flowers that unfurled were larger. If there was no rain for several days, the flowers were small.

In Yazhelbitsy in that fatal year 1950, when the Gypsy woman burdened me with her prophecies, my little house was awash in flowers. The heads of cosmos plants amazed me with the diversity of their sizes.

The standard and independence leapt out at me in 1953. Nasturtiums wound around my dacha, which was shaded by pine trees. Everything that I grew reached gigantic sizes. My nasturtiums had leaves the size of saucers. I don't recall and never will recall how and why I found myself at my neighbor's dacha or what academician Barannikov's widow and I talked about. I was looking at the nasturtiums. The path had lots of plants on both sides, and the plants near the dacha weren't too bad, but farther away they were pathetic. The leaves were about the size of a ten-kopeck coin. But the flowers on those plants did not at all correspond to the worsened growing conditions, as the stems and leaves on those unfortunate creatures did. The flowers maintained the standard size. And that wasn't all of it! The flowers on Barannikova's plants were not a bit smaller than my flowers, which were hidden from the human eye by leaves that exceeded the area of her neglected sprouts by a hundred times.

I thought that I was making a major discovery. The evolution of a flower is accompanied by the diminution of the number of parts—fewer petals, stamens, pistils, sepals, everything that makes up the flower. This is the general rule of progressive evolution—to diminish the number of parts performing one and the same job. More, different parts, each with its distinctive function, is the indispensable condition of progress. Every organ has to function being represented only once or as a pair of symmetrically located designs. Not only is economy thus ensured, but the accuracy of functioning as well. The elements of symmetry are involved in that process of diminution. The fewer identical petals a flower has, the fewer imaginary vertical planes divide it into equal sections. These imaginary vertical planes are elements of symmetry, and the general tendency of progressive evolution is to reduce their number to a single one.

The nasturtium flower is very different from the poppy. The latter is radially symmetrical, with four imaginary vertical planes that divide it into equal halves. The nasturtium has a single plane, like all freely moving animals, like you and me. Entire flourishing huge plant families—*Orchidacaeae, Scrophulariaceae*, the mint family, papilionaceous plants—have flowers provided by a single dividing plane. All the flowers that possess the symmetry of a mobile living thing are standardized within the limits of the species. I am a Darwinian and I do not recognize any mystical powers directing development along a path that leads to progress. There isn't anything that selection, the chance search for perfection, could not do. But, no matter how the organization of a living thing might organize the opportunities for that search, no matter how it might dictate the specter of arising changes, the moves remain a matter of chance. The field on which the drama of life is played out is not a chess board but a green cloth around which the playing figures dance. This is a commonly held opinion and I share it.

There is nothing more fascinating, pleasant, and risky than destroying a commonly held conviction and along with it your own. Standardized nasturtium flowers are not achieved in evolution because it's more profitable to be standardized. It does happen without selection. The advantage results in economy of construction. Standardization of size is a by-product of selection for economy not an adaptation in itself. It wasn't selection—my first thought—but the requirements of engineering artistry that led plant flowers down the evolutionary path.

All of my early speculations turned out to be a mistake. My eyes were opened to the truth at a talk by Boris Nikolayevich Shvanvich. Shvanvich, a professor of entomology, reported on the pollination of papilionaceous plants by bees and bumblebees. The insect doesn't wallow in the pollen of these plants, which are noble, if you consider the brilliance of their construction or, on the contrary, which are low if you consider their stinginess. The plant needs a winged carrier for its pollen—a bee or bumblebee. The plant is willing to pay for the transportation with a little nectar or a bit of pollen. Each plant species localizes a tiny portion of pollen in a specific spot on the insect's body. Shvanvich showed wonderful depictions of insects sucking nectar and taking pollen from clover, from bird's-foot, from trefoil, and from alfalfa blossoms. A little lump of pollen has to wind up in the spot on the insect's body that will touch the stigma of another flower before collecting a new portion of pollen. So the plant places its pollen in such a way that the bee's cleaning apparatus, the little brushes and combs on its legs, cannot completely clean them.

At Shvanvich's lecture, I realized that I hadn't made a major discovery and that my standardized flowers did not disclose laws of evolution that were independent of selection, like the laws of crystallization. I had provided an oblique proof of Schmalhausen's theory of stabilizing selection. The bees of a given species were standardized, their larvae developed in cells of standard size. An unbroken chain of events leads from the genes that govern a bee's building instinct to the standardized flowers of those plants whose pollen is transported by bees. You want to localize pollen on the body of a creature whose size is standardized? Please be so good as to be standardized. If your little lump of pollen misses that little spot with which the bee knocks against the flower's stigma, your genes will be cast out from life's arena to the chaos of selection. The vile principle of ends justifying means applies to insects and plants. Plants force the insect to adopt a certain pose, often an absurd and very painful one. The insects, trying to economize on effort, gnaw openings in the nectaries and cheat to get their nourishment, avoiding having to transport pollen. A flower's symmetry, with its single dividing plane, the growing together of parts, the diminution in the number of homologous parts, and finally, the standardization of flower size are the result of selection in favor of economizing the expenditure of pollen, the result of competition among creatures to whom much if not everything is permitted. Both plants and bees are guileful. The guile of nasturtiums shows through in every detail of their beautiful flowers. They are furnished with a "spur," as botanists call the horn-shaped container of nectar, a *cornucopia* full of most delicious drink. The three lower petals are closed together and form a landing strip where an insect can sit down and suck. A canopy made up of the upper petals that are grown together is furnished with indicator signs that say "Nectar here." All of this is a complete deceit. There is shamefully little nectar in the very long horn-shaped container. In order to suck it, the insect must extend its proboscis to full length, lean on its chest at the entrance to the spur, and lie flat with its legs outstretched on the flower's landing strip. A nasturtium flower draws a bumblebee's legs apart. The three component parts of the landing field separate under the insect's weight. The flower's construction, as well as its size, and the correlation of sizes of the flower's various parts are essential means for forcing a bumblebee first to touch the flower's stigma with its chest and surrender the alien pollen adhering to it, and then press its chest, the place where formerly there was alien pollen, firmly against the stamens and take on a new portion of pollen, intended for another flower.

The role of selection in the creation of standardized flower size was obvious. Everything had been made clear: who serviced the department of technical control and how its sorters worked. In transferring pollen from flower to flower, insects fulfill the role of calibraters. They chop off anything that diverges from the norm, since nonstandard flowers carried the lump of pollen past the sole point on the insect's body where it had to be.

All of this is the case if the size of the pollen carrier is standardized and the carriers are a single insect species. But if just anyone can pollinate the flowers, there will be no standardization. There is no reason for it to turn up. The poppy is a vivid model of that.

Where is the model of a husband!?! I wasn't joking or being spiteful when I called my husband high-minded and noble. In the battle for truth he is a fearless, irreproachable knight. He is brave and selfless. He is a talented scientist. His baseness is strictly limited to his relations with women.

After Elena Alexandrovna Timofeeva-Resovskaya met him, she told me that the blame for the split doubtless lay with me, because such a marvelous person as Kirpichnikov couldn't do anything bad. Sanctity itself spoke with her mouth. I didn't bother to disillusion her. Kirpichnikov had as his goal passing himself off as a highly ethical person, a model of morality, and the kind, trusting, and thoroughly virtuous Elena Alexandrovna, knowing Kirpichnikov as a selfless fighter for the freedom of science, and condemning me, demonstrated that he had quite achieved his goal.

My female friends from school, before whose eyes my family drama was played out, were of exactly the opposite opinion. When I said that Kirpichnikov was a first-class scientist who was fearless and unbending in the battle for truth, they were indignant, every single one of them. My high evaluation of him was evidence, in their eyes, of my blind feeling for a person who deserved nothing but condemnation, and from their point of view, it was shameful to nurse such feelings. But I could explain everything coherently to them. I was absolutely objective. Nothing that concerns sex is subject to control on the part of those categorical imperatives that regulate a person's social behavior.

The independence of moral criteria in various spheres of existence is related to the same category of phenomena as the independence of the size of a flower from the size of the plant as a whole. "Female matters" form their correlation pleiade of characteristics, and everything else goes into another pleiade. The inability to predict certain traits of character on the basis of others had done me a bad turn. I had married a highly moral man but found myself the wife of an unexpectedly negative person. The model of a husband that I put together with the help of flowers concretized the universal vague reasoning about the strange mixture of good and bad traits in every person's character.

Correlation pleiades have not played only a negative role in my life.

Having lost a husband because of them, I still don't feel that I have the right to regard them as unconditional proof of the world's imperfection. Correlation pleiades were a royal gift to me from a fate that has not always been kind. I have things for which to be grateful. From out of the most tragic events of my life, from rejection, humiliation, being doomed to loneliness—Eve expelled from Paradise without Adam—from the bottom of the abyss, my soul emerged at the end intact and unharmed thanks to creativity, the joy of learning, a step forward on the path of learning the laws of the natural universe. Correlation pleiades are the link in the chain of my spirit's victories over disaster after disaster.

The theses that I sent to the Tenth International Genetics Congress, Montreal, in 1958 were devoted to correlation pleiades. The independence of the dimensions of certain parts of an organism from other parts made an impression. I was elected chairman of the section on population genetics. Two journals offered to publish my articles, and I published in both of them. But I didn't have a chance to become a participant at the congress.

I was the only person at Leningrad State University to receive a personal invitation to the congress. A. D. Alexandrov, the chancellor of the university, took me under his wing. He tried very hard to obtain permission for me to go. He appealed to the Central Committee's Division of Science, which at that time was headed by V. N. Kirillin.* At Alexandrov's request, Kirillin arranged for me to be included in the delegation. I came very close to being a participant in the international congress. I'll tell you later when it was that I found out that the great scientist B. L. Astaurov, a single geneticist among the gang of Lysenkoist delegates to the congress, had refused to go, and how and why that happened, but I also refused.† Due to correlation pleiades, to the invitation to chair the section at the international congress, and especially because of my refusal to go abroad, Alexandrov took an interest in my fate.

Alexandrov and I met in Komarovo under a light, warm, barely sprinkling rain. "Don't you want to go to the Congress?" he asked me.

"Yes, I do. I'm dying to go, but not when I take all the circumstances

* Kirillin was replaced in that position in 1980. He fell into official disfavor at the same time that A. D. Sakharov was exiled to Gorky.

† In the USSR, scientists invited to go abroad to become members of scientific meetings have to apply for permission to the highest echelon of power. A delegation to the meeting is formed out of those who have proven their devotion to the party line.

into account. And why are you worrying about the congress? Participating in the congress is a cake, but I don't even have bread anymore."

He didn't understand. "You mean your part-time position?"

"No, the part-time position is fine with me. Otherwise I wouldn't be able to do any research. Teaching would eat up all my time. But I don't have the chance to work with flies. I need a laboratory." I was already working with flies, but not at the university. It was at the X-Ray Institute, without pay, and out of love for art. I'll tell about that later. But I hadn't had the chance to renew my population research.

The chancellor understood. "Write me an official note. I'll arrange to have the department examine the request."

A scandalous drama was played out at the Science Council when my project was discussed. Lobashev didn't show up. Zavadsky remained silent. B. P. Tokin, a bloody figure, a scientist of the new type, like Dubinin, used Schmalhausen's name as a synonym for a force hostile to true Michurinist science. He said "Schmalhausens are dumping garbage on our department." Tokin's use of the plural of Schmalhausen was not a slip of the tongue. He was using the language of newspaper tirades condemning the "enemies of the people," Bucharins, Kamenevs, and Pyatakovs. To increase the impact, the names were not even capitalized. Tokin's scornful intonation imitated this impressive style. He said that instead of granting money for the organization of a laboratory for me, the money should be granted to expand his laboratory. They said that I didn't have any publications, and the editors of the journals where my articles had been printed sat right there and didn't say a word. Gerbilsky said that I was a child prodigy who had failed to develop into anything. Dubinsky, who had gotten the laboratory of radiation genetics that was intended for me, said that I had a purely female attitude toward science; Pinevich, the deputy chancellor, shared with everyone gossip that he had received from Khakhina and Antonina Pavlovna from the Department of Darwinism.

The chancellor arrived in the middle of the meeting. He listened and listened, and then walked up to the podium and said one sentence, "I knew that monstrous things were going on in this department, but I had no idea that they were this monstrous." He left. The meeting continued as though nothing had happened.

The chancellor of the university really had learned some monstrous things going on in the Biology and Soil Department of his university, but he hadn't learned everything. Things were much worse than he imagined. If he had had a clear idea of the morals of the department, he would have informed *me* that, at his order, issued after he witnessed discussion at the Science Council I was to be given a laboratory, funds for the purchase of equipment and for paying lab assistants, and a full-time position as a senior scientific worker. But he neglected to tell me that. After the Department of Darwinism had received money to be used for organizing the laboratory, the money was given by Zavadsky to Gorobets, a true believer of Lysenko's, to create agricultural sorts of tomatoes while altering heredity with the help of suffering. The funds designated for me slipped away. I continued working completely dependent on my persecutors. Two lab assistants, Shirokov and Averyanova, allegedly were assigned to me, but they did nothing for me and did not even report for work. I did receive the salary due a senior science worker, and the children and I could finally eat our fill.

I learned the truth about the laboratory by accident. The departmental administration overdid things in attempting to cover up the traces of its crime. I was given a reprimand for the tardiness of one of my laboratory assistants, Yura Shirokov. Though I had spent my whole life long in university and academy laboratories, never had I heard of the head of a laboratory or a laboratory sub-unit being given a reprimand for an employee's tardiness. I didn't go to the administration to sort things out. I asked Yura what on earth was going on. And out of friendship he told me a secret that everyone knew except me: Gorobets was running the laboratory, the funds for equipping it were in his hands, my laboratory assistants were officially registered with his laboratory, they weren't obliged to do a thing for me, and so that I wouldn't guess what was going on I was given a reprimand for my alleged co-worker. If it hadn't been for that strange reprimand, I wouldn't have suspected a thing. The laboratory assistants weren't working, but that's in the nature of things. A lab assistant who works is a lucky accident. The laboratory assistant Averyanova filled the ranks of those who played their dirty tricks on me and destroyed fly strains. The only difference was that she knew which strains were the most valuable, and the others didn't. But Yura Shirokov didn't harm me or anyone. He was infinitely stupid. The only things that could have opened the university's doors to him were a party member card and a fencer's foil—he was an all-union champion.

Much later, Yura Shirokov came close to playing a fatal role in the history of our family. It was in 1970. He already occupied the high post of deputy dean for the Biology and Soil Department of the university and headed the entrance committee. By then I had long since left Leningrad State University. Masha, my youngest daughter, now a young married woman, was trying that year to enter the Zoology Division of the same department where Yura Shirokov was now a most important person. Nothing in her application gave any indication that she was my daughter—a dissident's daughter—since she had her husband's name. But her looks were highly inappropriate. When my pure Jewish genes combined with my former husband's Russo-Jewish ones, the results were two contrasting combinations. Blond, gray-eyed Liza is a Russian beauty. Brown-eyed Masha, with her chiseled nose and chestnut hair, is a Jewish beauty. If anyone should be flunked at the entrance examination in order to preserve the national purity ordered from above, then it was precisely she.

After she had given knowledgeable answers to all the questions on the examination ticket at the zoology exam, they began to ask her questions about botany. And they were such difficult ones that taciturn Masha—Manka, who wouldn't make a peep even if you tore her to pieces—protested. The chairman of the entrance committee is too lofty a figure to actually work during the examinations. But here there was conflict. He was called in. He decided to finish off the business with a single strike of his foil. But to the department's misfortune, he had once been my laboratory assistant, long, long ago, and Masha was my daughter. He asked her: "What is the difference between a population and a species?" She replied automatically: "A species is a system with a closed genotype, as opposed to a population, whose genotype is only relatively closed." Shirokov was defeated. Masha passed.

"They reproach me for having dirtied the department in regard to nationality while I've been dean," Zavadsky told me in one of his heart-to-heart conversations. "Kheysin, Olenov, and you." But he'd gotten things confused. I had been invited in 1954 by A. L. Takhtadzhan, a very skilled pilot who was capable of avoiding the most perfidious underwater rocks in the ideological ocean. It was he, not Zavadsky, who was the dean then. Zavadsky had profited from Takhtadzhan's diplomacy. In 1961, Takhtadzhan called me and said, "Submit your research on stabilizing selection to us at the Institute of Botany. That will serve as your doctoral dissertation and I'll be your opponent."* He pro-

* In the USSR there are two scientific degrees: a lower one, Candidate of Sciences, and a higher one, Doctor of Sciences. The latter gives the right to hold the position of full professor. The doctoral thesis defense procedure includes critical presentations by three authorities in the field of science explored by the pretender. They are called "opponents," and are summoned to evaluate the work before the scientific council of the institute where the defense is taking place. Members of the scientific council decide by ballot whether the pretender deserves the doctorate or not.

nounced the word *opponent* as though it had at least four *p*'s, not two.

I put together a volume. The usual difficulties that stand in the way of anyone who doesn't have access to free goods from his place of work turned into a blessing for me. I had to buy the typing paper at a stationery store. There was no paper there. The paper that I finally obtained was so thick that the typewriter wouldn't take four copies. Even the third copy was borderline acceptable. The volume had to be typed twice and I became the possessor of six copies. A dissertation goes through the approval of an office where the unfortunate victim of bureaucracy works. The volume came to Zavadsky for his judgment. More than a year passed—he was too busy to look at my dissertation. I decided to secure the evaluations of botanists. I called Takhtadzhan. (This was in February 1963, during the Symposium on the Multisystems Study of Art.)

"Please allow me to bring you a copy of my dissertation," I asked him. It turned out that he already had my dissertation. Zavadsky had given it to him and he could return it to me. I got the volume. It was the first copy, with remarks in it both by Zavadsky and Takhtadzhan. Zavadsky's temperament wouldn't allow him to limit himself to the margins. All up and down the text of the manuscript he wrote: "Nonsense!," "Absurd!" But he wrote with a pencil, not ink, although he may have failed to notice that it was an indelible pencil. He didn't touch on the heart of the matter. Takhtadzhan wrote in the margins thickly with a plain pencil. He exposed my botanical ignorance. He hadn't checked with specialists in groupings to see whether I was using the latest names for species or whether I was using outmoded terminology. He didn't trust my designations and the Science Council at the Institute of Botany wouldn't accept my dissertation. I should show the specimens of plants I studied to the specialists at the Institute of Botany and rely on their designations.

I can easily explain the nature of his objections with a single example. One of my research subjects was the dandelion. In the Leningrad region there is only one species of dandelion—*Taraxacum vulgare*. Formerly it was called *Taraxacum officinale*, but later it was renamed. I knew about that and Takhtadzhan didn't. He made fun of me while pointing out my error. But Takhtadzhan and Zavadsky are two very different people. At the last moment Takhtadzhan softened. I should check with a specialist in the flora of the Leningrad region at the Botany Department of Leningrad State University. He should go over my designations, and then we'd see what could be done. The specialist in the flora of the Leningrad region turned out to have been one of my fellow students at the university. He heard me out and then leaned across his desk and whispered, in reference to Takhtadzhan, "He's afraid." And there was reason to fear. A new catastrophe was approaching. Genetics wasn't in the clear. It had only one nostril above the water. Everyone knew that the blow would come at any second.

No mistakes in my designations were found. "What should I do about the dandelion?" I asked. "Use the old term. Takhtadzhan will see that you've been obedient." I changed the name and glued little strips of paper into all six copies of the dissertation. I didn't have to retype the first copy, the one that had been ruined by Zavadsky. "There was no good luck, so bad luck helped," goes a Russian saying. Because the shortage of writing paper made it necessary to type on thick paper, I had two first copies, thanks to the God of Critical Commodities.

As I was on my way to the Botany Department, I met Alexander Innokentyevich Tolmachev, the head of the department. He called me into his office and began talking about mechanisms of pollination. We talked and talked, and then I said to him: "I thought that I'd added something new, but after talking with you I see that even without me everything was quite clear." He said that it was precisely my works that he was basing himself on, but that he'd illustrated them with his own examples. He offered to write a preliminary evaluation for me. And he wrote it. When I saw what he'd written the world went black before my eyes.

On the last page, where the seal of the department's notary attested his signature, stood the following: "The author thanks the foreign scientists Dunn, Lerner, and Stebbins for the help that they gave her in publishing articles abroad. Can we award the degree in view of this circumstance? Wouldn't it be more correct to abstain from that?" I went to see him, showed him the place in the volume where I had excised the offending passage, and asked him to write the evaluation again. He did.

Do not condemn him for his vigilance. Do not condemn me for my obedience. Alexander Innokentyevich was subjected to terrible persecution in Stalin's time. He had spent many years in exile in Siberia before I met him. He returned to Leningrad from exile when Khrushchev sanctioned the rehabilitation of victims of Stalin's terror. During the last years of Stalin's life, a reference to the works of foreign scientists, let alone the expression of gratitude to them, could serve as a reason for arrest. Tolmachev carried his fear with him throughout the rest of his life. And I was ready to capitulate on such a minor matter as expressing gratitude to American scientists in a manuscript that none of them would ever see. I had expressed gratitude in order to show my opponents the high evaluation of my work on the part of genuinely great minds in world science.

Preliminary evaluations accumulated in the administration offices at the Institute of Botany, the dissertation was going around from hand to hand, but it had yet to be officially recommended as ready to be defended. The Department of Darwinism had absolutely no time to look at it. "Khodkov can't be expected to put off sowing his clover or Gorobets to delay planting his tomatoes because of your dissertation," Zavadsky told me. No meeting, no evaluation, no doctorate degree was the meaning of his words.

It was here that Vasily Sergeyevich Fedorov got involved. At that time he was the director of the Peterhof Institute of Biology. It was neither ambition nor toadying that elevated him to that post, but silent obedience. He broke his silence for the first time and raised his voice against those whom he had served faithfully up until then. He said that no departmental meeting or evaluation was necessary. I was a staff member of the institute and the institute would give me a recommendation and look at the dissertation. That shook up the Department of Darwinism. Averyanova handed me an evaluation signed by Zavadsky, the head of the department; Agayev, the party leader; and Khakhina, the union leader. I ought to know it by heart, but I only remember one sentence. It read: "Mendacious, twists facts in her favor." I made copies of that evaluation as souvenirs and sent the original to the chancellor of the university with a request that it be turned over to the university archives.

Zavadsky, Agayev, and Khakhina accused me of debauching young people. After reading that evaluation, I showed it to Ivan Ivanovich Kanayev, a well-known historian of science. Ivan Ivanovich presented me with an antique edition of Plato's *Socrates' Apology*.

Many people liked my work, but not everyone. At a lecture I was invited to give on stabilizing selection in the evolution of plants at the Biophysics Institute's Laboratory of Radiation Genetics, N. P. Dubinin, the organizer and head of the laboratory, was late for my presentation. When Dubinin came in, I handed him the manuscript of my latest article without interrupting my report. When I finished, the chairman of the gathering asked Dubinin to give us his opinion. He spoke in a lyrical tenor voice. "I was late. Raissa Lvovna gave me her manuscript. But I forgot my glasses. I borrowed Boris Nikolayevich Sidorov's. I took a look and thought 'How terribly this is written.' " He then fell silent and didn't say another word.

The year 1957 was a historic one in my life. After a ten-year interruption, I returned to my flies. I had been studying plants, and they quenched my thirst for knowledge. Returning to my flies meant re-opening an old wound and resurrecting the awareness of irreplaceable loss. I was afraid of flies. When I would go to Leon Abgarovich Orbeli's dacha to measure the flowers on fireweed, he would encourage me to study flies at home. He held up academician Kapitsa as an example for me. But I did nothing—until two people played a decisive role in my return to flies.

In 1957, Alexander Ilyich Klebanov called me. He was the husband of Natalya Vladimirovna Eltsina (the one who looked like Botticelli's Madonna) and had just been released from prison camp. He was a historian, a specialist in Old Russian religious art, the history of the church, and religious heresies. His first words were: "How are the flies?" I told him that I was studying plants, that in the Department of Darwinism where I was teaching there were no facilities for working with flies, that I was not planning to take up that work anew, and that I was afraid of irritating a wound that was gradually healing.

He couldn't believe his ears. "I was arrested while I was writing. The manuscript broke off in the middle of a word. When I returned, I put my pen on that same spot and continued writing."

In the spring of the same year, Samuel Naumovich Alexandrov, the head of the carcinogenic laboratory at the X-Ray Institute, invited me to participate in an expedition studying the influence of increased background radiation on living organisms. The study would be conducted in the northern Caucasus in the region of curative springs. These areas were noted for the high background of natural radiation. The frequency of occurrence of mutations would be studied. There would be many subjects, the fruit fly among them. No one knew how to do that. Anyone who had known how had forgotten. One of the expedition participants was to be Claudia Feodorovna Galkovskaya. Klavochka Galkovskaya! The same Klavochka whom the Lord had sent to adorn my life twenty years earlier when she and I had measured the frequency of occurrence of mutations in the city of Uman. At that time she had caught the very same flies among whom I had discovered the yellow males, the carriers of the mutation "yellow," my guiding star of Bethlehem. She was the co-author of my first article on population genetics. She was a merry woman, full of energy and goodwill, who did everything absolutely with the speed of lightning.

We met, and she laughed—and the flow of silver sounds rolled on and on, that's the way it had sounded back then, and that's how it sounded now, too. But she had forgotten how to measure the frequency of occurrence of mutations. So wouldn't I agree to take part in the expedition? The institute wouldn't allocate money for me, but they would pay my way to the northern Caucasus and back out of their own funds.

The connection between the frequency of occurrence of mutations and the dose of radiation is quite well-known. The founder of radiation genetics, my teacher and idol H. J. Muller, and after him many of his followers, among them Nikolay Vladimirovich Timofeev-Resovsky, studied this connection exhaustively and gave it a theoretical grounding. Knowing the radiation dose, one can predict the frequency of occurrence of mutations exactly. Knowing the difference in the doses, one can calculate the number of tests necessary for discovering the differences in their mutagenic action. The scale of the experiment is easy to calculate. I decided to determine whether the game was worth the candle, whether we could conduct experiments of the necessary scale. "But what's the background radiation there?" I asked Samuel Naumovich. He told me. "It's a lost cause," I said. "Even if there is any increased mutability with such a background, it's so insignificant that we won't find any differences. There's no point in wasting state money as well as your own." He implored me: "The topic has already been included in the plan. Let's go, and you'll do whatever you want, and we'll help you." I agreed, reluctantly.

Working with Samuel Naumovich and Klavochka and later with Natasha Pronina in Leningrad was sheer pleasure. We did research on the populations at Pyatigorsk and Inozemtsevo. The first year was spent laying the groundwork. We had to study the mutability of flies in their natural habitat and among their progeny raised in the laboratory. I had never encountered such low mutability as we found in Pyatigorsk, and particularly in Inozemtsevo. The research was carried out and when it was time to submit our scientific production we wrote an article comparing the mutability with my earlier data. We proposed several hypotheses regarding the low frequency of occurrence of mutations with increased background radiation. Some of the hypotheses did not link the frequency with the natural radiation level and some did. We sent the article, recommended by Schmalhausen, to *Reports of the Academy of Sciences*. An answer came immediately. The article was worthless and was written in laboratory jargon. I wrote the main editor of the *Reports* and asked him to deliver us from illiterate editors. That seemed to be the end of that.

But as soon as Khrushchev initiated atomic tests and the word *megatons*, which no one had ever heard of before, revealed the might of the Land of Soviets and, for those who were capable of hearing, the threat to people's health and lives, all the barriers to the publication of our article fell. We were urged to hurry. Our description of low mutability with increased background radiation had now acquired a political overtone, and that overtone turned out to be of a quite vile character. I don't even recall whether it was Alexandrov or I who said, "And will decent people be willing to shake hands with us after the article's publication?" It could have been either of us, because we were in complete agreement. We reclaimed the article from the editorial office.

Samuel Naumovich and Klavochka continued studying the mutability of flies in the northern Caucasus, while in 1958, having obtained funds from the chancellor's office at Leningrad State University, not the School of Darwinism, and having heard my fill of threats and abuse from the school's and the institute's administration, I went to Uman to study the fate of the golden knights. They had become a great rarity. Mutants had almost disappeared. Mutation occurred as infrequently as in 1946.

Just as in the late thirties, I observed a global pattern of mutability. Twenty years ago high mutability was inherent in every fly population studied at that time. Now the mutation rate was low everywhere.

I'm returning again to where I left off telling about what I was studying at the Department of Darwinism. In addition to activities not provided for by the plan—correlation pleiades and flies had been personal rather than departmental business until only recently—and fruitless attempts to defend my doctoral dissertation, I was giving genetics courses for doctors and writing a brochure together with Sergey Nikolayevich Davidenkov, a neuropathologist.

The fear of exceeding the measure of freedom ordained from on high, so as not to have even that freedom taken away from one, is a very important variety of fear. I had occasion to become acquainted with it and experience its pernicious effects when colleagues, under the influence of that category of fear, opposed the brochure's publication. It all began in 1957 or 1958 when I was asked by the *Medical Encyclopedia* to evaluate Sergey Nikolayevich's article. The article was entitled "Genetics."

Sergey Nikolayevich Davidenkov was a unique phenomenon in the Soviet medical world. At seventy-eight, he was an outstanding neuropathologist and a great specialist in the field of genetics—not just medical genetics, but general genetics as well. His *The Evolutionary-Genetic Bases of Neuropathology*, a book remarkable in all respects, appeared in 1947. He was a full member of the USSR Academy of Medical Sciences, a professor at the Institute for Advanced Medical Training for Doctors, and a consultant for the Kremlin Hospital. He was driven from all his administrative and teaching posts in 1948 and remained active only as a practicing doctor. The Kremlin did not reject his services.

The fact that the *Medical Encyclopedia* had commissioned an article from him and then chose me to review it was a symptom of "thaw." The *Encyclopedia* showed courage. In his article Davidenkov was bold enough as to use the word *gene* and not link medical genetics with racism. He wrote that medical genetics was developing rapidly abroad and in our country.

I wrote in the evaluation:

> An encyclopedia is not the place for polemics, nor is it a weapon of false information. Davidenkov is perfectly well acquainted with the fate of medical genetics in our country. He cannot help but know about the brilliant scientific and organizational successes in that field in the 1930s. In the whole world there was no establishment like the Institute for Medical Genetics in Moscow. Pavlov took it under his patronage. When Pavlov died in 1936, the institute was liquidated and its director, S. G. Levit, was physically destroyed. If the time has not come to speak up and, for reasons that have nothing to do with science, one has to be silent, then one should be silent.

I also pointed out the gaps in Davidenkov's description of contemporary achievements in Western medical genetics. These written evaluations, as is well-known, are undertaken anonymously. The reviewer and the author are unknown quantities to each other. But the *Medical Encyclopedia* did not observe that rule. The evaluation was sent to Davidenkov, and he called me to ask whether we could see each other and talk. He was ready to come see me wherever and whenever I wished.

There was not even the shadow of offense in his words. He had forgiven me my exposés. I grabbed the books that I would need in order to show him what he had missed and set off to his place. Thus was renewed our acquaintance, begun in the 1930s, when I attended a series of training lectures in statistics he had organized at the Advanced Training Institute.

In 1959, Davidenkov was asked to found an Institute of Medical Genetics within the framework of the Academy of Medical Sciences. He called Leningrad's geneticists together, and they hemmed and hawed harmoniously, saying that there were no people who knew genetics and that there was no one to work in such an institute. It would be better to wait until the Genetics Department at Leningrad State University could supply the personnel. They were all well set up, managing laboratories in biological establishments. They had all been traumatized, had endured persecution, and the fear of recidivism had turned genetics into a taboo subject for them. They were sitting here in Davidenkov's luxurious apartment on Revolution Square, eminent professors who were capable of producing hundreds of specialists, who were deprived of the opportunity to teach genetics and who were now rejecting that opportunity when it presented itself. I attempted to persuade them, and Davidenkov made his proposals, but it was obvious that he, too, was under the sway of this general passivity. Nonetheless, Davidenkov organized a laboratory of medical genetics which exists to this day. He died, however, without having had time to outline the laboratory's goals.

In 1958, Davidenkov was honored with a distinguished invitation from one other high source, one much higher even than the *Encyclopedia*. The puddles from the short Khrushchev thaw were continuing to grow larger. The Society Knowledge asked him to write a brochure entitled *Heredity and Hereditary Diseases in Man*.

The Society Knowledge is a very lofty organization. The Central Committee's Science Division stands directly behind its back. Anything that the society publishes has governmental sanction. The issuance of a popular brochure on human genetics would have signifed a great strategic victory. Davidenkov agreed, on the condition that I be the coauthor. The contract was signed. The brochure was ready by July 1961, and Davidenkov signed the typed manuscript. Two days later he died.

There are in all living organisms, including humans, several types of transmission of abnormal hereditary traits from affected parents to their offspring. A man suffering from achondroplasia, a stumpy dwarf, transmits the disease to half of his offspring, both to his sons and to his daughters. The trait is dominant. This means that only one gene inherited from either the father or mother is needed for its manifestation. This gene is transmitted by men and women and the probability of inheriting this single detrimental unit of hereditary information is one and the same for representatives of both sexes and it manifests itself in all of them with fatal inevitability. Transmission of hemophilia belongs to another type of gene that passes from parents to their offspring. A man suffering from hemophilia transmits his pathological gene only to his daughters. His sons are as safe as are sons of a healthy man. Daughters, unlucky to have the pathological gene, have, however, good luck in not manifesting the trait. Their pathogenic gene is recessive. Half of their sons are potential hemophiliacs and in them the trait manifests itself with the same fatal inevitability as achondroplasia. This is an example of a recessive sex-linked type of inheritance. There are hereditary diseases whose type of inheritance is recessive but not a sex-linked one. White tigers in the zoos all over the world (I have seen them three times, in Philadelphia, Miami, and New Delhi) have inherited the recessive gene from the single specimen captured in the wild.

Difference in prognosis of the probability of giving birth to an affected child depends on this diversity of types of inheritance. Davidenkov wasn't stingy. The brochure he wrote illustrated types of inheritance of diseases by family trees of sick people. Every family tree was based on tens of people whom Davidenkov himself had examined, and no fewer than three generations were under his observation. Such a compilation had never before been published.

The editorial board was satisfied. The process of getting evaluations began as well as the search for an editor, and it was at this point that there arose obstacles that could not have been foreseen.

Wonderful people of irreproachable boldness, people of scientific honor, refused to review the brochure on the basis that it was a bad piece of work, or else they wrote negative evaluations, saying that it was all outmoded, that what was needed was not types of hereditary transmission from generation to generation but the role of DNA, RNA, and proteins in the inheritance of human diseases. This seemed clearly untrue, given the brochure's purpose. I called one of the reviewers, a great friend of mine, and said: "Poorly written? That's the sort of thing that a decent person tells the author, but doesn't write it down for the editorial board. What's with you?" And he explained things to me.

The goal of the criticism was not to correct the brochure, but to prevent its publication. Now was not the time for publication of works on human genetics. A sword of Damocles—racism—is suspended over everything linked with human heredity. Surely you don't want genetics to be forbidden again? First it has to make its way into agriculture, then into the universities, and only then into medicine.

The manuscript lay at the editorial office. Bad times were ahead. The editor who had commissioned the brochure was removed from his post because of something unconnected with genetics. A new editor appeared—a young man by the last name of Soroka, who looked amazingly like the spiteful member of the Geographical Society at the Palace of Pioneers, a representative of the party District Committee, who was sent to watch my report on problems of cybernetics in biology. Soroka instructed me in the proper tone of contemporaneity, and I listened and heard. It was the gospel of accommodation. All the Russian sayings about the necessity of submitting to the caprices of the powers that be were used, and there are many of them. A single person in a field is not a warrior. You can't chop wood with a penknife. A slingshot isn't a cannon. A private isn't a general; and so on in that vein. And in addition, he said that although the book was a medical one, it couldn't be published without references to the gardener Michurin. Michurin had nothing to do with anything. He was a sort of fig leaf. What was implied was that questions of medical genetics were to be addressed from Lysenkoist positions. The genetic basis of certain human diseases was to be called into doubt. Absolutely all diseases were the result of the influence of external conditions. He was saying all this to a dead man's co-author.

Then he disappeared, as though he had dissolved, and I had dealings with a certain Romanova in the evaluation department. The years passed, the brochure lay there, and Romanova fell ill. Finally two evaluations were received from powerful figures in the Academy of Medical Sciences—Zhukov-Verezhnikov and Maysky—both of them opponents of Davidenkov. Maysky, thoroughly negative, wrote that the 300 types of fruit-bearing cultures derived by the great Michurin were not even mentioned in the brochure, and that it reeked through and through of medical opposition to Michurinism, which certainly was not permissible.

I spoke with Romanova once. "We need to educate the people," she said, "and in order to do that we have to inspire love for our national heroes Michurin and Lysenko and to propagandize their ideas about remaking nature."

"But you'll never educate anyone with lies," I told her, "quite the contrary."

"Why, what do you think you're telling me?" she said. "As though I didn't know. I'm a student of Karpechenko's myself, but there's nothing that anyone can do." Georgy Dmietryevich Karpechenko, the beautiful Karpechenko who seemed to be shot through with sunlight, had been a professor at Leningrad University and had created a productive cross-gender cabbage-radish hybrid. He was killed the same year as N. I. Vavilov.

I wrote to V. N. Kirillin, who was the chairman of the Society Knowledge, the head of the Central Committee's Science Division, and the vice-president of the USSR Academy of Sciences, and asked for his help. I sent along some of the junk that was being passed off as medical genetics at the time. "You need not be a specialist to be persuaded of the ignorance of these authors," I wrote, "while S. N. Davidenkov, a great authority in his field, has been condemned to silence. Hundreds of thousands of sick people need the correct diagnosis and prognosis for the inheritance of their diseases. Help them and help their doctors."

I wrote in vain. I received replies from high-ranking officials, Kirillin's assistants, on beautiful stationeries. They promised to summon me for a meeting, but never did. In 1964, after Khrushchev fell and genetics was revived, the rope restraining the sword of Damocles was replaced and had become thicker. I had been working at the USSR Academy of Sciences Institute of Cytology and Genetics in Novosibirsk, when my colleagues—geneticists, professors at Novosibirsk State University, and heads of laboratories at our institute, wrote evaluations and asked to have the brochure published. In vain.

I decided to take back the manuscript from the editorial board of the Society *Knowledge* and made an attempt to have the booklet published by the Science Publishing House of the Siberian Branch of the USSR Academy of Sciences. There was obstacle after obstacle. The brochure was included in the publisher's overall plan. Then it was thrown out. Included again.

Seven years had passed since it had been written. "You tell me that it will be a best-seller. It has Davidenkov's name. Hundreds of thousands of doctors will want it. But we're not interested," Falaleyev, the director of the publishing house, told me. "If our publications sell out, our production plan will be increased. Try to understand our position."

The year 1968 came and my chances for publishing the manuscript fell to zero. I was one of the "subscribers," as wags at the editorial office of one popular science journal called us. Forty-six people, I among them, spoke out on behalf of political prisoners, Ginzburg and Galanskov. I was forced by KGB, the secret police, to leave Novosibirsk. I took the manuscript back from the editorial office. In Leningrad, I asked S. N.

Davidenkov's widow for help. A neuropathologist herself who had formerly been one of Davidenkov's graduate students, she headed the Laboratory of Medical Genetics founded by him. At that time, or perhaps a little later, she was elected a corresponding member of the Academy of Medical Sciences. She was a party member and very well connected. She agreed politely to take a look at the manuscript. I gave her a copy. The mousetrap slammed shut. "You are in the anti-Soviet camp. You transmitted slanderous information to the West," she told me over the telephone. "I cannot permit Davidenkov's good name to be spoiled by co-authorship with you."

Her telephone may have been bugged. Mine couldn't have been, but she didn't know that. I lived in a communal apartment. I think that the indiscriminate bugging of telephones was limited to private telephones. I have indisputable evidence that my telephone, while it was mine alone, was bugged, and likewise, my telephone in Akademgorodok near Novosibirsk was the object of attention. However, wasting scarce tape to record the conversations of people living in communal apartments or assigning an employee of the KGB to listen to them would have been too costly for even the richest organization. It is simpler to appoint an informer from among the neighbors. My informer lived in the room next to the telephone, as I've mentioned earlier. I amused myself, reproducing in my imagination the tape recording of conversations captured from morning to night in the hallway of our communal apartment: "Cooked some noodles on a milk base. Dyed two sweaters, but I don't like the way they turned out—I'll do them again." All of that yelled out, without the slightest modulation, and for hours on end.

Davidenkova never returned the manuscript to me. She didn't limit herself to a telephone call. She wrote a brochure with the same title for the Society Knowledge. And her brochure was published. But I no longer had any hopes of publishing Sergey Nikolayevich's and my work with the Knowledge Publishing House. The brochure had turned into a small book and its size no longer corresponded to that of the brochures that the Society Knowledge issued.

I complained to one of my acquaintances about the book's pathetic fate. "I'll write a note for you to my friend. He'll help you. He has influence with Nauka, the publishing house in Leningrad." "I don't need any note," I told him. "Your friend is my friend too." And so I called him. It simply hadn't occurred to me to ask him to help. He helped. They accepted the book, included it in their publishing plan, and sent it to the USSR Academy of Sciences Publishing House Editorial Council in Moscow, and there it got lost. They searched for it in my presence. The chairman kissed my hand and told me that they wanted to find it but couldn't. "That means it's been sent back to Leningrad," they said in Moscow.

I found it by accident at the Science Publishing House in Moscow. I was there for another reason. But I thought I'd drop into the popular science division and ask. The head editor told me, "We'll look for it, but we won't publish it. We have too much work as it is. Don't even hope for anything. Maria Ivanovna, did we receive any book like that?"

"No," she said, "we didn't. But I'll ask the editors." And in a moment a girl was led in—she looked exactly like my daughter Liza and had a wonderful last name—Vyazemskaya.

"She has your book. How did you wind up with it?" the main editor asked her sternly.

"I took it from your desk to read," she said.

"But how did you dare?" he asked.

"I'm interested in genetics," she said.

"Well, all right, let's go," I said. "I'll collect the manuscript and that's the end of it." I went to her office.

She pulled out the manuscript, both copies, including the pictures. Everything was there, just as it should be. "But I won't give it back to you," she said. "I want us to publish it. Cut down the number of pictures, and do it as quickly as possible." I gave in, and sent the manuscript back to her. I waited and waited, but there was no reply. About another half year passed. I wrote her, citing Pushkin:

*I cast a living seed*
*Into enslaved furrows,*
*But I only lost time,*
*Noble thoughts and labors.*

Still no answer. And yet there was some movement, apparently independent of the quotation from Pushkin. The Moscow publisher forwarded the manuscript along with a resolution from the editorial council to the Leningrad branch of the publishing house.

The book was in the works. There was a problem with pictures. Dina Markovna, the head of the illustration section, directed me to the best photographer in Leningrad so that he could do the illustrations for me at an insanely high cost. Dina Markovna rejected half of them. It's difficult to understand why, but not impossible. I myself would never have guessed. A bright woman who has had more experience in publishing than I explained it all to me. "You have to pay for the illustrations again, but this time to the publishers," she said. "One of Dina Markovna's relatives is probably officially employed as a photographer by the publisher and earns his money, while the work is done by the best photographer in town. You'll have to pay or you'll have as much chance of seeing your book as your own ears."

And I called the main editor and told him that I had very poor illustrations and that he should subtract from my fees whatever it cost to fix them up. But then Dina Markovna was caught doing something illegal, and she was fired. They didn't subtract anything from my author's fees.

The book was still in the works. The editor and I had a very good relationship. And so now it was the end of 1971, ten years since the brochure had been written. I was on an expedition. I received a telegram from my daughter: "Fly to Leningrad. The publisher wants to see you immediately." I flew, and I took a taxi from the airport to the publisher. The main editor of the publishers' biology division, a very kind-hearted soul, told me that the book's publication was threatened. If I didn't cross out the references to Timofeev-Resovsky, the book wouldn't be printed.

I gasped. "Has he been arrested?"

"No," he said, "not so far as I know, but it's forbidden to refer to him. The order came from the Moscow branch, to which we are subordinate."

"I won't cross the references out," I said. "Such things didn't happen even under Stalin. If a person hadn't been arrested back then, it was permissible to refer to him. If he was arrested, the editorial staff didn't ask the author about it but simply crossed out the references. That's exactly what happened to me. I referred to Vavilov. The *Reports of the Academy of Science* threw out the reference. The passage in question had to do with the centers of origins of cultured plants. But that was then. Now, I guess there's been a mistake of some sort. Wait a bit. I'll find out and let you know."

I went straight from the publishers to the office of Boris Evseyevich Bukhovsky, who was the director of the Zoological Institute and the secretary of the Biology Division of the USSR Academy of Sciences. He would certainly know what had happened. And I was lucky; his secretary wasn't there and there was no one to tell me that I couldn't go in, that he was seeing Yugoslavs or Frenchmen or Germans. I just presented myself to him and explained what was the matter. He agreed that this had to be some sort of mistake and said that he would sort it out and call the publisher himself. This had to be sorted out in Moscow in person, not over the telephone—that was obvious. The prohibition against citing Timofeev-Resovsky turned out to be a mistake. The last barrier had fallen. The book came out in 1972.

# *The Political Economics of Socialism*

By the time that the last chapter of the story of the book *Heredity and Hereditary Diseases* was being played out, my political economics of socialism had already been thoroughly formulated. In creating it I didn't attempt to comprehend human nature, and certainly not its criminal manifestations, such as the actions of Dina Markovna and her hypothetical relative. Rather, I tried to comprehend the economic order produced by the interaction of human nature and the system imposed on man.

The social order is formed spontaneously, in accordance with human nature, no matter what conditions man may find himself in—whether he be an academician or a camp prisoner.

When *One Day in the Life of Ivan Denisovich* was published in the journal *New World* in 1962, the mathematics students to whom I was teaching genetics at that time told me that my understanding of the elemental processes at work in society coincided with what Solzhenitsyn wrote. They had read the novel, but I hadn't yet been able to. That issue of the journal had sold out immediately.

In the West there is the widespread and generally accepted notion that the Soviet Union's economic system is one of state capitalism. That notion is correct only if one narrows the definition of capitalism and if one recognizes that the Soviet economy is not a socialist one. Let's begin with the first clarification.

If we choose to define state capitalism as the exploitation of the majority by the ruling minority, then there is state capitalism in the Soviet Union. But capitalism has other traits besides exploitation. With good reason the state structure of Soviet Russia can be also denoted as state feudalism, which was temporarily ruined by the February Revolution in 1917, and subsequently restored by the October coup d'état.

Capitalism in Soviet Russia does not exist, because there is no free market with its market prices and its feedback system. Money does not play the role in the Soviet economy that it does in the economy of non-Communist countries. The regulating mechanism of the banking system, which levels gains, is entirely lacking. Prices are determined by political considerations and have nothing to do with the amount of work and material needed to produce a given product.* Prices also have nothing to do with laws of demand and supply, or with market capacity. Examples are the low prices of bread, gasoline, and state apartments. Salaries and the allocation of personnel are as dependent on politics and as independent of economy as are prices. The difference between the highest and the lowest salaries is enormously increased by privilege. No free market of manpower exists.

Laws are established to prevent the regeneration of capitalism. The free market of manpower itself is persecuted, not to mention the use of hired manpower. The products of labor, of your own labor, are not your possessions and you have no right to sell them.

To gain money outside the state system is practically impossible. It is forbidden to teach in one's own apartment, and it is also forbidden to sew for sale, to repair shoes, to create pieces of art. . . . The taxes paid by freelance workers are larger than their potential income. To shirk taxes is very dangerous and the penalties are severe. Nevertheless, an underground market does exist. It has its laws of supply and demand. The danger involved in selling the products of your own labor makes

* Some of the ironical situations resulting from production and price regulation became known during Khrushchev's de-Stalinization reforms in the 1950s, some are disclosed now under the aegis of Gorbachev's reconstructions. Thirty years ago we learned that peasants, members of collective farms deprived of any possibility of getting fodder for cows that they were permitted to own, bought bread in shops of nearby towns and used it to feed cattle. Measures were then taken to prevent that malicious practice. Cows were confiscated. What we've come to know more recently is even more comical. Collective and Soviet farms sell their production to the government. Governmental institutions sell it to individual purchasers. The purchase prices paid to collective and Soviet farms are higher than the retail prices paid by buyers. It is much more profitable for the sellers to buy the ordered amount of goods at the retail prices, instead of producing them, and selling them back to the government at a higher price. It has filtered into the press that this has been done habitually by some collective farm chairmen. Every cloud has a silver lining.

the demand exceed the supply. Those who dare want to be paid for the risk. The degree of danger is one of the determinators of price.

I am going to discuss another free market that provides a germ of capitalism in its most primitive form and has nothing to do with this basically underground personal manpower market. This new private economic system uses state institutions for personal gain. It by no means violates the law, and it certainly does not violate the plan that a planning office "drop" on every institution from "above." It is a very peculiar kind of free market because it is rooted in a planned economy, or to be more precise, it interferes with an economy proclaimed to be following a definite, worked-out plan. It feeds on the failure of the system to bring demand and supply into balance. The shortage of goods and services is a chronic aspect of the rigidly planned society. Nobody estimates what and how much is needed for the population. Hours spent in lines by working people do not worry anybody in the planning organizations. You have to stand in a line not only to buy meat or vegetables but to pay a bill for your apartment and utilities.

The new economy I am going to describe is based on the profit gained from the shortages of goods and from the waste of time in obtaining them. The institutions supplying personal needs—shops, restaurants, hotels, hospitals, schools, universities, courts, pharmacies, railroads, and airlines, all belong to the government. They exist to sell, to teach, to serve. . . . But there are too many to be served. The choice to sell or not to sell, to serve or not to serve is made by the staff of the institution. Without violating the law to the slightest degree, the employee chooses those pretenders who are useful to him (or her) in getting goods or services or medical help in exchange for goods or services he is able to offer.

Information that some kind of goods is on hand acquires value and the mere exchange of information becomes an economic factor of great importance. Instead of Marx's formula *product A → money → product B* another one comes into being: *product A → information that product A is present → information that product B is present → product B*. Americans accustomed to the ministrations of Madison Avenue are no doubt thinking "Ah, advertising!" The process I am discussing has nothing to do with advertising; it is in some respects its very opposite.

The attractive goods and services are hidden from the undesirable consumer. Information that these goods and services are present is sold to the favored consumer. It is not advertising. It is a process of channeling the distribution of goods and services with the maximum profit to the staff of the distributing organization. The most needed commodities are as a rule absent: cotton, thermometers, mustard plasters, and badly needed drugs are not to be found in ordinary pharmacies. No wild rose syrup, no cod liver oil. China teapots and cups, enamel pans and teapots are not available in china stores; no paper for a typewriter, no carbon paper, no typing ribbon in the stationery stores. To buy a towel, a toothbrush, shampoo, or laundry detergent can be an insoluble problem. To buy a book for which you have waited for years, you have to have an acquaintance in a bookstore. If the poet was first "forbidden" and then "permitted" as Mandelshtam was, it is easier for you to get his book from abroad than to buy it in the bookstore next door, assuming there is someone abroad who sends you books. Time spent in obtaining commodities is very short for those having a source of information. It is endless for those who do not have access to the free market of information exchange. But people are people. Sometimes you are served not because you offer something in exchange but because some human contact is established. I once asked for a thermometer in a pharmacy on Dzerzhinsky Street in Leningrad. This pharmacy is one of the best in town: people call it the "GPU pharmacy."

The Soviet secret police has changed its name several times. The first name, the Cheka, was replaced by the GPU, which was followed by the NKVD, and finally, the chain of renamings was capped by today's link, the KGB. The popular memory often retains what has been officially replaced: Leningrad is called Peter because of its original name Petersburg, the Kirov Theater of Opera and Ballet is called the Marinka, because it was called the Mariinsky Theater in honor of Empress Maria Feodorovna, the wife of Alexander III, and the frightening name GPU has been retained in the name of the pharmacy. It is located near the headquarters of the first intelligence agency.

"No thermometers," was the answer. I said: "Ah! My grandmother could measure my temperature and I can't do it for her great-granddaughter. My daughter is eight and she tells me that if Lenin were alive Sputnik would have been started long ago and that at the present time there would be thermometers as well." The drugstore saleswomen laughed and asked me to come by the following day. I got a thermometer.

At the same age of eight my Liza remarked on an advertisement. What was being advertised was . . . milk! Not the products of some giant state farm with the proud name "Red October" or "First Five-Year Plan," but plain old milk. The ad was on Lenin Prospect, the route for foreign tourists on their way from the airport to the hotels Astoria or Russia in the heart of Leningrad. Before them was the Soviet land flowing with milk and honey. "Drink milk!" proclaimed the advertisement. "Do you really need to persuade people to drink milk?" asked Liza. "It's so people won't be too lazy to stand in line to get milk," I said.

If there is no real shortage, it is fabricated so that information can be sold nevertheless. The value of information regarding the presence of some object depends on the availability of that object. Take potatoes, for example. No simple proportion exists between the amount of potatoes in all the vegetable stores of a town and the value of information about their availability. The value of the information that there are no potatoes and that there will be no potatoes in the near future is zero. And the information that potatoes are available in all stores is equally valueless. The maximum value coincides with a certain shortage, not a complete one.

Let us depict the curve showing the relation between the saturation of the market with some kind of goods (abscisse) and the value of information regarding the kinds of goods on hand (ordinate). The curve crosses the abscisse twice: At the very beginning, no goods of the particular kind are available, and at the 100 percent point. It seems to me that the curve depicting these probability phenomena coincides with the curve for entropy $H = -p \log p$. If so, the information has its maximum value when the saturation of the market is 35 percent, i.e., when about two-thirds of the people are not served. The question arises: Who are most oppressed by this system? The answer is: Those who have no information to exchange, such as blue-collar workers and scientists.

The process has not only economic but moral consequences as well. Everybody is a victim of this natural exchange system. It contributes its corrupting share to the depraving propaganda, with its thesis that "the ends justify the means," to the theory of class ethics, to the existence of the secret police, to the cruelty of punishments, including the death penalty.

Antiservice and anti-advertisement are on the surface. Service and information about the availability of goods are concealed. Hideous goods put on display are anti-advertisements, used by a salesperson to address an undesirable customer. A worthless item relieves the salesperson of the necessity of falling back on antiservice. "Take a look and clear out!" the goods say on their own, and the salesperson loafs.

Do you realize what antiservice means? No, you can't, unless you have lived in the Soviet Union. An undesirable customer not only has to be removed, but he has to be stung. Why? So that desirable customers will realize the value of the service that they are rendered. The more rude the rejection the undesirable customer receives, the greater the price of services. "Do you expect me to give you all meat so that other people only get bones?" That's how the butcher teaches you philanthropy as he short-weighs you, all in bones.

Like Gogol's nose in his surrealistic story "Nose," antiservice leaves the place where it fulfills its function and life becomes filled with mutual offenses, quarrels, and fights that have no function at all other than putting people in their place, asserting your superiority at least over someone, at least once in a while.

Service, detached from its profit-minded roots, wanders the streets of the USA. You're looking for a street sign, and right away you hear the familiar "Can I help you?" And one elderly gentleman added that he liked the pattern on my dress—lilies of the valley. The red stoplight gave him the chance to sing me a German song about *Maiglöckchen* (lily of the valley). Once in Moscow, I was standing on the Zemlyanoy Val, near the Belorussian railway station, on the square in front of the building where Efroimson, my idol, had lived. I'd gone there specially, just to take a look. An elderly man looked at me with a vague gaze and said, "Aw, go to hell, you old . . ."

The dietetic cafeteria in Pyatigorsk, where in 1957 I dined every day for the duration of my expedition there, was a hotbed of antiservice. Much of the time, by a happy concurrence of circumstances, the cafeteria was nearly empty. The waitresses loafed. It took a long time to catch their attention. But finally your order was taken. More or less; if you stumbled while listing the dishes you wanted to order, all you could do was yell the rest of the order at the waitress's back as she retreated into the distance. Then came a very long period of waiting. You had no time to waste sitting there, and you'd be dying of hunger. The waitress would sit down with her girl friends and chat. Is this, I thought, an attempt to make the expenditure of labor commensurate with a niggardly salary? But the labor expenditure is not the least bit greater if one first brings the food and *then* takes an extended break. To the torments of hunger and regrets about lost time is added irritation. That's very good. The overcoming of an obstacle gives birth to ideas. I would concentrate fiercely, trying to understand the logic of the waitress's actions. This concentration helped me while away the time. It even gave me joy. "Joy, O joy of suffering!/The pain of unknown wounds." That's Blok. The regularity with which the sufferings were repeated, day in and day out, excluded the possibility of chance or of the waitress's negligence.

There was an International Festival of Youth in Moscow in 1957. In the park near Krymsky Bridge crowds of Muscovites packed the pavilions at the art exhibit. There was a line in front of each of the many cafeterias. I got in line. The entrance to the cafeteria was blocked by a big bruiser, pardon me—a healthy-looking person—wearing a white gown. A robe, a dress coat, a cloak, ritual clothes, the rituals of scholarly meetings, courts, consultations—these are all things that allow certain people to interfere in the private lives of others because by meeting the feeling of inequality, they transform violence into order. The woman wearing the gown was watching to make certain that no one entered without her sanction. The cafeteria building was surrounded by an attractive fence, and the scene was played out at the gate. There were tables inside the building and outside on the veranda, behind columns. There was no service on the veranda. The chairs were turned upside down atop the tables. I was approximately the fifth in line. I said to the woman wearing the gown: "Lenin said that every cook must learn to manage the state. Why are you standing here instead of managing the state?" I said that cheerfully, as though I weren't standing in line with an empty stomach, but indulging my favorite pastime. The gown, looking confused, put itself on guard and didn't say anything. "I see that you don't want to take up managing the state. Then seat us at those outside tables and feed us."

"Go on in," the gown said to me without moving from the spot.

I turned to the people who were first in line: "It turns out that there are places. Go on in." A man and a young woman went in. "Excuse me," I said to the robe, "but I can't go out of turn." My turn came.

"Come sit with us," people whom I didn't even know at all called to me. "Why aren't you afraid to behave that way?"

"I'm fed up," I said.

In 1962, I was in Odessa with a whole brood of young people, my Masha among them. This expedition had as its goal the study of two fly populations. One of them was in Moldavia. It was the population of Tiraspol, which I had studied in 1946, when I'd caught the lowered mutability by the tail. That phenomenon had just occurred, and I was lucky enough to be a witness. The other population was in Armenia—the population of the apple orchards and domestic wineries in Dilizhan. We headed south and then turned east, following the itinerary Leningrad-Tiraspol-Odessa by land, and from there to Sukhumi by sea, and then again by land to Dilizhan.

There was a marvelous cafeteria in the center of Odessa. Its large veranda had many tables and in this case the visitors sitting at them were being served. But there were several empty tables, and there was a long line. People had to wait at least an hour and a half. Diners changed shifts very, very slowly. A group of young people came up—tourists. They were about to remove their backpacks. They stood a while and then left. "Let's go somewhere else. This is bad news," said Vika Gorbunova, a participant in my expedition.

"Let's stay in line," I said. "It's a field trip for a political economics course." Zavadsky, Agayev, and Khakhina weren't so very far from the truth when they wrote in my recommendation that I corrupted young people.

We finally replaced those who had been eating. We walked past the untouchable tables. TABLE FOR CHESS PLAYERS read the sign on each of them. We ordered. We had stood for our honest hour and a half. No chess players had appeared. But as soon as we were seated comfortably, two chess players sat down at one of the tables. They were brawny fellows of an obviously proletarian sort. The waitress fluttered over to them at once. And immediately two bottles of Stolichnaya and two bottles of beer appeared on the table. With quite practiced gestures, as though Stanislavsky himself had trained them before putting them on the stage of the Moscow Art Theater, the fellows were putting the bottles in their pants pockets. Four pockets, four bottles. After the chess players had paid, they left.

"And have swarms of chess players come to Odessa?" I asked the waitress. She realized that I understood how the restaurant administration and personnel were killing two birds with one stone—reserving tables for suitable customers and sending a stream of scarce goods to the black market. The manager of the shoe warehouse for a department store does not look any different than a chess player. Just try to find out whether he's a chess player or not. His services are evidence of the fact that cheap imported shoes have arrived at the warehouse, and that he has already put aside a box with a pair of shoes just the size that the dining-hall manager's wife wears, or the very waitress to whom I addressed the tricky question. He will be fed instantly and he won't have to wait a second. Tables for the participants in a chess tournament are always free, since there's no chess tournament—the tables themselves are reserved for the event—and again, just try to prove that there isn't one. The function of the fellows who sat down at the table for participants in the nonexistent chess tournament is a different one. They honestly paid for alcohol at the high restaurant price and left. They had no intention of dining. No matter how much the drinks may have cost them, you can be certain that the price on the black market would be even higher. The waitress had sensed the danger of exposure and decided to buy us off. We were given first-class service.

Odessa had found a much better way of limiting the influx of undesirable customers and reserving a place for the desirable ones than chairs turned upside down. But we did see chairs in that sacramental position in Odessa, too. It was at the Beacon Restaurant. It's simply called the "Beacon," but everyone understands that it's not just any old beacon, but the beacon of the future, the beacon of socialism. The most innocent question—"Whom are you waiting for? What delegation?"—is often sufficient to get the management, if you should ever be so lucky as to come into contact with the manager, to arrange a table for you right away. But the question must be asked in the appropriate tone of voice. What kind? I leave that to the reader's imagination. There are many variations. Only one tone of voice is excluded: that of a person interested in eating, the tone of a person who gives bribes, rather than that of one to whom bribes are given.

I don't recall what year it was. It was at the airport hotel in Tashkent. There were no spaces available. You could put a five-ruble note in your passport and hand the passport and the bill to the manager. You could show the manager your Doctor of Science diploma. Many people do that, and they told me that it helped. I didn't try. I was standing and waiting.

A family came up. A grandmother, mother, and two children. They were going to have to spend forty-eight hours at the airport. The manager told them that there were places available only for people whose flight was the next day or who had been held up by bad weather or for mechanical reasons. They would have to seek accommodations in town, or they could apply to the airport manager, and if he issued an order, they'd be given a place. The entire group of them was about to head off to find the manager. I stopped them. "There's no reason for all of you to go. Let grandmother and the children stand over there by the wall"—there was no place to sit, the airport was crowded as could be, and the weather was unusually bad—"and let the young woman go to see the manager. I'll go with her and tell him what I witnessed." They were given a room.

A huge Uzbek woman arrived carrying a boy of about ten on her back. They refused her a place, too. She put the boy on the floor and was preparing to go see the manager. I offered to go instead of her. The hotel manager lost her patience. She gave both the Uzbek woman and her son and me places, on different floors. "There aren't any rooms," the manager half growled and half hissed venomously, "you'll have to take a cot in the hall." "Well, that doesn't frighten me in the least!" It turned out that a cot in the hall was a malicious exaggeration. The hotel's lounges were partitioned off from the halls by very luxurious curtains. There were no fewer than twenty cleanly made-up beds, none of them occupied, in the quarters where I was to spend the night. But good fortune shined on me just then. As I was registering with the hall attendant on my floor, two pilots checked out and I was given their room. There was a knock at the door just as soon as I had got into bed. The room was a double. I assumed that some high-ranking woman had got the second bed. But from behind the door I heard a very lively male voice: "Open up, Granny."

That form of address—"granny"—had nothing to do with my age. The person with the lively voice didn't know my age. The only information that he had concerned my sex. He chose the politest form of address that occurred to him. The form of address for strangers in the Soviet Union is a sore subject. Even the press devotes space to it. In Leningrad the only old form of address that has been retained is the word *baryshnya*, which is roughly the equivalent of the English "miss" and is used in addressing a young woman. "Mister," "Mrs.," "master," and "mistress" have all turned into terms of abuse. The officially accepted "comrade" and "citizen" are not good for anything: "Comrade" doesn't work when addressing a woman, and "citizen" or "citizeness" sound wrong except in the mouths of the police, judges, and investigators. The accused wouldn't dare call them "comrades." To him they are "citizens," just as he is to them. The broad masses have returned to the ancient forms of address that imitate kinship in order to show friendliness. A middle-aged woman is addressed as "Mama," and an elderly woman is called "granny." Children use the terms "uncle" and "aunty." A young woman is addressed as "girl." This is all when people wish to be polite. If they don't wish to, they say "woman" or "man." I always found the latter terms especially repulsive. On the lips of the man behind the door, "granny" sounded almost affectionate.

After a few clarifications, made with the door closed, it turned out that the other bed in the room was for the wife of the man with the lively voice, not for him. I opened the door. Three people walked in: papa, mama, and their son. The woman was ill. She had a sore throat and a fever. They wouldn't let her into the medical station at the train depot. He would make do somehow, but the wife and son might as well sleep together. He would sit out in the hall. "That's no good," I said. I was very averse to the prospect of spending the night in close company with someone who had tonsillitis. Twice a shot of penicillin in the throat had saved me from certain death. And the boy shouldn't sleep in the same bed with his sick mother. "Take the boy," I said to the father, "and go see the manager again. Get her to give you a place. Say that you're a single parent. She surely doesn't remember that she already gave the boy a place."

"No," objected the father, "the boy is listed on his mother's ticket. And he's wearing a red suit. It won't work."

"Do you mean to say there isn't a blue suit in the suitcase? Change his clothes. And let me have your ticket." He gave it to me, and I wrote in the boy's name. They left and didn't come back. Good fortune continued to spoil me. I didn't come down with tonsillitis.

In an airport hotel, where places are made available after showing your ticket and passport, you cannot pretend complete lack of personal interest (as you must at a cafeteria). It is obvious that you need a room; otherwise, why are you there? However, you may show interest in the fate of others and readiness to support their complaints against the management, no matter how high and mighty.

Likhachev, Maletsky, and I were deprived of that opportunity at the Turachak airport, where it was plane tickets that we were trying to get, not a room in a hotel, since there is none there. But would we take interest in the fate of other people at Turachak? You must be kidding. "Others" were quite out of the question. All we cared about was flying out of there. Complain to the management? The management was Lord knows how many miles away. You'd need an airplane ticket to reach the manager!

Do you think that the head of the Turachak airport sold tickets with the help of a sort of underground auction—to the highest bidder? You are only partially right. The situation was much more complicated than that.

Your suffering is the most important component of the system. In causing you to suffer, neither the head of the airport, nor the cafeteria manager, nor the waitress breaks the law or stains his or her conscience. They aren't extorting a bribe. They are enacting the only possible method for the co-existence of people under the yoke of socialism. The waitress's time pattern in the cafeteria in Pyatigorsk, the tables for the chess players in Odessa, and the incident at the Turachak airport were all related to the economic basis. They served the ultimate goal of increasing one's real earnings; and you have to increase them, otherwise you can't get by. The man's coarse swearing which interrupted my delightful observation on the Moscow square was related to the superstructure. He was singing me his own song about *Maiglöckchen*. Not by bread alone. . . .

The assertion that the economic system in the Land of the Soviets is one of state capitalism assumes that the country's economy is not a socialist one. That is a harmful and dangerous error. Distinguishing state capitalism from socialism is based on a romantic faith in socialism, in a socioeconomic order that corresponds to true human nature better than any other. I note parenthetically that, according to the Marxist doctrine, there is an economic order that corresponds to human nature even better than socialism. This idea system, communism, was achieved by our distant ancestors, primitive men. They enjoyed the benefits of the primitive communal system, the advantage of primeval communism. We recent humans are still only growing up in the process of building communism. The holdovers of capitalism first have to be eradicated from our consciousness.

The greatest calamity on the path to remaking society on an allegedly just foundation was the universal collectivization of agriculture. The land that the peasant had at last acquired as a result of the October Revolution in 1917 went, in 1929, for the permanent collective use of the members of the collective farm. In working the land, collective farmers fulfill a plan imposed upon them by the district party organization, or more accurately, the hierarchy of republic, district, and regional party organizations. The harvest becomes state property. Collective farmers receive wages based on the numbers of days they have worked. The payment depends on the extent to which the plan has been overfulfilled. If that payment were the only source of maintenance for collective farmers, they would all die of starvation. In order to avert that undesirable phenomenon and simultaneously to retain the opportunity to appropriate the harvest from the collective farmland, another sort of payment has been introduced. Every family has its own hut and attached to it is a tiny little personal plot of land. Whatever is grown on that land as a result of the whole family's work is the peasants' property. He can sell surplus produce at the market known as the collective farm market. What may be grown on personal plots and what may not is strictly regulated. And the same thing goes for livestock. It is forbidden to grow grain cultures, just as it is to keep horses. The size of the plots excludes the use of that cumbersome technology that is used and misused on collective farm land. All forms of cooperation are forbidden. Authorities assert that a third of the agricultural production of the USSR, a third of what the citizenry consumes, is raised by hand on private plots.

I hear from many people in the USA that soon there will be a socialist revolution in America, and American socialism will be of the true kind and society will be transformed along rational bases. Americans who dream of such a socialist revolution ought to know about collective agriculture. I'm picturing in my imagination socialism in the USA. And I see these people standing in line at the collective farm market to buy potatoes, where the leftovers of what people have raised with their own hands in their garden plots are sold. And one of them looks at the other and sings him his song, in his own way, as best he can. . . . Russia's socialism is the real and only possible one. There is no other sort of socialism in nature, nor was there ever, nor will there ever be.

Is this brand of socialism an isolated phenomenon? Yes, it's isolated, but the shadow that the Earth casts on the Moon is also an isolated phenomenon. Aristotle knew, however, that the Earth is a ball, and he came to that conclusion by watching its shadow on the Moon. Every circle is a model of infinity. Nearly 280 million inhabitants of the Land of the Soviets are not an infinity, but they're not a small number either. There are isolated phenomena the accuracy of whose evidence is no less than the accuracy of a statistically saturated sample of phenomena. Their statisticality is enclosed within themselves; they are integrated multitudes, like a circle, like a sphere; like the 280 million Soviet citizens who are doomed to suffering.

Now, at the time when communications about reforms of the new Soviet ruler Gorbachev flood the media and are on everybody's lips, it is worth a glance into the future to evaluate the fate of the information market in the USSR. The spontaneous process of channeling the distribution of goods and services, and the ruling mechanisms of powers that be, are supported by one and the same pillars: by shortages of goods and services and by the privileges in obtaining the deficient supply. These pillars are as vital for the information market as they are for the ruling force. They are unavoidable and thus indestructible.

In the face of abundance, when the price of the information that there are potatoes in all supermarkets equals zero, the information market is doomed to extinction. But a powerful remedy of control practiced by the rulers escapes. Possibilities to create inequality in consumption vanish. To buy souls of subjects becomes puzzling, if not impossible. Shortages and privileges are heads and tails of one and the same coin. They can't be exterminated without harm for both the information market and the police state.

In some respect the interests of both contradict each other. The state wants to arrange strata according to its interests, and here comes the information market making a salesman in a street beer stall a millionaire and a persona grata. The penitentiary system is helpless in face of these sources of enrichment. Theft and bribery can be punished, and they are punished in the process of Gorbachev's purges. Bargain partners of the information exchange market can't be made liable to punishment; they are outside the competence of court. They don't violate laws, they fulfill the output plan "dropped from above" (according to colloquial Soviet expression). To nose out the channeled distribution is, in the vast majority of cases, impossible. A furniture dealer putting a sign SOLD on a suite waiting for a needed customer can sleep the sleep of the just. The sign in a cafeteria TABLE FOR CHESS PLAYERS, or a more simple sign RESERVED, serves the same goals: to remove the undesirable customer and to avoid the vigilance of inspection.

The mimetic display of deficit sometimes does not work properly and the affair comes to light. The pitfalls are the lack of desired customers and inability to hide the attractive goods and services. Railroad and theater tickets belong to this category of wares. Vacant seats are on display.

Elena Sokolova in the newspaper *Novoe Russkoe Slovo*, published in New York, refers to an article that appeared recently in one of the most prestigious Soviet newspapers. The Soviet author of this inveighing pamphlet points to people who enter an empty car after spending nights in a line at the railroad agency or station to get a ticket. Cashiers are endangered.

Elena Sokolova does not relate the small amount of sold tickets to the general rule of Soviet economics, to the mimetic display of scarcity, to failure in awaiting a needed customer. She refers to a too-huge amount of tickets reserved for nomenclature, the amount the privileged persons did not make use of. She writes that in the pre-Gorbachev era, such denouncements did not occur. She is right, the disapprovals have become much stronger, but sharp pamphlets with public condemnations also slipped into the press from time to time in a pre-Gorbachev era. The purpose of their publication was exactly the same as it is now, to pass the whole for a part, to camouflage the brave, truthful disclosure of negative features of social life.

I remember one essay published in the sixties in the newspaper *Leningradskaya Pravda* or in *Vecherniy Leningrad.* It was devoted to the same question as the article referred to by Elena Sokolova, to the vacant seats in railroad cars, and to the endless lines at the railroad stations. It was entitled "The Voyage of the Invisibles."

To eliminate the arbitrariness in ticket sales and at the same time preserve the privileges of the elite is possible. In the Kirov Opera and Ballet Theater in Leningrad tickets are sold to foreigners for currency. I have never seen any empty seats at that theater. But in the Leningrad Philharmonia I observed plenty of them. With great difficulties I got tickets for a concert. *Sinfonia Robusta* of Tishchenko was performed. Nearly half the seats were vacant, although lots of people wanted to have tickets.

Vacant seats bringing the affair to light are an exceptional phenomenon. The vast majority of goods and services do not let their distributors down. To get it better, to control the artificially created, the exaggerated scarcity of these silent things is absolutely impossible.

"Turning the screws tight" is the Russian cliché for strengthening the penitentiary control. This is what is going on under Gorbachev. He uses tightening the screws as a measure to force the populace to serve the regime honestly, not as a remedy to prevent the impoverishment due to theft and bribery. No attempt is made to raise the living standard of those who feed, cherish, entertain the elite, who defend these minions of fortune by increasing the war potential of their country.

The new ruling force has to accept the fact of existence of spontaneous processes going on in the society. The information exchange market belongs to these uncontrollable forces of nature. The rulers will accept it easily, the more so as they have no choice.

# The Unvanquished

Now, trying my readers' patience, I abandon temporarily the narration about my own relatively safe fate. I start resolutely to speak about people whom I knew, my pupils, co-workers, colleagues; my friends and idols, whose tragic fates tell a tale in a much more significant way than anything that happened to me. The tale is about moral resistance to manipulation—manipulation by means of shadowing and terror, by means of mendacious propaganda—and resistance to attempts to subordinate the very souls of the whole populace. My book is not about me; it is about a time I had the opportunity to witness. Deprived of the right to protest the most outrageous crimes of the regime, perpetrated before my eyes, I was silent. Let me now tell what I wanted to tell at the time the events I describe happened.

Great words were once spoken under circumstances that were not predisposed to such an utterance: "Forgive them, Lord. They know not what they do." But they did know what they were doing, those who were destroying science and its best representatives; those who killed Vavilov, Karpechenko, Levitsky, and Levit; those who forced the persecuted Chetverikov and Rozanova to die a slow, lingering death, since they were doomed to unemployment, and who would have inflicted the same fate on Koltsov, had he not suddenly died himself; those who sent Efroimson to Dzhezkazgan and Romashov to prison; who drove Sakharov, Rapoport, Svetlov, Kerkis, Nikoro, Prokofyeva-Belgovskaya, Sidorov, Rokitsky, Sokolov, Davidenkov, Kryshova, Kanayev, and many, many others out of genetics.

The new "teaching," the crowned scion of a gendarme and a quack woman, was a child born of post-revolutionary reaction. Treachery and terror stood by its cradle. Fear rocked it. In order to succeed, to occupy a post, one didn't need scientific merit or a knowledge of the truth, but unconditioned readiness to betray the truth. That's the way it was and that's the way it remains to this day. Zhores Medvedev's fate is proof of that. A geneticist at work on the most advanced edge of contemporary science, he combines the gifts of a scientific experimenter and a historian. He wrote a history of Lysenkoism. He sent the first copy of that superb book to the government. Only then did the book become the property of Samizdat. Medvedev is now in England as an exile; he has been denied the right to return home. While Lysenko, unharmed, remained a full member of three academies—the USSR Academy of Sciences, the LAAAS, and the Ukrainian Academy of Sciences—up to his death in November 1976, and was buried with full honors.

None of Lysenko's underlings could plead ignorance: They knew what they were doing in supporting Lysenkoism. Their justification was the necessity of subordinating themselves to a higher force—fear for their own trembling skins.

There are three ways of saving oneself from one's guilty conscience, of saving one's face before oneself, of justifying lies and lawlessness: 1) a high goal, an end that justifies the means; 2) the sad but unavoidable necessity of undesirable, but inescapable side effects—eggs get broken when you make an omelet. . . . and 3) We're little people. Conscience is only a prerogative of those more important than we who have freedom of choice. Faith, oblivion, and thousands of autopsychiatric subterfuges come in handy.

The final outcome was no accident. The chain of events was strictly determined even at the very beginning. At the outset, an ignoramus was elevated onto a pedestal. From there on, the size of the salary and the height of the pedestal became criteria of truth. It couldn't be that a full member of the USSR Academy of Sciences, elected by the Biology branch, receiving a salary twenty times that of a janitress (and that's not counting the benefits that double his income), didn't know biology. Absurdity upon absurdity received the approbation of his rank. The bolder the projects, the more rapid the movement upward, and the higher the rank, the more ridiculous the project.

In Russia there is a scientist, Vladimir Pavlovich Efroimson, who, in 1956, estimated the loss in rubles that Lysenko had caused agriculture. He took his volume to the prosecutor general of the USSR and proposed that criminal proceedings be instigated against the academician. His file was accepted. As everyone knows, no action was brought against Lysenko. The latter continued as the director of his institute, all the while running things in three academies. Vladimir Pavlovich was twice imprisoned and spent years in the camps. His years of exile are reckoned in decades. He was tortured in Stalin's torture chambers.

There are many reasons why the study of genetics was not ultimately crushed by the heel of Lysenkoism. One of them was the fearlessness in the face of death of such people as V. P. Efroimson, I. A. Rapoport, B. L. Astaurov, Z. S. Nikoro, M. L. Belgovsky, V. S. Kirpichnikov, V. V. Sakharov, and A. A. Malinovsky. Death was waiting for them so as to multiply the unknown graves in which lay their friends, teachers, and those worth modeling oneself on. But graves, even lost ones, and perhaps even most of all the lost graves, possess a great power. They are a barrier on the path to universal demoralization. Genetics did not perish in the Soviet Union, because Vavilov, Karpechenko, Levit, Levitsky, Agol, and many others died for it in torture chambers, because many people took the martyr's crown for it, rejected glory for its sake, and were doomed to an ignoramus's existence.

Brave people endangered not only themselves, but their friends, acquaintances, all the numerous groups of which they were members—all the representatives of their place of work, their field, their nationality, and their people. A bold person jeopardizes all of them, turns them into potential enemies of the people—and always of the entire people, no more and no less—and transforms them into an object of prophylactic terror, and bloody vigilance.

The genetics of fear is a favorite topic of contemporary animal genetics: the inheritance of timidity and bravery, the inheritance of the tendency and ability to induce terror. Much is already known. The scale of defensive and aggressive types of behavior has a wide range. The diversity is created by group selection—diversity is useful to the group as a whole. Those who win in life's competition—winning means surviving—are the groups whose members are the most diverse. Management, the rational use of resources, is one of the strategies in the struggle for life, and it requires coordination. The feeling of fear in some of the group's members, and the ability of others to induce it, consolidates the members of the community and increases its chances for survival.

Individual selection supports leaders; the strong survive. It is opposed by a potent force—group selection. Group selection supports groups that assure there will be both oppressed and oppressors. The heritability of leadership is slight. Cowards occasionally give birth to fearless offspring. But leaders, conjugal pairs of leaders (for they select each other on the basis of similarity, not contrast), give birth both to shy, retiring types and to those capable of inspiring dread.

One can be quite certain that the inheritance of timidity, fearlessness, and leadership qualities in man does not differ from the relay of group diversity in animals. That's good and bad. Watch out that you, with your revulsion at the very fact of the rat hierarchy's existence among humans, don't give birth to a rat king. . . . That's bad. What's good is that the combinations of hereditary tendencies toward bravery, altruism, and self-sacrifice are ineradicable. Nothing can prevent the birth of brave people or avert the beneficial influence of sacrifices made and graves lost.

"Why don't scientists fight for the freedom of science?" I was asked in 1968 by students at the evening branch of Novosibirsk University. "After all, writers fought. Some of them are even in prison. That means that it's all right," said a female student with a troubled look. Think about those words: "They're in prison—that means it's all right."

Student protests in Stalin's time were a great rarity. I know only of two such instances. In 1940, when Professor G. D. Karpechenko, the chairman of the Department of Plant Genetics at Leningrad University, was arrested, one female student, as a sign of protest, asked to withdraw from the university. In 1948, after the August Session of the LAAAS (Lenin All-Union Academy of Agricultural Sciences), a student at Leningrad's Institute of Medicine protested the dismissal of Professor Ivan Ivanovich Kanayev, a well-known geneticist and historian of science who taught at the Department of General Biology. The student was arrested, given ten years in the camps, rehabilitated in the Khrushchev era, and became a geneticist—a specialist in the genetics of mental illnesses. He considers arrest a psychic trauma capable of inducing illness. He is alive and, for understandable reasons, I cannot state his name.

The female student got away with her boldness. She withdrew and that was that. She and her mother died of cold and starvation two years later in the blockade. I can tell about her. Her name was Edna Borisovna Brissenden. She was my student. In 1937, she was brought to me at the Department of Genetics and Experimental Zoology at Leningrad University. They wanted me to teach her genetics. She was in the eighth grade, fifteen years old, and a student at the University for Secondary School Students. There was such a thing. Any student who wanted to could attend evening courses there. It turned out that the girl spoke

Russian correctly, but with a strong accent. Her first sentence was: "I need mice."

"Why?"

"I want to study mouse genetics."

"No one in our department works with mice, but we could get some for you."

"Can you get pure strains?"

I opened my eyes wide. The question showed that the child knew genetics well and knew what she wanted. I suggested that she study fruit flies under my guidance, and she agreed. My first work on population genetics, published in the *Journal of General Biology* in 1941, was written in 1938 in co-authorship with her. Not at all afraid of spoiling the child with praise, I told her, "There will come a time when co-authorship with you will be a source of great pride for me."

I *am* proud of having been a co-author with her, but the reason for my present source of pride is not the one I had in mind back then. I was predicting a future as a great scientist for her. That didn't come to pass. Her greatness was in her fearlessness.

All her actions, as well as her mother's, were rooted in humanitarianism. They were vegetarians. I asked her much later why she had needed the mice. She opened Sinnot and Denn's genetics textbook and showed me a picture: It was the result of crossbreeding gray mice and the white albino ones. In the second generation of that cross, albino mice turn up among the offspring of the first generation gray hybrids. But only one-quarter of them are albinos. Three-quarters are gray. "I wanted to help the white mice transmit their traits to their offspring," said Edna.

The beautiful little girl with large gray eyes was an American. Her parents had separated long before the events being described. Edna was brought to Leningrad along with her mother by Nikolay Ivanovich Vavilov. Her mother, an American, had been a specialist in Russian language in the USA. A member of the Communist party, she participated in a protest against the conviction and execution of Sacco and Vanzetti, and lost her job. Vavilov brought her to Leningrad as a secretary. He said that she was a genius at that business. Mother and daughter lived in a tiny little apartment in the Stroganov Palace on Nevsky Prospect at the corner of the Moyka Canal. The building belonged to the Institute of Plant Breeding, the director of which was Vavilov. They lived right under the roof. The palace servants had probably lived there. But for the 1930s that was luxury living—it wasn't in a communal apartment, after all. The windows looked out on Nevsky Prospect. I went to visit them once. There was a wool rug hanging on the wall, with a large red swastika portrayed on it. "Take that rug down right now," I said, "before someone comes and sees it and denounces you." Stalin did not as yet make friends with Hitler. Nazism was the main fright of the official propaganda. "No," said Edna, "Mama won't take the rug down. It's Indian, not Nazi, and we're against the oppression of minorities."

In 1937, I wanted to take her with me on an expedition. The head of the expedition agreed to include her without pay, though he also agreed to pretend it was with pay, but I'd provide the money for her expenses. Helping her was no easy matter. Her pride was excessive. She viewed any help as a handout. Her jacket was a leather one, not at all appropriate for the Leningrad winter. We in the department decided to dress her in my thick Iranian wool pullover. We begged her to take it just for a while. She wouldn't hear of it. The gang and I—Muretov, Gratsiansky, Rosenshtein (they all died during the war) forcibly dressed her in her jacket and my pullover. We joked and laughed, but acted decisively. She stepped over to the door, whipped off her jacket in a flash, tossed off the pullover, and left.

The head of the expedition and I decided to fool her. Things turned out differently, however.

It turned out that as far as the chief was considered, this really was a deceit, but of another sort. He never had intended to take Edna, and he made the promise in order to ensure that I would be part of his expedition. And the only reason that I was going was so as to help Edna. I was involved in radiation genetics under Muller at that time and wasn't interested in populations.

The head told Edna that his funds had been cut and that he wouldn't be able to finance her participation in the expedition. "How could you have believed him?" the fifteen-year-old girl asked me later. "It was obvious at first glance that he's a liar." Of course she knew nothing about my understanding with him. I traveled to the expedition in Uman separately. I'd been in a holiday home in the South. I was disappointed to find out that Edna wasn't on the expedition.

Don't judge that chief too harshly. No sensible person (I've never belonged to that group) would have risked taking an American citizen on an expedition. It was 1937, and that says it all. When we arrived at the fruit wine plant in Uman and asked the director for permission to catch fruit flies in the fermentation department, he lifted the phone receiver without answering us and called the NKVD (the equivalent of the KGB in those days). He was asking what he should do. He flinched the whole time, as though he had St. Vitus's dance, and one of his eyes twitched. After the phone call he gave us permission to catch flies.

I became a population geneticist beginning with that expedition. And now, in St. Louis, Missouri, USA, I'm doing exactly what I did in 1937. The global splashes of the mutation process, the first surge of which I had the good fortune to observe in Uman back then in 1937, permanently seduced me.

I returned to Leningrad with huge amounts of material. Edna Brissenden took part in processing it and carried out an important aspect of the research. After the article was written she said, "Raissa Lyubovna"—she couldn't pronounce Lvovna—"you write well, but Muller writes better." I couldn't argue with her.

She didn't waste any time over the summer. She went directly from the eighth grade to the tenth. For her literature examination, she wrote an essay in English entitled "Mayakovsky and Whitman." The teacher suggested that she publish it. He was willing to do the translation himself. She refused. I asked her to let me see the essay. She wouldn't give it to me. She liked me, but sometimes treated me like a dog.

At the beginning of 1939 (I was still a graduate student in Leningrad), she came to the department and said that she and her mother couldn't get their visas extended, they'd been denied citizenship, and apparently would have to go back to America. She was in despair. I told her that it was better for them to leave. She, proud Edna Brissenden, buried her face against the wall and said in a constrained voice, "Go to hell! Here I'd be able to study at the university, but over there I'll be a dishwasher." "No," I said, "over there you'll be alive, while if you stay here there won't be even dust or ashes left of you."

She entered the university in 1939. Before entering it she had worked as a laboratory assistant at the Institute of Zoology. Now that she'd become a university student, she lost the opportunity to earn money. She wasn't given a stipend. Vavilov was no longer the institute director and he couldn't afford a secretary. Edna's mother worked as a librarian in the foreign language section of the library at the Institute of Plant Breeding and earned forty rubles a month. That was the salary level for parents, and initially their children were not eligible for a stipend. I managed to arrange a stipend for Edna.

In 1940, Vavilov was arrested, as was Karpechenko. Edna withdrew from the university as a sign of protest. She said that in America, in her hateful America, not a single student would have stayed. I asked her what she was doing with herself. She refused to answer. She said that there were things more important than science. I was living in Moscow at that time. She was supposed to come to visit me in the summer of 1941. That didn't happen. The war began. Word of the circumstances of her and her mother's death didn't reach me until almost forty years later, when American botanist N. S. Forest became interested in the fate of the two American citizens whose traces had been lost in blockaded Leningrad. He found out from neighbors and from her mother's co-workers that Edna died of starvation there. A neighbor had offered her bouillon made out of glue, but Edna refused it. Glue was an animal product, and she and her mother were vegetarians. Her mother was evacuated, but she died of malnutrition. Forest made Edna the heroine of an enchanting short novel, where he gave her the name Edith.

Arrest is far from the only method for remaking the nature of the Soviet intelligentsia. Sometimes there was no need for murder—people died all on their own.

Roza Andreyevna Mazing was an assistant at the university when I was an undergraduate. The Department of Genetics at Leningrad University was organized in 1925 by Yuri Alexandrovich Filipchenko. Roza Andreyevna was his student.

Yu. A. Filipchenko was an encyclopedia biologist. Biometrics and the theory of evolution, genetics and systematics were all equally his specialties. He named the department that he created the Department of Genetics and Experimental Zoology. He didn't limit himself to zoology, however. He studied the genetics and selection of quantitative traits in wheat and ducks. He created a variety of wheat for the Leningrad region called "Peterhof wheat." He was interested in the inheritance of giftedness. On the basis of an analysis of the genealogies of outstanding Russians, he attempted to bring to light the relative role of heredity and upbringing in the formation of the creative person. His conclusions were of a most humane and democratic character. He had many, many disciples, and they idolized him. He was forty-eight years old when he died in the summer of 1930.

When I arrived at the department in the fall of 1930 Filipchenko was no longer alive. His students and the new chairman, Alexander Petrovich Vladimirsky, were sitting there with wooden faces and the bleached-out eyes of rotting fish, sullen and silent. The newcomers, party young people, were trampling the name of their teacher and idol in the mud—the obscurantist, racist representative of the bourgeois intelligentsia, the only just deceased Yuri Alexandrovich Filipchenko. Such was my first impression of the university and genetics. It was a roaring smithy where future traitors were being forged . . . as well as the heroes, the devotees of science. A polarization was underway.

I don't recall seeing Roza Andreyevna at these meetings. She would sooner have put her head on the block than speak ill of the late Filipchenko. She remained at the department until Vladimirsky's death in 1939 and worked with the fruit fly. She discovered increased viability in flies that contained a hidden lethal mutation—the phenomenon of super-dominance. That was a very important discovery. It allows one to understand the genetic reasons for hybrid vigor.

That work was the object of Lysenko's profanation. At a general meeting of the Academy of Sciences, speaking before academicians of all specialties, he permitted himself such rude behavior that those who knew what had happened could only make do with hints in the presence of ladies, while they couldn't even do that to poor Roza Andreyevna. When it reached her that the quintessence of Lysenko's dirty joke was coarse swearing, she was very upset to learn that such language existed. Up until then she hadn't known about it. She was from an intelligentsia family. Her ancestor had been the children's doctor for Pushkin's family. All her brothers were professors. She took up the question of the origin of the mother oath and found out from folklorists that it dated from the time of the Tatar Yoke (a political euphemism). Still, she had decided to find out what was concealed behind the strange silences. She invited me to come visit her, and when it was just the two of us, she asked me what was going on. And so I enlightened her.

The flies with which Roza Andreyevna worked were black, the color of ebony. The strictness of the experiment demanded the presence of a marking trait. The marker was the black color of the body—ebony, as it's called by the international English nomenclature. (Latin is the international language of zoologists and botanists, while it's English for geneticists and drosophilists. Hundreds of mutations of that famous fly bear English names.) "You write that virgin females come from the ebony strain," I said to Roza Andreyevna. "So here's what happened: Lysenko said, 'Why how could they be virgins if they're . . .' and then he pronounced the English name with a Russian accent."*

"But if you read the name for the females in Russian, it doesn't mean

* Lysenko's "witticism" would thus read: "Why, how could they be virgins if they've been f———ed?"—*Translator's note.*

anything," Roza Andreyevna said. "I asked my acquaintances, and even wrote it down for them on a slip of paper, and no one knows the word. Here, I'll write it out for you."

"Why, they're lying," I said. "They're too embarrassed to tell you. How could it be that I know and they don't?" She took a piece of paper and wrote out *yevopu*. That's how the English word "ebony" would sound if the English letters were read as Cyrillic ones. And there really is no such Russian word. "Roza Andreyevna," I said, dying from laughter (and I'd already begun to die when she told me about writing the word out for people), "haven't you ever heard carters swear when they urge the horses on?"

"I've heard three funny words, but I couldn't figure them out," said Roza Andreyevna.*

I told my father about Lysenko's prank. He said that he refused to believe it. But the infinite sadness with which he said that showed that he did believe it. That was in 1939. Twenty years later, in 1959, I became the victim of a similar attack on the part of Lysenko. Without naming names, he quoted a sentence from an article of mine that treated the genetic bases of evolution. The sentence was a quite usual one, written in professional language. He read it as an example of abracadabra, which for him it was. His choice of the sentence was determined by an expression I had used. The expression was "genetic *dreyf*." A *dreyf*, "drift," a generally known concept, is a change in a ratio of normal and mutant genes in the population. "Drift" is applied to describe accidental shifts occurring without interference of selection. If 35 percent of a given generation on some closed human population have brown eyes, and in the next generation 40 percent do, a shift, a drift, the *dreyf* has occurred. In colloquial Russian *dreyf* means "fear, panicked flight." *Sdreyfit*, is to show the white feather. Now, genetic *dreyf*. All Lysenko had to do was to pronounce those words, and everyone understood that he was referring to the fear that he had put into geneticists. Vavilov's de facto murderer was flaunting his ability to inspire fear. He himself had recovered by then from the terror that he experienced when Stalin, his protector and the great leader of all the peoples of the world, died. In 1959, there didn't seem to be the slightest chance that he would be "demoted *to* full membership in the academy." This is what the wits called the very harmless fall of Lysenko in 1964.

*The equivalent of "I have known your mother carnally," the standard Russian curse.

In 1950, Lysenko paid Orbeli back for the support that he had given geneticists when he was vice-president of the USSR Academy of Sciences and director of the academy institute. L. A. Orbeli was exposed at a session of the USSR Academy of Medical Sciences and removed from his position as director. When Roza Andreyevna learned of that, she suffered an attack of angina pectoris and died.

M. E. Lobashev was a co-worker of Orbeli's at the time. In August 1948, he had been driven from his post as head of the Genetics Department by N. V. Turbin, a fierce Lysenkoist and thug who is now not just a geneticist, but the president of the N. I. Vavilov Society of Geneticists and Selectionists. He has transformed himself right before everyone's eyes, but no one in the Soviet Union even remembers what it means to be surprised by anything anymore. M. E. Lobashev said at Roza Andreyevna's funeral, "Her life and death are an example of how lonely a Soviet person can be." He told me that himself. He got into great trouble because of it. He, a party member, ought to have known that a Soviet person can never be lonely under any circumstances.

Roza Andreyevna's doctoral dissertation was destroyed when one of her brothers, a professor, threw it into the stove in a fit of terror.

At the meeting of the Biology Division of the Academy of Sciences when Orbeli was removed from his post as institute director, Orbeli tried to control his fate. He was being dismissed because of his overestimation of the role of higher nervous activity in the physiological functions of the human organism, but now said that he would write a disavowal of his former position, since at this very session he had been persuaded by the example of his colleagues that the stomach exerts a powerful influence on the brain. But his colleagues refused to accept that sort of a declaration.

One hundred forty years prior to the August Session of the LAAAS, the great encyclopedist Lamarck created a theory of evolution.

In August 1948 the least sturdy underpinnings of the magnificent edifice erected by Lamarck—the adaptation to local conditions through acquiring new adaptive characteristics and the transmission of these acquired traits to offspring—was proclaimed the cornerstone of contemporary biology. *Pravda* trumpeted with great fanfare that for the first time in history the inheritance of traits acquired in the course of individual development had become a state doctrine.

And in spite of that, Lysenkoism does not in the slightest degree deserve the name Lamarckism. Lamarckism is a noble teaching. The first theory of evolution in the history of science was Lamarck's. It proposed that progress is a primary trait of living creatures. That ability to perfect oneself is an integral attribute of living things. Lamarck called it the "gradation principle." Changeability flowing along fixed channels outlined by the internal organization of a living thing—just as the form of crystals is determined by the organization of the crystallizing matter—and the harmonization of the vegetable, animal, and mineral elements of the landscape are the moving forces of progressive evolution. Adaptation to the local conditions of existence, according to Lamarck, is a perversion of the great principle of gradation.

I knew many Lamarckians very well. My father was among them. Alexander Alexandrovich Lyubishchev and Pavel Grigoriyevich Svetlov were my friends. I know Boris Mikhailovich Kuzin only from reading about him. Not one of them took up the banner of Lysenkoism. That's understandable. They were anti-Darwinians in an era of Darwinian dictates. They remained scientists in an antiscientific era. The victory of Lysenkoism not only spelled the demise of Darwinism, but of Lamarckism as well.

In 1922, my father wrote a book entitled *Nomogenesis or Evolution Determined by Law*. He contrasted nomogenesis to tychogenesis, as he called Darwin's theory, producing the name from the Greek word *tyche*—"accident, chance." At the end of the 1930s he foresaw the advent of social Darwinism and wanted to prevent it. He told me that Darwin's theory was harmful for humanity, since as applied to man the theory sanctioned warfare as a factor of progress. It seemed necessary to counter this antihumanitarian conception with the theory of mutual assistance. I said that it wasn't the theory that should be disclaimed, but its application in a sphere where it is inapplicable, and that Darwin understood the struggle for existence not in the literal sense, but as competition. One can compete for mutual assistance, just as species in fact do. The ones who survive are those where mutual aid, including concern for future generations, is perfected to the greatest degree. Mutual assistance is a means of achieving victory in the struggle. My father wouldn't even hear of such a thing and just grew angry. He showed most convincingly that evolution is a process based on law. His argumentation was purely materialistic. He rejected the idea of supernatural forces manifesting themselves in living beings and forcing them to evolve: "The history of science has taught us that vital force, as a hypothesis, is valueless: it has in nowise aided us in making any progress in the interpretation of facts," he wrote. "We are enabled to work fruitfully in the field of natural science only by the aid of forces recognized in physics, and every naturalist should as far as possible endeavor to interpret nature by mechanical means . . ."*

But in the 1920s Darwin was not to be criticized. Darwin had been canonized and figured as a small icon on the official altar of Marxism-Leninism. A hailstorm of attacks fell on my father. Prezent was beside himself with rage. Label after label was hung on the guilty party. Berg was an obscurantist. His teaching was priestliness. His denial of the role of chance in evolution was camouflaged preaching of the biblical myth. The critic didn't look for errors of logic or fact, he was tearing the mask from a class enemy, exposing him as a counterrevolutionary, revealing the real motives for Berg's actions, which were aimed at the restoration of capitalism, feudalism, and tsarism. Don't forget what sort of a period that was. Today an exposure speech, and tomorrow the police wagon was already calling at your door. But fate was kind to Lev Semyonovich. He was not only a theoretical biologist, but a major zoologist, as well—a specialist in fish. In 1927, he was elected an associate member of the USSR Academy of Sciences Division of Biology. In 1939, L. S. Berg was nominated for election as a full member of the Biological Division of the USSR Academy of Sciences. And then an article appeared in *Pravda* under the heading "No Room in the Academy for Pseudo-Scientists." The other pseudo-scientist was Nikolay Konstantinovich Koltsov.

Koltsov was an associate member of the USSR Academy of Sciences, a geneticist, cytologist, a scientist of world renown, a brilliant organizer and pedagogue. The power of N. K. Koltsov's thought was striking. He was several decades ahead of his time, the first to penetrate the principle of self-reproduction of a chromosome. He realized that a chromosome is a single molecule. Hence, it did not grow, it did not divide, but reproduced itself by building an exact copy of itself and pushing it away. Koltsov knew better than anyone else how to link the results of genetic research with microscopic pictures and chart new paths in science. It was under his guidance that I. A. Rapoport began his studies of chemical mutagenesis, and they brought him worldwide fame. But there was no room in the Academy of Science for these two pseudo-scientists, Berg and Koltsov.

* *Nomogenesis or Evolution Determined by Law* (Cambridge, Mass. and London, England: M.I.T. Press, 1969), p. 6. First published in English by Constable and Co., Ltd., London, 1926.

The same article also cleared the way for Dubinin to replace the pseudo-scientist Koltsov as director of the Institute of Experimental Biology and for Lysenko to become a full member of the academy. Dubinin's plans fell through. But Lysenko was elected and for the first time in the more than 200 years of the academy's existence, curses were heard from the podium at a general meeting. It goes without saying that Berg and Koltsov were not elected to the academy. I think that they probably were not concerned about the election. They were happy not to have been arrested. L. S. Berg didn't occupy any high posts, so there was no place from which to dismiss him. N. K. Koltsov was driven from his post as director of the Institute of Experimental Biology that he had founded, and he died suddenly and unexpectedly in December 1940. His wife committed suicide. They were buried together. I remember very clearly N. P. Dubinin standing by their open coffins and vowing eternal loyalty to his teacher's memory. In a plangent tenor he said slowly: "We will forever preserve the memory of this life and this death—a memory which teaches us how to live." Somehow or other it was a poor teacher for that traitor. Read his book *Permanent Movement*, a political denunciation of the dead. Koltsov comes in for special attention in it.

The article in *Pravda* attacking Koltsov and my father was signed by academicians Bakh and Keller, and by party functionaries at the academy, Koshtoyants, Nuzhdin, Kosikov, and Dozortseva. Who wrote it is anyone's guess. One of the signers? Prezent? It doesn't make any difference. The article contained monstrous accusations that had absolutely nothing to do with reality. Father was accused of sympathizing with Hitlerism. He began to age noticeably. His teeth began to fall out.

All the same, he was elected an academician seven years later. That miracle occurred quite by chance. He was nominated both in 1943 and in 1946, but both times his chances for election were zero. But in 1946, two candidates were nominated for a single place in the Geography Division. The nominees were Berg and Baransky, and it was at this point that something unheard of happened. Baransky refused to be elected. "No one can be an academician if Berg isn't one," that remarkable man wrote to the Presidium of the Academy of Sciences. Berg was elected.

Soon after the August Session of the LAAAS there occurred a meeting of two academicians—Berg and Prezent. Prezent was at the zenith of his glory. He chaired the Department of Darwinism at Leningrad University and ran things at Moscow University and the USSR Academy of Sciences Institute of Genetics. He was a full member of the LAAAS, a philosopher, professor, and agronomist. They met in an international car of the Leningrad-Moscow train. Lev Semyonovich was on his way to a meeting at the Academy of Sciences. Academicians are required to be present at general meetings of the academy. They travel first class in "international cars" and their expenses are paid by the academy. A compartment in an "international car" has two sleeping places and a toilet, as opposed to the four-person compartments in the next lowest category of first-class cars. To judge by the dress coats that they wore back then, Lev Semyonovich's neighbor was a high-ranking railroad official.

No sooner had they left Leningrad, than a short sharp-nosed gentleman entered the compartment. "Hello, Lev Semyonovich! Do you often have occasion to travel?" the gentleman asked politely, with animation.

"Yes," said my father, "although I try to avoid it."

The gentleman said, "I spend almost half of my nights on the train—I work three days a week in Moscow and three days in Leningrad."

"How awful! How do you feed yourself? At cafeterias?"

"No, I have a permanent room in a Moscow hotel, and they bring me food from the restaurant."

"But, it's all cold."

"I heat it up on a hot plate," said the gentleman. And that was the extent of their evening conversation. In the morning the gentleman came again. "Good morning to you, Lev Semyonovich!" And again a conversation began. The stranger asked how Lev Semyonovich felt about field-protective afforestation. Lev Semyonovich said that rivers and their upper waters needed to be protected, as prominent experts had recommended long ago. The gentleman asked what Lev Semyonovich's attitude was toward fishing on the Volga, but just then they entered Moscow and the stranger hurried off. Then the high-ranking railroad official turned to Lev Semyonovich with great piety and exclaimed in amazement, "So, that's what sort of friends you have!" Father didn't have that eagle look that turned up in some of his portraits, but my friend, a geneticist, Professor V. V. Sakharov said of him, "Even if you're not on the university embankment of the Neva you can tell three miles off that that's a professor walking along."

"Who was that?" Lev Semyonovich asked.

"Why, that was Prezent himself!" exclaimed the official.

When Father told this story he cited a quotation from Goethe, the words that Mephistopheles speaks after his conversation with God: " 'Es ist so nett von so einem hohen Herren so menschlich selbst mit dem

Teufel zu sprechen' [How kind it is for such a high Lord to speak in such a human way with the Devil himself] God forbid, that you should tell that to anyone," Father said. But who could restrain herself? And I told it to a physiologist, E. Sh. Ayrapetyanets, a professor from among the promotees, who had been the community prosecutor at the Mordukhay-Boltovskoy trial. I'm ashamed now to have been cordial to him. He said that this was an extremely interesting case for learning about the physiology of higher nervous activity (that's what human psychology was called in the USSR). "Lev Semyonovich couldn't help but have known Prezent before," Ayrapetyanets explained. "They were at meetings together for the Greater Science Council at Leningrad University. But he didn't notice him, and if he did, he forgot, because Prezent represented a threat of destruction. That phenomenon is called 'protective inhibition' in scientific language. And no matter how many times your father meets Prezent, he'll never recognize him."

I told Ayrapetyanets that Father had asked me not to tell anyone. A few days later Father and I were walking down the street and ran into Ayrapetyanets. "Lev Semyonovich," he exclaimed, smiling that radiant smile of his, "tell me about how you and Prezent met. I want to include it in a lecture to my students, but I won't mention your name, of course. Raissa told me about it." He was obviously giving me away, but Father didn't notice that and was about to tell him the story, when suddenly it turned out, in complete accord with Ayrapetyanet's analysis, that Lev Semyonovich couldn't remember a thing about it. "I've forgotten," he said. "Let Raissa tell it. She remembers everything."

In 1971, I told this story to Viktor Amazaspovich Ambartsumyan, an astronomer, a member of the USSR Academy of Sciences, and president of the Armenian SSR Academy of Sciences. He began his scientific career as a professor at Leningrad University and had been a member of the Science Council along with Berg and Prezent. Viktor Amazaspovich said that once, when Prezent was speaking, Lev Semyonovich asked him, "Who's that?"

There was one other instance of forgetfulness on Lev Semyonovich's part. In 1945, during the anniversary celebration of the academy, academicians and associate members received medals. Lev Semyonovich was an associate member and his rank entitled him to a medal, not as lofty a one as the academicians were being awarded, but still a medal. When R. L. Dozortseva, the secretary of the Presidium of the academy, congratulated him on receiving a medal, he said to her: "But what do our medals mean in comparison with your military honors?" It could not possibly have occurred to him what a devilish ambiguity his words contained.

R. L. Dozortseva, in the past a co-worker with N. I. Vavilov and H. J. Muller, had deserted to Lysenko, and was one of those who in 1939 had signed the article "No Room in the Academy for Pseudo-Scientists." Her participation in the Great Fatherland War was limited exclusively to her shameful, panicked flight from Moscow in a state car during the panic of October 16, 1941. But the medals that she had received for loyal service to antiscience really were earned in battle. Father told me about his reply to the woman who was Secretary of the Presidium, but I was not the one who opened his eyes to who she was. And that's just the way she remained for him—that traitress, all decked out for show in her thirty pieces of silver, the Joan of Arc of the Great Fatherland War.

Protective inhibition, if one is to speak simply, means the inability to remember something. It took Stalinism spiced with Lysenkoism to do any damage to the memory and ability to recognize that my father possessed. He was not only a zoologist but a paleontologist as well—he could identify a fish by a single bone. His memory was legendary.

And now Berg, the Lamarckian Berg, was an associate member of one branch of the Academy of Sciences and an academician in another. Darwinism was no longer a state doctrine. Falsified Lamarckism under the name of creative Darwinism had won the government's recognition.

In 1946, the first volume of the collected works of the great leader of the people of the world, I. V. Stalin, was published, and we, the staff of the Department of Darwinism at Moscow State University, were horrified to read that in his early youth, barely out of the seminary, i.e., having received an education that wasn't even the equivalent of a high school diploma, Stalin had had an opinion about the theory of evolution. He wrote that the argument between Darwinism and Lamarckism was not yet finished, and that in his opinion, Lamarckism would triumph. We realized that the end had come for us, Schmalhausen's co-workers, and for Schmalhausen himself. Lysenko had received a mandate for the liquidation of those who didn't consider his sorcery great achievements. In 1948, genetics was liquidated. A shameful caricature of science took the place of true science. But certain of the tenets of the insane constructs that had received governmental sanction coincided with what had been written in *Nomogenesis* more than a quarter of a century earlier.

These included some false statements—the rejection of chance and the inheritance of acquired characters.

Election to the academy offered Lev Semyonovich the broadest opportunities to propagandize his theory. But the author of *Nomogenesis* never felt or showed any solidarity with the victors. Their victory had been achieved with bloody hands. The Darwinists with whom Berg polemicized were his friends—Vavilov, Filipchenko, Schmalhausen. The difference in their views did not at all get in the way of friendship. They were murdered or cast into the dust and deprived not even of freedom of speech, but speech itself. Their names were scrubbed from books and it was forbidden to mention them. Freedom of science, that freedom which he valued more highly than anything else in the world, no longer existed. Nor did the people who believed as he did. There was no one before whom he could defend his rightness. Raising his voice now meant committing treason in regard to the dead and those in prison. Life became hateful to him. Not long before his death he said: "I fought against the idea of natural selection. Now I see that I was mistaken. The selection of scoundrels is occurring."

My father was elected president of the Geographical Society in 1940. His predecessor was Nikolay Ivanovich Vavilov, who was the fourth president since the founding of the Geographical Society. When Vavilov was arrested, Berg was elected president.

In 1947, the Geographical Society celebrated its hundredth anniversary. Lev Semyonovich wrote a history of the society. He devoted a special chapter to the great geographer, botanist, and agronomist Vavilov. Censorship demanded that the chapter be excised. Lev Semyonovich refused, saying that in that case there would be no book. The book was published, with the Vavilov material intact. That was unheard of boldness and a great victory in the Stalin era. No one knew what had happened to Vavilov. They knew that he'd been arrested. But then what? Had he died, or was he serving a sentence? Lev Semyonovich Berg didn't know either. The author of such a book was in danger of being arrested at any second.

As long as Berg was president, Vavilov's name was not crossed out of the books that were kept in the Geographic Society's library, and Vavilov's works were not destroyed. But when Father died in 1950, the library was closed and the order was given to blot out the name of the anathematized president with India ink.

I was often in the Geographic Society's archives, and as I walked by the glass doors of the library I saw some sort of strange activity. A. G. Grumm-Grzhimaylo, a historian, the son of a famous explorer, and the library's director explained to me what was going on. He was completely covered with some sort of boils, the result of nervous shock. I remember one very nice old lady librarian who was completely gray. At the same time that Grumm-Grzhimaylo broke out in boils, she had become semi-transparent, and in place of eyelids there were watery swellings. She somehow seemed broken. She had known Vavilov personally, and it was impossible to know him and not love him. It was her Orwellian task to cross out the name of the best of all people in all the books where it was mentioned. What would have happened to Father had he still been alive?

Vavilov's arrest was a turning point in my life. I had known him personally. Spending time with him raised one above everyday reality and transcended the boundaries of existence, though he was infinitely down to earth. The whole world knew him and loved him. He won hearts, and that love was transferred to the Soviet Union. People loved Soviet Russia because they loved Vavilov. Muller told me that Vavilov was the Peter the Great of the twentieth century. That was both true and not true. Muller had only the tsar's positive traits in mind.

In 1942, Vavilov was elected a foreign member of the Royal Society of England. The Royal Society limits the number of foreign members, elected from all countries and in all fields, to fifty. Sir Henry Dale, the president of the Royal Society, told Julian Huxley that the news of Vavilov's arrest and death did not reach the society until 1945. Repeated requests for information about the place and time of Vavilov's death, sent to the Soviet Union through all possible channels, remained unanswered. Huxley tells about Vavilov in his book *Heredity East and West: Lysenko and World Science*, which was published in 1949. He concludes his narration with the words, "Such was the unfortunate fate of one of the best scientists that Russia ever produced."

To murder Vavilov meant to spit in the face of world public opinion, to drive off one's well-wishers, to reveal shameless tyranny. Through that arrest a great number of things become apparent: the friendship pact with Hitler; the occupation of Latvia, Lithuania, and Estonia; the knife in Poland's back; tanks on Czechoslovak soil; the murder of Mikhoels, Meyerhold, and Mandelshtam; the pillorying of Akhmatova, Zoshchenko, Pasternak, Solzhenitsyn, Sakharov; the trials of writers; and the arrests, arrests, and arrests. Stalin, Khrushchev, Brezhnev. The arrest and destruction of Vavilov indicated that the Soviet Union did not need solidarity with the progressive forces of the world, and that it didn't reckon on the effects of its mendacious propaganda. Diplomacy was cast aside as unnecessary. Or is this the higher diplomacy of routs? "We'll destroy those whom you loved, whom worldwide fame made invulnerable in your opinion, and you, the representatives of world public opinion, won't dare even make a peep. And the greater the people who become our victims, the more loyally you will come to worship us." History shows that evil has been correct in its reckoning. These are thugs who are interested in having witnesses, who are happy to commit crimes before the eyes of a crowd grown silent.

One of those who had grown silent was my father. He lived on for ten years after Vavilov's arrest. The moral countenance of the country was saved for a while by the incalculable suffering brought by the war and the forced alliance with the Western democracies. Things grew progressively worse after the war. A victory is demoralizing, even a victory over hellish forces. The creation of the atomic bomb radically changed the moral countenance of the entire planet. I remember Father saying: "The atom bomb couldn't be created by our country. The outlays required are too great." Idealism itself was speaking with his lips. In order to create an atomic bomb you don't need wealth; what you need is authority that knows how to concentrate great resources within its own hands, no matter what the circumstances and in spite of the nation's poverty. The Soviet atomic bomb was created in his lifetime.

The victorious nation was clamped in an iron vice. Arrests followed arrests. Anyone who had been a German prisoner of war or who had been driven off to Germany for forced labor was sent to the camps. Censors cracked down with a vengeance. Anything done in any research institution was declared to be secret information. Humor was declared seditious. Shostakovich was lambasted for a second time—his music spoke the truth in language that was all too clear. His post-war symphony did not "resound with the thunder of victory"; instead, it depicted forte-fortissimo the feverish activity of plunderers who were making a living off the nation's blood. The passacaglia of that immortal work is a lament for those who remained among the living. The August Session of the LAAAS was an ordinary episode for those times. Whom didn't they destroy?

Honors and privileges became Judas' thirty pieces of silver. Those who didn't lose their minds (and there were such cases, but I won't give any names) died. Father died before the Affair of the Kremlin Doctors, before that bloody anti-Semitic rampage. How many times I thought: "What good fortune that Father died. What would be happening to him if he were alive?" At his funeral I heard someone say: "The president of the Geographic Society, yet no one from the City Soviet spoke at the mourning assembly. Why, I wonder?" "He's a Jew," someone else replied. That was one of the reasons. The main reason why no representative of the authorities spoke at the huge mourning assembly was that Father was an honest person. There was a complete rift between the high position in the hierarchy of lies and the categorical imperative of morality to which the soul of the deceased president was subordinate.

The people who could preserve spiritual harmony were those who could reject honors, people who had a profound faith, who were religious by their very nature, who found themselves at the very bottom of existence, but who had their religion. They were the playthings of fate, not of the authorities. One of those fortunate people was the Lamarckist, Lyubishchev. He was not merely a follower of Lamarck, but a Platonist. He considered the material world an incarnation of a great and good idea. It goes without saying that there was no room for the creative role of natural selection in his noble system of ideas.

Alexander Alexandrovich Lyubishchev, a zoologist and specialist in plant protection, was a friend of my father's and thought like him. Just like my father, he placed the means for achieving a goal higher than the goal itself. Lysenko's ignorance and his bloody actions were not expiated in Lyubishchev's eyes by the recognition of the inheritance of acquired traits.

Unlike Lev Semyonovich, Lyubishchev rushed into battle. He wrote volumes criticizing Lysenko and Michurin. By a miracle he remained at freedom. Out of his enormous number of works perhaps only a tenth were published. His criticism of Lysenko, I need not tell you, was not among them.

His courage was infinite. He was persecuted and lived in exile in Ulyanovsk, on a pension. He protested against the unfreedom of science. He didn't worry at all about his own fate, didn't care whether he went abroad, and was happy when he managed to arrange to have his works published, but he wrote them absolutely freely and without the slightest hope for their publication.

He was delightfully wily. Once, on a flight from Novosibirsk to Leningrad, I was reading his essay on Saint-Exupéry. It was a first-class model of *samizdat*. At the most interesting place, where the topic of the essay on the great humanist was the similarity between Stalinism and Hitlerism, I fell asleep. The manuscript lay open on my knees. I was sitting by the window, and I had only one neighbor. The airplane set down in Sverdlovsk for a transit stop. I awoke when the passengers were preparing to exit. My neighbor, on leaving, said, "I saw what you were reading."

"That means that you spent your time usefully," I said, and then shuddered. But neither in Sverdlovsk, nor at the Leningrad airport did they come for me.

Lyubishchev's protests were not limited to science. He wrote a letter to the Moscow Soviet against raising a monument to Prince Yuri Dolgoruky, allegedly the founder of Moscow, in the center of Moscow in front of the building that houses the Magistracy of the USSR. Erecting this monument was a mendacious gift of tribute to Russian chauvinism on the part of the government—that same government's criminal actions had produced a cavern where there had formerly been the revolutionary ideals of proletarian solidarity and internationalism. Lyubishchev provided historical information about the prince and suggested that the monument be removed. He received a reply: The Moscow Soviet agreed with his arguments and the monument would be moved to another place and the place was named. Lyubishchev sent copies of that letter through the mail to all his friends. That was great boldness at a time when people's mail was read. The monument remained where it had been—a monument to a lie, as Joseph Brodsky entitled one of his poems.

Lyubishchev was enormously charming. He was the embodiment of freedom. When he made visits to Leningrad as late as the 1970s, and gave papers at the Mathematics Department of Leningrad State University, hordes of young people trailed after him. His last presentation in Leningrad was devoted to my father's work, *Nomogenesis*. I had the role of discussant and treated the questions from a Darwinian point of view.

Darwin's theory and Lamarck's do not at all exclude each other. Systems theory makes a synthesis of Lamarckism and Darwinism possible, and allows us to combine the idea of natural harmony with the idea of selection. I treated selection as one of the regularities of evolution, as a process of regulation within a system of an order higher than the organism, and underlined the factors that limit it; in particular, the collaboration of different elements within the system. There is no struggle between different organs of one and the same organism. Parts of the organism cooperate to provide the survival of the integral system.

Parts of a population, individuals, are involved by virtue of their equality with each other, by the identity of their needs and requirements, into an endless competition. Victory or defeat is natural selection. There are, however, many functions that individuals have to achieve in combined action with their counterparts. Perfection of these actions based on cooperation is the target of natural selection. Differences, not similarities, between individuals sometimes ensure the most successful outcome. Reproduction and parental care are examples of activities based not on struggle but on cooperation of different representatives of the species—males and females in this particular case. Survival of groups whose members are diverse and cooperative limit the intragroup competition. I showed that if there was such a thing as the inheritance of traits acquired in the process of the development of an individual, it would limit the freedom for the realization of individual properties of an organism and would inhibit evolution.

After my presentation, Lyubishchev said, "If that's what you call Darwinism, then I'm a Darwinian." It was the moral aspect of the question that was most important as far as he was concerned.

Long before Darwin's *Origin of the Species* was published, Goethe guessed through poetic intuition the secret of Nature's game of chance. Nature stands at the playing table, and exclaiming "*Au double*," doubles the stakes. She gambles boldly, happily, and passionately. She stakes everything on prize—animals, plants. . . . "And isn't man himself just a stake on a higher goal?" asked the poet. The question was an illumination. It is an idea born of nobility itself.

Still and all, at the highest level of calculation, selection and the survival of the fittest, and the struggle for existence are Nature's Darwinian pillories of shame. This was expressed best of all by Osip Emiliyevich Mandelshtam in his poem "Lamarck": "Who is the dueler for Nature's honor?/Undoubtedly, the fiery Lamarck." In rejecting selection as a moving force of evolution, the adherents of Lamarckism interceded on behalf of Nature's honor. They refused to believe in the triumph of adaptation, in the rights of the strong, in the beneficial results of struggle. Lyubishchev, Berg, Svetlov, Kuzin. Let us mention their names with reverence. I have only told of two of them. But each of them, besides fighting for Nature's honor, defended human dignity by not joining Lysenko and his distorted Lamarckism.

# *Vanished Victors*

The rulers' attitudes toward genetics fluctuated. At the end of his reign, Stalin started his attack against Arakcheyev's regime in linguistics. There were inconstant signs that the next victim of his struggle against false authorities would be Lysenko. Khrushchev's criticism of Lysenko's agricultural projects turned at the end of his reign into complete subordination to Lysenko. Dismissal of Khrushchev heralded the restoration of genetics.

During all these ups and downs Lysenkoism was far from extermination. Its ideology; its demagogic type of manipulating of the local party and Soviet ruling centers, universities, academies, and the press; its readiness to submit to every desire of the highest echelon of power provided Lysenko and his gang with a shelter, paved them the way to the highest ranks in the scientific hierarchy. From time to time it was less easy to persecute genetic principles, but no change occurred in oppressing the adherents of those principles. During the struggle against the Arakcheyev regime, it was clear that support of Lysenko's quackery was not spreading to that regime, while defending genetics was. Efroimson, Malinovsky, Bakhteyev, Kirpichnikov, Z. A. Medvedev, V. V. Sakharov, Timofeev-Resovsky, Astaurov, and myself were permanent victims. How that came to pass is the tale of this chapter.

On March 19, 1957, a debate took place at the Department of Genetics at Leningrad State University. Two events had occurred not long before that. V. P. Efroimson was released from Dzhezkazkan, and N. V. Turbin published a collection of works by the Department of Genetics at Leningrad State University covering the nine years that he had "ruled with glory." That collection, entitled *Questions of the Biology of Fertilization* (Leningrad, 1954), attempted to return science to its prehistoric roots. Its central idea was that fertilization is the mutual assimilation of male and female sex cells. To use the words of Ayzenshtat, one of the authors: "The degree of transmittal of traits of the parents is largely determined by the power of assimilation of each of the cells. The offspring will develop in the direction of the parent whose sexual cell assimilates the sexual cell of the other to the greater degree. We link the individual strength of the sexual cells in the period of the act of fertilization, or mutual assimilation, with their age." An old egg cell "in a state of fading life activity, possesses fewer opportunities in the process of mutual assimilation" and is unable to transmit to the offspring the traits of the maternal organism, including sex. Weaken the egg cell and males will predominate among the offspring. Weaken the spermatozoa and the offspring will consist primarily of females. The shattering of heredity via suffering and the subsequent remaking of nature by means of the forced assimilation of external conditions are at the center of the attention of N. V. Turbin's colleagues.

With the help of V. I. Tsalkin, a true friend of genetics and the deputy editor of the *Bulletin of the Moscow Society of Naturalists*, Efroimson published in that *Bulletin* an article that criticized the collection. In regard to the change in the ratio of males and females in the offspring of parents subjected to one influence or another, Efroimson calmly pointed out that the subject under discussion in Turbin's book was deviation from numerical equality of sexes in one experiment or another, but no one had even tried to explain the fact of equality itself. And it is precisely that fact that is explained by the chromosome theory of heredity.

Alexander Danilovich Alexandrov, the chancellor of Leningrad State University, suggested that the department react to the article. Turbin was no longer the head of the department. In recognition of his great services in the field of Michurinist biology, each "service" more shameful than the one before, he had been elected a full member of the Byelorussian Academy of Sciences and had moved to Minsk to savor the joys of a privileged and thoroughly bourgeois existence. M. Y. Lobashev, the new department head who had been driven out of that same position by Turbin in 1948, organized the debate. He invited the editor of Turbin's collection and asked Efroimson to be the opponent. Turbin didn't show. He left his leaderless followers to fend for themselves. The debate took place. It was a marvelous play. The geneticists adduced scientific arguments and the Lysenkoists babbled some sort of pathetic nonsense, some of them swearing loyalty to the Lysenkoist credo. Georgy Alexandrovich Novikov—George—the faculty dean, threatened the geneticists with the liquidation of that pathetic freedom which the post-Stalin period had granted them. The emaciated Efroimson shone.

M. Y. Lobashev said, "If the geneticists had put bars of gold on the table in 1948, even that would have been in vain." In vain! What an understatement!

On August 7, 1948, Yuri Zhdanov, A. A. Zhdanov's son and the head of the Central Committee's Science Section, wrote a repentant letter addressed to Stalin. It was printed in *Pravda*. On the closing day of the August Session of the LAAAS, Yuri Zhdanov repented of the support that he had given geneticists. That hereditary member of the intelligentsia found a marvelous formulation to describe the achievements of genetics: "These are Greeks bearing gifts, a Trojan horse."

One of the gifts hidden inside the Trojan horse, according to *Pravda*'s publications, was a highly productive variety of buckwheat produced by my friend Vladimir Vladimirovich Sakharov. Russia's food, buckwheat porridge, no longer existed, but that wasn't what alarmed the opponents of genetics—it was the ideological damage that might result from harvests of buckwheat created by a geneticist through purely genetic methods. All the great achievements of world and Soviet agronomical science were under a prohibition: the method of double interstrain hybridization of corn; high mountain seed production and hybridization of sugar beets; no polyploid varieties of potato, herbs, and radishes; the method for evaluating of sires by their offspring and many, many other things. Everything that might feed the country and prevent famine, avert the shameful necessity of importing wheat, and deliver the nation from general poverty, was rejected. All of this contradicted the principles of dialectical materialism, and the hair from a tail of a dwarf Pomeranian, a fairy's magic wand that created economically valuable types of plants, continued to be the irreplaceable implement for discovering the laws of nature.* Sakharov's buckwheat didn't suit the tastes of Lysenkoists.

* I have to remind you of my meeting with Vasily Sergeyevich Fyodorov, a lecturer at the same department where the great controversy between Efroimson and Turbin's personnel took place. Fyodorov told me secretly and after being sure no one was eavesdropping that suffering produced at the Department of Genetics by artificial fertilization of tomato flowers with an insignificant amount of pollen "shatters heredity" and results in variations beneficial to agriculture. Tomato flower pollen for artificial fertilization was transferred by means of the tiny hair from the tail of a dwarf Pomeranian dog.

V. V. Sakharov began his scientific career under the guidance of N. K. Koltsov. His first works were devoted to the genogeography of humans. He studied the goiter in mountaineers. The disease seemed to be transmitted from generation to generation. Whole villages populated by blood relatives suffered from it to the last man. V. V. Sakharov showed that this was a mistaken notion. The reason for the disease was external—the absence of iodine in the drinking water.

When medical genetics was prohibited, in spite of the practical significance of its results, Sakharov worked in the field of chemical mutagenesis. The very first discoveries in that field in the whole world were his. His subject was the fruit fly. Even while Koltsov was still alive, Sakharov decided to work with buckwheat. He wanted to be useful to his country immediately. A new type of buckwheat was created by doubling the number of chromosomes and with the help of selection.

Sakharov stayed in Moscow during the war and continued perfecting this variety of buckwheat under very difficult conditions. In 1944, his head of hair that had turned white early could be seen outlined against the background of a flowering field of buckwheat. I'll never forget it. He sowed it on the banks of the Oka at his institute's biostation in Kropotovo. In 1948, he was expelled from everywhere. He lived with his mother and his unmarried sister. The sister became the family's sole breadwinner. She died of a heart attack on the stairs. He was without work for several years. He lived in a communal apartment where the neighbors were abhorrent and one could only fear them.

Apparently he grew very thin at that time. He told me that when the times had changed, he was again working at the academy and was not exactly heavy, but full. He spoke with great love of life and humor and preferred a full figure to a thin one. He loved to eat. I would bring him little smoked fish from Leningrad. You couldn't get them in Moscow. I brought very tiny ones once. "You were probably chatting right and left in line and not watching what the salesman put on the scales," he said, and he was right. I like to talk to people in lines. "Liking lines is a Russian trait," I used to tell people who were standing with me. "Catholics sit down in church, as do Lutherans, as do Jews in the synagogues and Mohammedans in their mosques, but the Orthodox stand."

He didn't like Roquefort, the blue cheese. He would describe in a very funny way what Roquefort was, and how his sister had served Roquefort, while their cat—by the name Kassirsha—Cashier Woman—stood behind her and sniffed disapprovingly. And when a piece of the silver wrapping fell—not even a piece of the cheese itself—the cat walked up to it and tried to bury it.

His Russian was magnificent and he knew the history of Russia in detail. A Muscovite, he would walk around Leningrad and tell the history behind each palace, every building that was even the least bit famous. He was a bachelor. In the small room where he received guests at a large dining table, there was a photograph of a young woman on the wall. She possessed that sort of outstanding charm and refined beauty which exert their influence even from the beyond. His solitude and that photograph seemed to me to be links in a single chain of events that was obviously tragic. I didn't ask.

Everyone loved him. If you asked Mr. A whether Sakharov had even a single fault, Mr. A would reply unhesitatingly, "Of course, he receives Mr. B in his home," and Mr. B would say exactly the same thing vis-à-vis Mr. A. V. V. Sakharov really did receive D. F. Petrov, one of those vile people who deserted from Lysenkoism to genetics any number of times, always trimming his sails to the wind. But Vladimir Vladimirovich wasn't picky about his attachments. People showed their best side to him and he believed.

Vladimir Vladimirovich pursued neither glory nor rank. The degree of Doctor of Biological Sciences was awarded to him without his defending a dissertation. Sixteen outstanding geneticists were awarded that honor. That happened in 1965, after Khrushchev had been retired and Lysenko lowered in rank and removed from the position of director of the Academy of Sciences Institute of Genetics. Just as he had been a full member of three academies, so Lysenko remained one. According to the constitution of the academy, a director of an academy institute may be removed from his post but cannot be stripped of membership in the academy. Vladimir Vladimirovich joked about Lysenko's being lowered in position and said that Lysenko had been demoted to academician. Congratulating Vladimir Vladimirovich, on his degree, I wrote: "It sometimes happens that the awarding of a degree does honor to the degree itself, raising its worth. This is just such a case."

I was with him and Vladimir Pavlovich Efroimson at the French Exhibit in Moscow in 1961. In the Science Hall a young Frenchman who spoke Russian superbly was overseeing a display of photographs—at that time the talk of the whole world—of chromosome patterns of people suffering from so-called chromosomal diseases. Each violation of the number of chromosomes corresponds to a specific set of pathological symptoms, including mental retardation. Photographs of the patients were on display as well. I drew the young man's attention to the fact that the names of the diseases were mixed up and that the labels needed to be changed.

"Yes, I noticed that," said the young man, "but it's difficult to remove them without doing damage to the display. You're the first person who pointed this out. You deserve an 'A.' "

"Thank you," I said, "but life itself has already given me an 'A.' "

"What do you mean?"

"I mean that I've been teaching a course on evolutionary genetics at Leningrad State University for seven years now and I use Lejeune's work. And I compare man with the fruit fly, talking about the evolution of sex chromosomes."

"But I was told that it's forbidden in your country to talk about chromosomes," he said.

"Not completely," I said, and extracted from my briefcase an issue of a journal, a chemistry journal published by the Mendeleyev Chemistry Society. I showed the young man Efroimson's article about chromosome diseases in man with the same photographs from Lejeune's work that were on display at this booth.

Just then a group of young people spoke up. They wanted to attend my lectures. "I'm a Leningrader," I told them, "I'm only passing through Moscow. But this is the author of the article, the geneticist Efroimson and his friend, the geneticist Sakharov. Both of them are Muscovites."

"I would like to speak to you alone," the young French fellow said. "Let's go outside and sit on the bench in front of the pavilion."

We sat down and he asked what I thought about Lysenko and whether he had really made any contribution to the theory and practice of agriculture. I didn't even have a chance to open my mouth when Sakharov and Efroimson came running up.

"Excuse us please, but Madame is late for a train, we have to see her off to the station." They thought that they were saving me from mortal danger.

On my next visit to Moscow, Vladimir Vladimrovich, laughing his marvelous laugh and smiling his marvelous smile, told me that he'd visited the exhibit again. At the chromosome stand he'd heard the following conversation: "Why are you displaying this nonsense? Academician Lysenko has shown that there are no such things as chromosomes." "Well, I happen to know that in the Soviet Union courses on chromosomes are taught and articles about them are published." "And do you suppose that it's only in your country that idiots like that are born? What, you think we have a shortage of idiots?"

During Stalin's time, another martyr, A. R. Zhebrak, helped Sakharov to become an assistant professor at the Department of Medicinal Plants at the Pharmaceutical Institute. He lectured on botany for the pharmacists. Then a new social category of people appeared—Sakharov's birdlets. Vladimir Vladimirovich taught the pharmacists genetics at home. Many of today's young geneticists came from that nest, from that underground university whose only professor was a lecturer at the Department of Medicinal Plants at the Pharmaceutical Institute.

In 1944 or 1945, when Sakharov was still in the Academy of Sciences, working at the institute created on the ruins of Koltsov's institute, I went to see L. A. Orbeli, the vice-president of the USSR Academy of Sciences, and academician G. M. Krzhizhanovsky. I took with me a plant of buckwheat that was much taller than my tall stature and three or four times taller than the average height of a buckwheat plant, and a box of seeds from the buckwheat. I asked for their support. Orbeli said sadly, "If it had been created by Michurinist methods your Sakharov would be made a comrade of the president of the Republic, but as it is, there's no hope."

Gleb Maximilyanovich Krzhizhanovsky said, "Fedula's great, but a fool," referring to the Russian proverb "Velika Fedula, da dura," meaning that being big does not mean that one is smart. He meant that he was powerless to do anything. He got around the buckwheat with allegories, saying that the impossibility of introducing a new strain of it was nothing compared to what he had witnessed and against which he could only exert passive resistance.

In 1943, Vavilov was already dead and lying in a common grave in Saratov. He died January 26, 1943. But we thought that he might still be alive, and I asked Gleb Maximilyanovich to talk to Stalin about it. Gleb Maximilyanovich said straight out that he was powerless. On that occasion, because he understood the importance of my appeal, he told me how it happened that he had lost any possibility of intervening in Stalin's policy.

In 1930, Krzhizhanovsky had been assigned to evaluate a technical project by Ramzin. The project had been declared the work of a saboteur and legal proceedings instigated against Ramzin. He was threatened with the death sentence. Krzhizhanovsky gave the project a positive review. At the meeting, Stalin only said three words, "Look it over," and smirked sardonically, according to Gleb Maximilyanovich. After that Krzhizhanovsky was expelled from the government and Ramzin sentenced to be shot. There had been no sabotage at all—it had been invented by Stalin as a convenient way of making short work of the intelligentsia, using them as a scapegoat for all the innumerable disasters and the poverty of those glorious years. Ramzin and his alleged co-conspirators were condemned to death both by a court and at the demand of millions of workers who voted for the death sentence at meetings.

Ramzin wasn't shot, however. The sentence had an educational significance for the millions of workers. Ramzin was later given a Stalin Prize. That was announced in the newspaper without any commentary. Krzhizhanovsky was not returned to government spheres.

He was clowning about the buckwheat. There is no doubt that he would have helped had there been anything that he could have done. But Swift's characters were still coming to powers. And it was still two or three years away from 1948, when research in genetics and its application in agriculture and medicine were proclaimed criminal.

During the five-year period that fate measured off to V. V. Sakharov after genetics had its rights restored, he tried to introduce his variety of buckwheat into production. Barrier after barrier arose. The collective farms around the Kropotovo Biostation had been sowing Sakharov's buckwheat since the war. Testing of new varieties was entirely in the hand of the Lysenkoists. It was absolutely no problem for them to blackball a strain if they wanted to. As far as I know, Sakharov's buckwheat never received sanction.

When he died, I experienced a feeling that I had never experienced with regard to any other death. How could he have done that, I thought. He was only thinking about himself; he didn't care about us. Two white wings sprouted over his shoulders and carried him away, and he is happy, and we are left behind.

Vladimir Vladimirovich died January 9, 1969.

Could it really be that great Russian science, the science of Vavilov, Zavarzin, Vernadsky, Dogel, Koltsov, Serebrovsky, three generations of Severtsovs, D. P. Filatov, Astaurov, Schmalhausen, Sukachev, Zenkevich, and Beklemishev didn't create anything great during that sick, gloomy period, nothing that would have illuminated the same questions that Lysenkoism was trading on? No, even under the most difficult conditions science continued its brilliant development. There are fewer achievements than there might have been, but they were significant. You can't put a prohibition on science.

N. K. Koltsov's ideas found their embodiment in I. A. Rapoport's work on chemical mutagenesis and B. L. Astaurov's studies, about which there will be more later. Vavilov's law of homologous series was developed in B. M. Zhukovsky's work with plants and studies of the fruit fly by a number of geneticists. Vavilov's and Serebrovsky's genogeography was not muffled. The role of nonhereditary variation in evolution was especially lucky in that regard.

Can it really be that the individual, carrier of individuality, endowed with a huge number of means of reacting to changes in the environment, the most complex and differentiated of all living systems, is but a nameless digit in the statistical mass that comprises a population, a pawn on the chessboard of the game of evolution? And if not, then what can one say about the role in evolution of those changes that an organism undergoes in its individual development? This was the question that A. N. Severtsov pondered. Now Kirpichnikov, Lukin, Gause, Krushinsky, and, finally, Schmalhausen had provided an answer to that question without resorting to the aid of the theory of the inheritance of characters acquired in individual development. Schmalhausen's mind was working at the most advanced edge of contemporary science. The question of the integrity of an organism despite the autonomy of its different parts during its growth and development was at the center of his attention.

Under unthinkable conditions Orbeli, Mazing, Promptov, and Krushinsky worked out the genetic basis of animal behavior. They all had to camouflage themselves. Krushinsky, for instance, called thought "reflex," so that it would appear that he was developing Pavlov's "teaching." Pavlov, just like Darwin, had been assigned a small icon on the altar of official materialism.

The magnificent school of Russian forest specialists and the great authorities on the life of plant communites found their followers in the person of V. N. Sukachev and his colleagues.

The nearly total liquidation of malaria in the Soviet Union was achieved by V. N. Beklemishev, a zoologist, embryologist, and ecologist, on the basis of a profound knowledge of the life of plant and animal communities. The question of the levels of integration of living systems was worked out by him and V. N. Sukachev. Sukachev studied systems that included the organism as a subordinate part; the forest was his province. Beklemishev illuminated the organizing principles of living systems at all levels, beginning with the cell and finishing with the multitude of living organisms that inhabit the planet with what Vernadsky called the living component of the biosphere.

L. V. Krushinsky, D. P. Filatov, T. A. Detlaf, M. S. Gilyarov, A. A. Malinovsky, and V. P. Efroimson inscribed brilliant pages in the development of Soviet science, and one could write a book about each of them.

The name of V. I. Vernadsky, a persecuted and half-proscribed titan, will eventually be placed in a single row with the names of Galileo, Newton, Lamarck, and Einstein. Vernadsky created a new branch of science and coined its name, biogeochemistry. Biology, geology, and chemistry were mobilized to study distribution on a global scale of chemical elements, their accumulation and dispersion, and the role of living beings in this process. Life depicted by this new science emerged as a mighty global geological force. Its evolution was considered a most essential component of the evolution of the universe.

Vernadsky found a brilliant follower in the person of N. V. Timofeev-Resovsky, who switched over from genetics to biogeochemistry under the most tragic circumstances of his life. The shift was caused by arrest and imprisonment in a *sharashka,* a punitory establishment, where scientists were gathered to raise the military potential of the socialist state. There was no such power in the universe to force Timofeev-Resovsky to work for war. He had chosen the most peaceful area of research to find out the role of aquatic plants and animals in absorbing radioactive elements, thus preventing the pollution of reservoirs.

As a geneticist, Timofeev-Resovsky played a major role in establishing the contact between genetics and physics. It was he who attracted the attention of the most prominent physicists of his time—Schrödinger, Bohr, Delbrück—to the molecules representing the code of hereditary information, to the giant molecules transmitted in the process of reproduction from parents to offspring. Erwin Schrödinger's book *What Is Life*, first published in 1944 and reprinted since then again and again, is inspired by Timofeev-Resovsky. Many pages of this epoch-making book are devoted to Timofeev-Resovsky's research and theorizing.*

I had imagined Nikolay Vladimirovich Timofeev-Resovsky to be a quiet Russian intellectual. He lived in Germany, I had never seen him, and I had only read his works, written in German. We met in Moscow

* The book *What Is Life: The Physical Aspect of the Living Cell* was translated by A. A. Malinovsky and was published in Moscow in 1947 by the State Publishing House of Foreign Literature. At the August Session of the LAAAS in 1948, Malinovsky along with Timofeev-Resovsky and Schrödinger were the main targets of condemnation. Malinovsky paid for his interest in world science with years of unemployment and starvation.

in 1956. Thanks to Khrushchev's de-Stalinization policy Timofeev-Resovsky had just been released from a Siberian camp. He was half blind. Nevertheless, it turned out that he had a mighty Russian nature, full of wild energy, a powerful voice; extraordinary liveliness, and an exceptional sense of artistry. He immediately became the center of attention and devotion in any company. And he wasn't against a drink or three either. I asked him why he had seemed to me to be such a quiet Russian intellectual when I read his works. He said, "Because using the German language, you can't be anything except a quiet Russian intellectual."

He said that there was no such thing as class antagonism: The aristocrat and the proletarian were at one pole and the petite bourgeoisie at the other. He combined the proletarian and the aristocrat in himself. Fear, an attribute of the petite bourgeoisie, was absolutely alien to him. He had a brilliant mind, phenomenal strength and discipline, incredible patience, and a first-class memory.

When released from incarceration, he first lived in Sverdlovsk, then in Obninsk, not far from Moscow, and occasionally he came to Leningrad. He didn't use public transportation. He either walked or took taxis. He didn't walk the way everyone else did. Elena Alexandrovna, his wife, and I would walk along normally, and he would run on ahead, stop, turn around, and run back to us, then run on ahead again, and so on. In a taxi he would say in his unimaginably colorful, springy voice: "Lelka, let me have a twenty-fiver." Twenty-five rubles when the meter showed seven. That was before the money reform, when rubles were exchanged, ten for one. And Lelka would give it to him, and not because they were wealthy, because they certainly weren't. He received pathetically little money. He had no Soviet degrees, and Soviet bookkeepers didn't take into consideration his membership in two European academies—the German and Italian—when they calculated his salary. But it was just his nature to give things away right and left. Leningrad taxi drivers (a clan of working aristocracy) adored him; the older ones, of course, remembered him from the very first ride, and some of them would even recognize me as his companion and ask about him. He was a "certain professor" for them. "You and a certain professor rode with me once. How is he doing?"

At the age of twenty-six he won a Rockefeller stipend and left for Berlin's Institute for Study of the Brain. He became the director of the institute and worldwide fame came to him, though he did not chase after it. He remained in Germany when Hitler came to power. His elder son was apprehended for participating in resistance to Nazism. The parents could only guess about his death. They didn't know anything, but hoped and waited. They just couldn't leave. Nikolay Vladimirovich himself wasn't touched.

The well-known Swedish geneticist Arne Müntzing, who visited Leningrad as part of some sort of delegation, told me that he was in Germany at a conference in 1936 or 1937. The meeting was interrupted. A speech by Hitler was being broadcast. Everyone was supposed to stand and listen in silence. Everyone rose, and amid the universal silence Nikolay Vladimirovich's voice thundered out, "Wann wird denn dieser Wahnsinn endlich aufhören?" (When will this madness finally cease?) People like that aren't imprisoned. You'd only have trouble with them.

When Berlin fell in 1945, Nikolay Vladimirovich's institute turned out to be in the Soviet sector. He could have fled to the West, but he was blameless before Russia, had never lost his Russian citizenship. His son's fate was still unknown, and he decided to wait for his own people. Besides which he had never fled anything—he wasn't of that breed. He wasn't touched for a year. In 1946, N. I. Nuzhdin presented himself to Nikolay Vladimirovich. Nuzhdin, if you recall, was one of the vilest figures in the panopticon of Lysenkoism. A co-worker of Vavilov's in the past, he was, in 1946, one of Lysenko's colleagues. He took Nikolay Vladimirovich's strains of fruit flies, asked that the boxes full of vials be wrapped in wrapping paper, not newspaper, and left. After that, Nikolay Vladimirovich was arrested. He spent some time in prison and then was sent to a labor camp. Solzhenitsyn was his wardmate in prison. In his memoirs, *Gulag Archipelago*, several pages are devoted to their common stay. Solzhenitsyn speaks about his cellmate with great reverence. Nikolay Vladimirovich came close to dying of pellagra in the camp and had nearly gone blind. In my presence Nikolay Vladimirovich told V. P. Efroimson what pellagra was like: "If a teaspoon with three tea leaves is poured into you *per os*, it will immediately come out of you *per rectum*, together with the three leaves," he said with extraordinary vivacity. I don't know how it happened, but after two years in camp he was sought out at the order of influential prison officials and sent to a *sharashka* where he could do scientific work. It was noninfluential officials who had brought him half dead from a Siberian camp to Moscow, gave him horrible medical treatment, so that it seems as though his blindness was not so much consequence of the illness as of the treatment. His wife and their younger son came from Berlin to where he was in Siberia. His wife had worked with Nachtsheim at the University of

Berlin. They spent eight years in incarceration. Elena Alexandrovna brought stocks of fruit flies to Siberia. But they were destroyed in 1948 upon a strict order from above.

Nikolay Vladimirovich was "rehabilitated" in 1956. I attended his lecture in Moscow. He spoke about the results of the research that he had done in the *sharashka*. He had studied the radiation stimulation of plants. At freedom, in the "larger prison zone," his scientific activity achieved a grand scale. He based himself first in Sverdlovsk. His biostation, "Miassovo," in the southern Urals in a fabulously beautiful area, was an oasis in a desert of frightened people. The oasis was populated almost entirely by former camp prisoners. The biostation became a place of pilgrimage for scientists of all specializations. Work went on there from morning until late at night. A scientific seminar met every evening. Later N. V. Timofeev-Resovsky moved to Obnisk in the Kaluga region and organized a laboratory of radiation genetics at the Institute of Medical Radiology. His laboratory, which functioned marvelously, was extraordinarily differentiated. It was the best institute of genetics I have ever seen.

One of the most reassuring signs of Khrushchev's reforms was the posthumous rehabilitation of Vavilov. A book of collected articles in genetics and selection was planned as a memorial to him by the Academy of Sciences Publishing House.

The first article Timofeev-Resovsky wrote after his return from the "small zone" to the "big zone" was written for this issue devoted to his friend. I was his co-author. The editorial board rejected it. The name of Timofeev-Resovsky was too offensive, too jeopardizing.*

M. Y. Lobashev, who had come to head the Department of Genetics at Leningrad University, organized an All-Union Conference On Experimental Genetics in 1958. The conference was under the protection of Stoletov himself, a Lysenkoist and great diplomat who was already then beginning to reorient himself. As the deputy minister of Higher Education, he ran things at universities and headed the Department of Genetics at Moscow University. With his permission, Lobashev invited everyone. Lysenkoists were represented in abundance. Only that which was comically shameful was rejected; what was only simply shameful was accepted. But the geneticists too, including the fighters, were rep-

* Our article didn't perish, however. It is published in the fifth volume of a non-periodical journal founded by A. A. Lyapunov: *Problems of Cybernetics*. Its title is "On the Trends of Evolution of the Genotype." My part, "Evolution of Mutability," is a small one. Timofeev-Resovsky put his name after mine, however.

resented for all the world to gape at—Rapoport, Kershner, Efroimson, and Timofeev-Resovsky. Abstracts of the reports were printed. On the very eve of the conference, it was learned that the conference had been forbidden by a decision of the Central Committee. The participants who had gathered from all over meekly went home.

Nikolay Vladimirovich's friends persuaded him to defend his doctoral dissertation. The USSR Academy of Sciences Institute of Botany accepted it for defense, and the defense took place. That was in 1960, I think, when he was still living in Sverdlovsk. The dissertation was on the topic of "Vernadskyology," as he called it. But the Supreme Attestation Commission didn't plan to award him the degree. Khrushchev was in power, Lysenkoism was flexing its muscles again, and he would never have seen the degree of Doctor had it not been for the Little October Revolution in 1964, as we called Khrushchev's ouster. Now, just as in 1953 when Stalin died and a sharp turn of history was expected, the Lysenkoists started to tremble. Genetics was on the rise. Article after article in its defense appeared in the newspapers. The turn was much, much more abrupt than it had been ten years earlier. The years of Khrushchev's reign, a time when the reins of power were loosened, gave scientists in other fields outside of genetics—physicists, chemists, mathematicians, and cyberneticists—the chance to speak out in defense of genetics. And now, finally, the order to resurrect genetics was dropped on everyone from above.

But the fright in the Lysenko camp was less than what it had been back then. Ten years before they had expected a bloodletting. Now mores had softened, and besides, they knew how to act in order to hang on to power. They needed to change course and give immediate and solid evidence of their loyalty. The Supreme Attestation Commission, permeated through and through with the metastasis of Lysenkoism, awarded the degree of Doctor of Biological Sciences to Nikolay Vladimirovich Timofeev-Resovsky, a full member of two European academies.

In 1966 I was working at the USSR Academy of Sciences Institute of Cytology and Genetics in Novosibirsk and was a member of its Science Council. At one of the council's meetings we proposed candidates for the elections to the academy. When the genetics candidates were named, I said that we ought to nominate Timofeev-Resovsky. Kerkis agreed that he should be nominated, but for biophysics, not genetics. And so the matter reached the biophysicists. Lovely Ninel

Borisovna Khristolyubova* said that she would be the one to nominate Timofeev-Resovsky. But then Galina Andreyevna Stakan spoke up and said that she thought he had committed unpatriotic acts. And Olga Ivanovna Moystrenko said that he had lived in Nazi Germany. And Yuly Oskarovich Raushenbakh stood up and said that he refused to nominate such a person to the academy, and Deputy Director Privalov shared his view.

Then Zoya Sofronyevna Nikoro started up. She said that yes, Nikolay Vladimirovich had lived in Germany and worked in a Rockefeller Fund Institute and had not returned when his fellowship expired. "But I'd like to see whether any of the people who have just spoken would have returned either. If he had, we wouldn't have a thing to talk about right now. It is one hundred percent certain that he would have been destroyed. It was Bukharin who could make up his mind to go to Paris, walk around on free soil, breathe the free air, and return, knowing that he would die as an enemy of the Fatherland, instead of dying *for* it. Nikolay Vladimirovich didn't return, but he didn't betray the Fatherland. He was a luminary of its science, and he served it. The people who betrayed the Fatherland were those who betrayed its science, who ruined genetics in the name of Michurinist biology, who wrote denunciations of its best people, who signed false statements about sabotage. Timofeev-Resovsky's services to science make him worthy of being an academician, and I support his nomination."

I have reproduced Zoya Sovronyevna's speech word for word, with one exception. She did not name Bukharin's last name. She called him Nikolay Ivanovich, but people understood perfectly well whom she meant. Speaking about the betrayal of the Fatherland by those who took part in the destruction of intelligentsia in Stalin's time, Zoya Sofronyevna did not state Raushenbakh's name, either, but he was exactly the person she had in mind in mentioning people who had given false testimony, calling innocent people wreckers. At the end of the 1930s, when a group of veterinarians was tried for poisoning horses with the goal of undermining the country's military potential, the science expert who signed the findings of sabotage was Raushenbakh.

Reaction to Zoya Sofronyevna's words on the part of the party or-

* The first name of Ninel Khristolyubova is Lenin, read backwards. This is in contrast to her last name, which means "Christlover." The odious combination points to a father, a scion of a priestly family, eager to demonstrate its loyalty to the Soviet regime.

ganization was immediate. "It's very bad that speeches such as that should be heard at our Science Council," Shkvarnikov, a member of the party, said quietly and threateningly.

Raushenbakh took the floor and said with sincere bewilderment: "How could it be that adherents of Michurinist teaching betrayed the Fatherland, when it was approved by the Central Committee of the party? But Timofeev-Resovsky betrayed it, that's as clear as anything—he lived in Nazi Germany."

Then another deputy director, R. I. Salganik, a very educated and talented person, spoke up passionately and persuasively. He had received Nikolay Vladimirovich at his home and was no fool. But he was a Jew, and a Jew's sacrifice to the altar of true loyalty has to be not only rich, but something very close to his own heart. Salganik said: "Timofeev-Resovsky could and should have returned long before 1937, before there was any danger. But he went to pro-Nazi Germany and stayed there. Fascism apparently was more to his liking than the land of burgeoning socialism."

Ninel Borisovna cut him off. She said that she was retracting her nomination, since she didn't want to subject Nikolay Vladimirovich to attacks of that sort.

It was at that point that I entered the fray. I said that when Nikolay Vladimirovich had left in 1926, he was not on his way to pro-Nazi Germany. He was going to a country that was on the eve of a Communist revolution, a country whose Communist party was the strongest Communist party in Europe, and that on a worldly scale it had ranked second only to China's. The Soviet Union was not entirely blameless in regard to the triumph of Nazism in Germany. The fatal role was played by the fact that the Social Democrats were declared the chief enemies of communism, not the Nazis. In preventing the formation of a united front against Nazism, Stalin had facilitated Hitler's triumph. It was absurd to raise the question of Nikolay Vladimirovich's patriotism. He lost a son in Nazi Germany, a son who was a member of the resistance. And if Ninel Borisovna retracts her nomination, then I would nominate him.

D. K. Belayev, the institute director, had been silent up until that time. He always played the role of offstage director of a puppet theater. His abilities at diplomacy were developed beyond belief. Now he said that in view of the difference of opinion that had arisen, he thought that we ought not to nominate anyone for the Biophysics Division. That motion was passed by the majority. And Zoya Sofronyevna was "worked over" at a special meeting. They didn't touch me, but I don't know why.

Here, I must abandon for a short time my narration about Timofeev-Resovsky and devote a page or two to Zoya Sofronyevna, the remarkable woman who was so bald as to defend not only the rehabilitated Timofeev-Resovsky but also Bukharin, who was shot in 1938, a most innocent victim of Stalin's terror. His rehabilitation was absolutely unthinkable at the time when the meeting at the Institute of Cytology and Genetics took place. I do not know how it happened that Zoya Sofronyevna's parents, Russian through and through, lived in Rumania and why it was in that country that Zoya was born. Because of her Communist activity she had opportunity to be acquainted with Rumania's torture chambers. After her release, she got over the border to the Soviet Union, crossing in sight of Rumanian borderguards and escaping by chance before the bullets hit their mark. At the time of her Scientific Council escapade she headed the Laboratory of Cattle Selection. Her speech at the council, after which she was worked over, was not the first one she'd made there.

In 1963, before Khrushchev's ouster, when there was a motion on the floor to create Communist labor brigades, Zoya Sofronyevna had told us we were unworthy of that standing, that there was hunger in the land, not enough bread, and the reason for it was ignorant planning for agriculture, while we—specialists in agriculture—were keeping silent. She said we knew that obvious mistakes were being made and we weren't even attempting to correct them.

She spoke, and not all of the audience, but perhaps three-quarters of them, applauded. Then D. K. Belayev leapt to the platform and said everything that it would have been suggested he say if he had had time to consult with the Regional Party Committee. The words "demagogic speech with the goal of winning applause" figured in his remarks. He was very handsome, lean, and his neck was slender. It jerked in a nervous tic from the corner of his mouth to his collarbone. After, Zoya Sofronyevna was supposed to be "worked over." But Yu. P. Miryuta, a staff member at the institute and a party member, who had known Zoya Sofronyevna at Gorky University where they had worked together and from where she had been dismissed in 1948, said that he had always considered her a wrecker and an immoral person. "Do the people present know that she used to play the piano in a restaurant orchestra at night?" And she really had played there; that had been her only source of income after she was driven out of Gorky University. She had three children of her own and several other foster children.

At that point, those present forgot about their party duty to concern themselves with Zoya Sofronyevna's re-education, pounced on Miryuta, and he had a bad time of it. And with that, the shameful business came to an end.

The scheduling and canceling of official anniversaries and birthday celebrations is overseen by the institute's party organization along with the Trade Union Committee and administration. They subsequently decided to cancel Zoya Sofronyevna's birthday jubilee. In the service of justice in the Soviet Union one finds confusion and lack of agreement among the various parts of the unbelievably huge bureaucratic machine. That was what I took advantage of. I announced Zoya Sofronyevna's sixtieth birthday to the party organizations to which the party organization of the Institute of Cytology and Genetics of the Siberian Division of the Academy of Sciences is subordinate. I signed my own name. Just let them try to prove that I was not the chair of a committee for conducting the celebration. We, her colleagues and students, arranged a celebration such as the officials had never even dreamed of. Congratulations came from the very highest party offices. Zoya Sofronyevna did not know the source of information to her powerful well-wishers and was at a loss as to who it could be. The administration and party organization of the institute kept quiet. Confusion is a great thing! You just need to know how to take advantage of it.

In 1970, Nikolay Vladimirovich turned seventy, and the Moscow Society of Naturalists celebrated the occasion. The Academy of Sciences announced a celebration, too. Invitations were sent out, the speakers named, and the titles of their talks printed. But then his jubilee was canceled. I can imagine what the denunciations contained that served as the reason for the cancellation.

In 1971, Nikolai Vladimirovich became jobless. The wonderful genetics lab at the Institute of Medical Radiology was broken up at the direct instruction of the KGB. No seemly pretext was even needed. But this time world public opinion was not silent. Delbrück, a Nobel Prize winner—Nikolay Vladimirovich had worked with him in Germany—came to the Soviet Union from the USA to talk about Timofeev-Resovsky with M. V. Keldysh, the president of the USSR Academy of Sciences. Nikolay Vladimirovich was enlisted as a consultant for the USSR Academy of Medical Sciences Institute of Medico-Biological Problems. His laboratory was not reestablished.

The last time that I saw Nikolay Vladimirovich was at Astaurov's funeral. Boris Lvovich Astaurov was a classic figure in the natural sciences. He worked with the fruit fly, too, and did first-class research.

But his main subject was the silkworm. His practical work was organically linked with theoretical biology. He accomplished great things in the area of theory, too. One of the most elegant proofs of the chromosome theory of heredity belongs to him. Within a single egg cell he was able to combine the plasma of one species with the nucleus of another. Larvae were hatched from eggs deposited by a female silkworm *Bombix mori*. They ate, shed the way they were supposed to, and formed cocoons. Out of the cocoons came butterflies of another species of silkworm, *Bombix mandarina*. Before being crossed with male *Bombix mandarina*, the females *Bombix mori* had been subjected to X-rays so as to kill the nuclei of their egg cells. Then they were crossed with males *Bombix mandarina*. During fertilization, several spermatozoa wind up in the egg cell—two are sufficient for producing a normal nucleus of the paternal species. The fertilized eggs were activated by high temperature. These peculiar hybrid larvae lacked sufficient strength to hatch out of the egg of the maternal species. They were helped with a needle. Neither the larvae, nor the cocoons, nor the butterflies produced in this way looked the least bit like the maternal species from the eggs of which they were hatched and in the plasma of which their genes had carried out their formative action. They belonged to the paternal species. And they were all males. They were males because two spermatozoa of the parental species had merged, introduced two identical sex chromosomes, and created the chromosomal situation that determines the male sex in butterflies. And if there is a gene with the sex chromosome of the parental species that causes an unusual coloration of the wings or antennae, it will definitely show up in these bastard sons. The role of chromosomes in the transmittal of traits from generation to generation is shown by this example with extreme persuasiveness. And, if you please, a model is provided for the creation of one species within the bosom of another, since butterflies, which manifest the traits of their species with exact precision, are gotten from eggs deposited by a female of another species.

I have already had the opportunity to describe this most shameful fabrication of the official Soviet science, a foundation stone of the evolutionary theory created by Lysenko together with another quack, Lepeshinskaya. The origin of a species, this theory says, roots in an individual exposed to unusual external conditions. There are particular grains, *krupinki*, in the living matter of the individual. In an unaccustomed environment these grains change to become grains of another species. The unusual environment is not properly digested and assimilated. A new species is nature's belching. The tree of one species starts to produce besides its own old branches, leaves, flowers, fruits, and seeds, all belonging to another species. Hornbeam grows branches of hazelnut. Out of a spruce tree grows a pine tree. A cuckoo youngling hatches out of the egg deposited by a chiffchaff. Astaurov's model ironically seemed to demonstrate something of that kind. However, Lepeshinskaya's grains and Lysenko's belching have absolutely nothing to do with it, just as the methods of the alchemists have nothing to do with splitting the uranium atom.

Boris Lvovich didn't lose his job in 1948, but he was forbidden to study the silkworm. Fish were to become his subject. "*You* could probably study the silkworm now," he wrote to a specialist in the genetics of fish.

His fate was surprising. In a world of moral values perverted in accordance with the postulates of the materialist dialectic, he was able to combine service to truth with movement upward through the ranks, with what one bright woman, my school friend, called "elite success."

As a rule, geneticists did not participate in the democratic movement. Their signatures were not to be found under petitions to the party and the government for less harsh rule, less severe punishments. That was if genetics wasn't involved. When it was, they wrote. There were 300 signatures attached to the 1955 petition to the government asking that genetics be reinstated.

Being a geneticist—a teacher, researcher, selectioner, doctor—*was* their democratic movement. By teaching Mendel's laws or implementing genetic methods in selection, they were on guard for democracy. Reason and good sense told them to keep away from politics, lest they bring down upon themselves the wrath of those whose good fortune included the right to wrath, lest they lose the chance to teach Mendel's laws and implement genetic methods in selection.

Boris Lvovich was an exception.

In 1958, he was invited to the Tenth International Genetics Congress in Montreal. Many people had invitations, but Boris Lvovich was the only geneticist to receive permission to go, not including about eighteen Lysenkoists. He wrote the Central Committee that he couldn't go to the congress as part of a delegation all of whose members were adherents of an anti-scientific doctrine. His reputation as a scientist would be stained. "Not wishing to have recourse to strong expressions in this document," he wrote, "I cannot describe their views in any other way than as approaching the absurd." In order to avert the fatal conflict that might arise if he were to receive an order to go, he adduced a second reason for his refusal—his father was very ill at the time. Anyone else would have paid dearly for such boldness.

The "delegation" abandoned by Astaurov was just the "delegation" I had become a member of thanks to the appeal on my behalf of A. D. Alexandrov, the chancellor of Leningrad State University, to the Central Committee of the Communist party. The delegates to the congress were a bloody gang including Nuzhdin, one of six people who attacked my father and Koltsov in a lampoon: "No Room in the Academy for Pseudo-scientists." Its publication paved the way for Lysenko to become a full member of the Academy of Sciences and led to Koltsov's dismissal from the post of director of the Institute of Experimental Biology, which he had founded and headed for twenty-three years. Other members of the delegation were of the same type. Astaurov did not want to be a sole scientist of that delegation. I did not want to either. As soon as I learned that Boris Lvovich had refused to attend the congress, I canceled my application. Boris Lvovich was allowed to go abroad even after that, and he won love and respect outside of the Fatherland's borders.

In 1963, Boris Lvovich supported my hopeless efforts to dispute the accusation of a political prisoner. He opened the door of his apartment to me and to a person who, according to our estimation, could help—Dudintsev. A mad boldness was an indispensable condition of such a deed. No! You cannot believe me. Comparison is a tool of cognition. My best friends were afraid to type my manuscripts; the used typewriter ribbon could betray them to the KGB. Some were afraid to speak in my apartment, being sure that listening devices had been secretly installed.

It was during Khrushchev's reign, when the poet Brodsky, then just a young boy, was convicted of parasitism and sentenced to five years of exile in the Archangelsk region and forced labor at timber cutting, I appealed to Dudintsev for help. He was the author of the much-talked-of book *Not By Bread Alone*, which had lifted the curtain a little on the dominance of evil and the impassable swamps of bureaucracy in the USSR. Although the book lifted the curtain only the slightest bit, it was a sensation, a life-giving spring of truth in a waterless desert. This freedom-loving author was a public prosecutor, and I sought him out because it seemed to me that Brodsky's sentence could be contested, that one could prove the flimsiness of the accusations from a juridical point of view and show that there were procedural errors. Perhaps, perhaps, I thought, "they," the all-powerful "they," had realized that they had gone too far, that an international outcry was possible—Brodsky had translated English, Spanish, and Polish poets—and would agree to dump the errors on the judge and pardon the innocent.

Both Astaurov and Dudintsev were Muscovites. I came with my troubles from Novosibirsk to Moscow.

The conversation transpired in Boris Lvovich's luxurious apartment. Astaurov moved into that apartment at no. 3 Dmitry Ulyanov Street not long before his death. For many years he had lived in desperate circumstances in a communal apartment.

"Why should I believe you?" the prosecutor asked quite logically.

I placed before him the shorthand record of Brodsky's trial. The writer Vigdorova had made it.

"Why should I believe this report?" the prosecutor asked just as logically.

I placed before him my own verbatim record, written out in hand, that I had kept during Brodsky's trial. "Compare them," I said. "Everything that coincides is the truth."

"But I've seen his poetry," said the prosecutor. "You're the one who showed it to me when I was at your place in Leningrad. He wrote against the people, against the party."

I placed before him the very same verses that I had shown him in Leningrad. It was a marvelous poem, and there were no politics in it.

"That isn't the same poem," the prosecutor yelled, "why should I believe you? Great people do great things, but he's just dangling between their legs; he's grown a beard, goes on drinking sprees, and chases after women." Brodsky didn't drink and didn't even smoke, and he was suffering through his tragic love, about which he has since written in his marvelous cycle *Mary Stuart*.

Dudintsev kept on yelling and yelling, but by then I'd stopped listening to him. I turned to Boris Lvovich and said, "Freedom is measured out from on high. Everyone wants to be its sole herald, take advantage of his little piece of it, and prevent anyone else from doing so, lest he exceed the measure and he be forbidden and everyone forbidden and the last little grain taken away. That's what's going on."

Boris Lvovich said that I was quite right. His daughter, who was present for all of this, wept. She was seventeen then. Nothing came of my efforts, and I wasn't able to help Brodsky.

Boris Lvovich was elected an academician in 1966, the same year that we didn't manage even to nominate Timofeev-Resovsky. He organized a new institute at the Academy of Sciences and became its director. He was the chairman of the N. I. Vavilov All-Union Society of Geneticists and Selectionists. He refused to fire the people at his institute who participated in the democratic movement.

When Stalin died, his successors, one after the other, loosened the reigns of authority: Malenkov released criminals; Khrushchev opened the prison doors for most, if not all, of those sentenced for political crimes; the new collective rule headed by Brezhnev removed the prohibition from genetics in particular and tightened up on Lysenko a bit. After the granting of certain freedoms came a phase of tightening the reigns of power. In 1967, there came a time when they had to change the criminal code so that public demonstrations, reading *samizdat* material, and political jokes could be punished without recourse to Article 70. That article, with a sentence of up to seven years and subsequent loss of rights, was a little too ominous; besides which, Khrushchev had said that we had no political criminals, and the Criminal Code, strictly interpreted, turned the entire nation into real or potential criminals. And so some additions were made to Article 190, the result of which was to deprive citizens of their basic civil rights. It carries a punishment of up to three years in prison. Fourteen academicians sent a collective protest to the highest spheres. Among them was Boris Lvovich Astaurov.

Boris Lvovich was not completely immune to his enemies' venom, however. In 1974, one of the staff members of his institute went to a conference in Venice and didn't return. The powers that be hadn't wanted to let him go in the first place, but Boris Lvovich had gone to the Regional Party Committee and pleaded his staff member's case. They made use of that.

The fatal roles were played by Yu. Ovchinnikov, a young biophysicist and vice-president of the USSR Academy of Sciences, and academician N. P. Dubinin, a geneticist. Both were capable of committing those crimes that the government found desirable. At a regularly scheduled meeting at the academy Presidium, Boris Lvovich was given to understand that they didn't consider the flight of his staff member a chance occurrence, and the very existence of his institute was challenged. After Boris Lvovich returned home from that meeting, he suddenly died.

The subordination of the life and death of every Soviet citizen to the ruling force is nowhere seen more clearly than at burials. The open coffin of an elite person is displayed in some luxurious hall of some luxurious building—at the headquarters of the Union of Writers, if the deceased was a writer, or at the Presidium of the academy, if an academician is buried. The hall is chosen according to the number of people expected to come to pay last respects to the dead. In that hall the mourning assembly takes place. An orchestra performs, ceasing only long enough to allow the speakers to deliver their eulogies. An honor guard, four people, stands near the coffin, mourning bands round their sleeves. There are strict rules for positioning those who form the foursome, and rules on changing the guard. Several designated persons put the mourning bands on the arms of volunteers. An appointed person leads the members of the guard to their places. There are rules which arm—left or right—bears the band. The order of red and black stripes of the band is fixed. The time each foursome stands depends on the number of volunteers. The first honor guard takes position before the mourning meeting begins; it is solemnly replaced every few minutes during the speeches; and the cycle is continued even after the meeting's conclusion, during the procession of mourning past the coffin. The mourning guard is not removed until the time comes for the relatives to take their leave of the dead.

Boris Lvovich was an academician. His funeral meeting was held at the Palace of Scientists on Kropotkin Street in Moscow. I arrived as the ceremony had just begun and the first men of the honor guard had taken their places—Ovchinnikov, Turbin, Belayev, and a man whom I did not know. I looked for the line where people were waiting their turn as guards of honor and got in at the end of the line. The line went into the assembly hall, a separate room, not the hall where the open coffin was placed. The line made a loop there, and came out again. The leading end and the tail end met at the widely opened folding-door. At first, I didn't see anything and didn't notice who was standing nearby, but then I did—it was Andrey Dmitriyevich Sakharov. On that very day Moscow had seen Alexander Arkadyevich Galich, a satirist poet, off into exile—to Norway, thank God, not to Siberia, and now Andrey Dmitriyevich had come to Astaurov's funeral from the airport. Fatikh Khafisovich Bakhteyev came up to me—he had been a colleague of Vavilov's and was now his biographer. Bakhteyev was with Vavilov on that last expedition when Vavilov had been seized and taken away to prison. Then one other person came up—he was of that group who are born in order to become martyrs. I introduced Bakhteyev and the future ascetic to Andrey Dmitriyevich, and we formed a foursome for the honor guard. We got to the door leading into the assembly hall and stood face to face next to the foursome on whose arms mourning bands were being placed. Closest of all stood Iosif Abramovich Rapoport. I introduced him to

Sakharov, they shook hands, and the foursome moved on. The plump, curly-haired young man, who had led the foursome to their places, heard the names of the people whom I had introduced to each other. And suddenly the line began to melt away, there was no one left around us, and it was announced that the meeting of mourning had begun and that the honor guard had been removed. The last foursome had taken their places at the coffin—Ovchinnikov, Belayev, Turbin, and the man whom I didn't recognize—the same people who were there when I came. Sakharov at Astaurov's coffin . . . that was not at all in the plans prepared by the behind-the-scenes funeral directors.

In the eulogies that followed, without naming names, the speakers hurled thunder and lightning at N. P. Dubinin's head. An academician's absence at the funeral of a colleague in the same field is an extraordinary phenomenon. "No one and nothing has been forgotten," D. K. Belayev said with the voice of an iron tribune. That was exactly what the higher-ups wanted from him. Honor had been observed, and capital acquired. Though Dubinin's name wasn't mentioned, everyone understood that Belayev had him in mind. Not that Dubinin had cause for alarm, though. He would be awarded for his services. And the price for his services would be all the higher the more humiliating were the anonymous reproaches in his direction. But Dubinin could have attended Astaurov's funeral; he could have stood in the same honor guard with Ovchinnikov, Turbin, and Belayev.

Dubinin was not the only one absent. Boris Lvovich's wife, Natalya Sergeyevna, didn't come to the funeral at the House of Scientists. In her eyes the honors could not right the fatal harm done to her husband by those who took part in the ceremonial. Sakharov stood for a while, moved away quietly to the exit, shook Rapoport's hand, and left. Bakhteyev stood some time alone near the coffin and wept.

B. L. Astaurov is buried in the cemetery of the former Novodevichiy Monastery, near the grave of N. S. Khrushchev. This is a very prestigious graveyard, second in esteem only to the cemetery just outside the Kremlin wall near the Lenin Mausoleum. Burial at the cemetery near the Novodevichiy Monastery is allowed only by permission of the Executive Committee of the Moscow Soviet of Working People's Deputies. M. V. Keldysh, at that time the president of the Academy of Sciences, and academician V. A. Engelgart obtained this permission. Engelgart, more to be able to continue his struggle for genetics, had publicly condemned A. D. Sakharov. This had seemed to him the only way to preserve his relatively influential position. Now he used his influence, bought for this high price, to get an honorable grave for a geneticist.

The moral code of the Academy is somewhat worse than that of the nomads of Genghis-Khan. Changing banners was not considered treason by the Mongols. Personal harm done to the former leader was taken by the Mongols, but not by the Soviet powers that be, as a crime. Astaurov refused or perhaps was even not asked to sign the condemnation of Sakharov. He had to die, but a grave of high rank was permitted. The customs of Genghis-Khan included just the same observance for titled enemies: the murder without bloodshed, and honorable execution.

Eight people, all of the same height, carried the coffin from the Monastery gates first to a place where there was another mourning assembly, and then to the grave. I remember that Efroimson was among those eight—he looked very miserable.

Now at the churchyard monastery near the open grave, the rules of protocol gave way, and those in disgrace were allowed to speak. Out in the open air before a crowd of hundreds, his thunderous voice scarcely audible, Timofeev-Resovsky said, "He perished for us, he risked his own defenseless chest and his own vulnerable, tender human heart in our defense."

I think that had Boris Lvovich been asked where he would like his final resting place to be, he would have said without hesitation that he preferred to be in the common grave where Vavilov lay. A prestigious grave at the Novodevichiy Cemetery placed his spotless honor under suspicion.

The cemetery orchestra, appropriate for one of such a high rank, accompanied Boris Lvovich from the monastery gates to the grave, playing something that was martial, official, and monstrously out of tune. At least, that's the way it seemed to me.

# *The Ballet* Sleeping Beauty *Performed by Flies*

The healing nature of oblivion will now be the subject of my narration. Here are the words which Pushkin, a poet, a sufferer, spoke to ease the grief of all who suffer:

*The desert of the world, so sad, and so immense,*
*Mysteriously is broken by three streams.*
*The spring of youth, rebellious, hot and fast.*
*Its boiling jets don't stop to purl and gleam.*
*The Castal fount, the spring of inspiration, gives golden dreams*
*To the exiled, the banished whose fate is hard.*
*The cold stream of oblivion, the final stream,*
*Gives coolest calm to burning wounds of heart.**

In the drama *Boris Godunov*, Pushkin portrays the ability to forget as a way to prosper during history's tight turns. The cunning courtier Count Shuisky, a backstage aspirant to the throne, reveals his intention and his right to be crowned, and later fails to become tsar. (Let me note that he reveals this as a character in a play—if the characters were silent about their feelings and intentions there would be no play. Silence and forgetfulness are twins, both shields and arms in the struggle for existence.) "The time to remember is gone. Now and then it's useful to forget. Take my advice," Shuisky tells his former chargé d'affaires.

The episode that I want to relate as an illustration of the healing

* Translated by R. L. Berg.

power of oblivion begins in early 1946 and concludes in early 1976.

*Act I.* I was working at Moscow University in the Department of Darwinism, which was headed by I. I. Schmalhausen. It was located in the old university quarters by Manege Square across from the Kremlin. Just across the stairwell was M. M. Zavadovsky's Department of the Dynamics of Development, and next to us, A. S. Serebrovsky's Department of Genetics. We were separated from the Department of Genetics by a door, and the door was covered by a bookcase. Repairs were being made. The bookcase was moved aside, the door opened, and I was walking around both laboratories. Some department members called me over and one woman asked, "Do you want to have some fun?"

"Sure. Why not?" I answered.

"Here's a jar full of flies for you and an etherizer. Put these flies to sleep." These were house flies—big ones, not fruit flies. I etherized the giants. The flies were lying on a porcelain plate.

"You see," they said, "some of them are lying quietly, but others are jerking their legs. Sort them out."

No sooner said than done.

"Count," they said.

I counted them: 3:1! Three-quarters of the flies were sleeping peacefully, and one-quarter were nervous.

"There you have it," they said. "Mendelism. Did you enjoy it? Put the flies back in the jar."

The women in the Department of Genetics knew who I was. They had no intention of teaching me genetics. All that they wanted to do was to demonstrate to me the applicability of Mendel's laws to the pathological behavior of a sleeping fly. At the basis of the laws formulated by Mendel lies that very same segregation with a ratio of 3:1. The grandmother or grandfather of the flies counted by me had had a nervous disease. The hospitable women had crossed a sick fly with a healthy one. The children were healthy—they slept quietly and didn't jerk their legs. These banal, normal flies became the parents of a mixture of ill and healthy children. This mixture now slept on the porcelain plate. The hidden malady manifested itself in only one-quarter of them. Three-quarters either had not inherited the pathological gene nesting in their parents or else inherited a healthy gene in addition to the pathological one. And these three-quarters had no reason to wave their legs in their sleep. One-quarter of the children received diseased genes from their parents and became their victims, and I received pleasure on their account.

Mendel observed segregation in the ratio of 3:1 while studying the inheritance of traits in peas. With his brilliant mind he perceived that at the basis of inheritance lay the transmittal from generation to generation of some particles capable of combining with each other while remaining unaltered. Mendel did not coin the name for these particles. The name—gene—was introduced much later.

The pathological gene of the flies that slept poorly was inherited by them from healthy parents. It was not influenced by the healthy gene, and the health of its carriers was not reflected in it. When two pathological genes met in that first cell out of which the sick fly developed, there was nothing else that fly could do but jerk its legs in its sleep. The remarkable stability of the gene, its capacity to remain unchanged irrespective of whether it has fulfilled its function or was suppressed by another gene or by environment—a cure, for instance—is a scientific truth. The stability of the gene indicates its molecular nature. The idea of changeability of the hereditary substance in accordance with environment is pure alchemy.

Two years later, in 1948, Lysenko came to power, and the traditions and aspirations of the alchemists were resurrected in Soviet science under the name of Michurinist biology. Mendel became the main enemy of pseudo-science.

The year 1946 was a time of extremely sharp antagonism between science and obscurantism. On the side of science was the forced alliance with the Western democracies which had just assured the victory over Hitler's Germany, and on the side of obscurantism was the betrayal of that alliance by Stalin's continually increasing isolationism and forging of the Iron Curtain.

With their diversity, the flies sleeping on the porcelain plate strengthened science's position against obscurantism. In a single word—Mendelism—the women expressed their understanding of this antagonism and of my position. In offering this refined show, they knew whose side I was on.

*Act II.* Ten years later—1956. I was working at Leningrad State University's Department of Darwinism under Zavadsky. "There's an article here on genetics for the *Great Encyclopedia* that they sent to me for an evaluation," Zavadsky told me. "Won't you write the review? It will bear your signature."

A reviewer receives money from a publisher. I didn't receive any money. Zavadsky had appropriated it. That was a very significant fact, one that was politically important and, moreover, positive. It indicated that Zavadsky had grown so bold that he considered it possible to place his signature under my devastating criticism of Lysenkoism. The article was Lysenkoist, and in the battle with Lysenkoism a review of an article in the *Encyclopedia* was a very important weapon. Neither money nor authorship had the slightest meaning for me.

I wrote the evaluation. The "thaw" had begun. Stalin was dead. Khrushchev had not yet fallen completely under Lysenko's influence. *Pravda* wrote that Khrushchev had condemned the Lysenkoist innovations and said publicly to Lysenko, "You and your experiments can go to the moon."

The article to be reviewed did at least bear the clear stamp of progress. It did not contain Lysenko's theory of the origins of the species according to which the representative of a new species arose in the body of a representative of the old species, involving grains of the living matter called *krupinki*.

It did not seem to matter to the Lysenkoists that this theory contradicted the entire elegant edifice of so-called Michurnist biology, i.e., Lysenkoism. The graft and the stock are supposed to become similar to each other and remake each other, since they have common nourishment. But according to the theory of the origin of species, all of a sudden, without rhyme or reason, within the bosom of one species, the grains of another species are formed, and out of a well-behaved spruce tree a pine tree grows up that is in no way different from a pine tree that makes do with its own roots. Oats gave rise to wild oats, wheat to rye, and in some mystical way even peas were derived from wheat. These grains, *krupinki,* responsible for the sudden emergence of a new type of metabolism, clearly fell out of the context and ruined the dynamic picture of Michurnist biology, a picture that vibrated in response to the slightest change in external conditions. And the article, the one sent to Zavadsky for evaluation, didn't contain these grains, a self-contradictory element of Lysenkoism. The contradictory part of the "teaching" was cut off.

I wrote that there were two conceptions and laid out in detail the postulates of each and the methods by which results are attained. I pointed out the lack of correspondence of Lysenko's entire theory with reality. The internal consistency of absurdities does not move scientific constructs toward the truth; rather, it distances them from it. I gave one copy of the review to Zavadsky and sent the other, along with a letter, to academician Vinogradov, the director of the editorial board for the

*Great Encyclopedia*. I wrote him that the time had come to return science, which had been driven off the pages of scientific literature, to its proper place.

Instead of throwing out the Lysenkoist nonsense and ordering an article from a geneticist, the *Encyclopedia* made a decision that was typical for the times. Two articles were printed, one written by a geneticist and the other by an adherent of Michurnist biology. The article issuing from the pen of an obscurantist looked very much like that part of my evaluation in which I described how clearly all the parts of that pseudo-teaching agree with each other.

*Act III*. Erevan, Armenia, 1961. I was still working at Leningrad State University in the department headed by Zavadsky. I had gone to Erevan on expedition, where I was studying population genetics. I was sitting at the university there, counting flies—this time my own fruit flies—and suddenly I was invited to a scientific serviculture station near Erevan so that I could learn about its work. They were sending a car for me. I went, wondering what it was all about. It turned out that Martiros Khristoforovich Sarkisyan, the head of the station, wanted to acquaint me with his graduate student so that I could give my opinion about his dissertation for the degree of Candidate of Sciences. The student's last name was Sarkisyan, too. They would show me not only the finished dissertation, but the animal subjects and those remarkable forms that had made the conclusions possible. And the conclusions were of extraordinary importance—they overthrew the chromosomal theory of heredity.

The head of the station, an Armenian who spoke Russian fluently, but not faultlessly, hurried to fill me in on the situation. It sounded like this: "Here, look—you see this female silkworm? She has a trait that she could only have if she were male. Well, what you say now about genetic bases of sex determination—about the citadel of chromosome theory of heredity? Not at all agree with what Boris Lvovich Astaurov he write. And whole world know his work. Look, you look these caterpillars and butterfly and write evaluation that Sarkisyan he right, Astaurov make mistake, chromosome theory false theory."

And along with all that, an excellent breakfast was served. Their calculations were obvious: I was a drosophilist and didn't understand anything about silkworms. Besides which, I was a woman and was probably—no, not probably, certainly—stupid, and would take the bait.

I really don't understand anything about silkworms but I had a friend,

V. P. Efroimson, who understood a great deal about them. He had told me that one can often observe mosaicism in the silkworm, and mosaicism had led astray even such an outstanding geneticist as Goldschmidt when he formulated his theory of sex determination in the silkworm.

Now, let me explain what mosaicism is. It is a mixture of parts differing by their genetic makeup within one and the same organism. These differences in the hereditary equipment of different cells of the organism show up during its development. They are caused by mutations arising in some cells of the growing organism. One very frequent type of mutation is the loss of a whole chromosome. A male silkworm has in his genotype two sex chromosomes, while in the female genotype there is only one representative of that pair. A male having a hidden gene—say, for abnormal coloration—in one of his sex chromosomes, looks normal. Now, suppose that in a cell that gives rise during development of such a male to a wing, the chromosome having the gene providing normal coloration is lost. All the cells derived from the mutant one now have a mutant deficient genotype. A mosaic creature results. One of the wings of this creature is normal and manifests male characteristics, small dimensions among them. The other wing has a mutant female appearance. It is big and abnormally colored.

So, with Efroimson in mind, I said: "This female—here—with the traits of a male is not a female, but a male, a male-mosaic. His female parts are the result of loss of one of the chromosomes. Where a chromosome is lost, one group of traits appears, and where it hasn't been lost, a different group turns up. And everything is just as it should be. Complete correspondence with the chromosome theory of heredity and even more than that—an elegant proof of it. Let's look through the tables of crossbreeding and find out where and which chromosome was lost. That will be a very good project. There's lots of material, and the young man has worked so hard."

But the Sarkisyans weren't about to show me the tables of crossbreeding. We were already having breakfast and a polemic had started up. The young people were silent, while the director of the station defended the principles of Michurinist biology: the inheritance of traits acquired in individual development. "Imagine to yourself a gene. We'll indicate it with the letter *N*." He took a piece of paper and wrote out an elegant Latin capital *N*. "And next to it," he went on, "there occurs a change in metabolism. What will be the result?"

"The result," I said, "will have absolutely no relation to the inheritance of acquired traits. You're like alchemists who stand beside a cyclotron and exclaim: 'Our side won! The transformation of elements has been achieved.' Yes, it's been achieved, but alchemy has nothing to do with it."

The young people reacted demonstratively, approving my polemical lunges. The elder Sarkisyan shifted the front: "You're saying all of this because you don't know all the facts on which Michurinist genetics are based."

"No," I replied, "I know them and I'll prove that to you right now." And I told him the previous episode, including the remarkable similarity between my review and the Lysenkoist article.

He broke into a smile. No one has the command of the art of smiling that the Armenians do. Radiant, he exclaimed, "I'm the one who wrote the article about Michurinist genetics."

"Well, that means that you know that I'm familiar with the fundamentals of your teaching. The universal character of Mendel's laws," I said, "is shown in two ways. They are applicable to all organisms that have achieved a certain stage of organization of their hereditary apparatus in the course of their evolutionary development, that is, where there is an alteration of the haplophases and diplophases in the chromosome cycle. That's the first way. They manifest themselves in regard to all traits: structural and physiological, biochemical and behavioral. That's the second way. And the theory of the physical basis of heredity is true for all traits of all organisms, without exception. Mendelism is its subordinate part, just as Newton's mechanics are part of the theory of relativity and just as Euclidean geometry is part of Lobachevsky's geometry. And to prove to you that the inheritance of all traits occurs according to Mendel's laws, I'll tell you a story." And I told him the story of Act I.

He broke into a smile and said, "That was my graduate work. I did it under the guidance of Alexander Sergeyevich Serebrovsky. I produced mutations with the help of X-rays. What they showed you was just a tiny part of what I did. I had strains of flies that slept peacefully if it [the room] was quiet, but that jerked their little paws if it was noisy. Certain strains only reacted to a sound of a particular pitch, and for different strains the pitch was different. I used to lay out flies of different genotypes in pretty rows and play the xylophone. First one row and then another would come to life. It was a *corps de ballet* of sleeping flies. Different alleles of the same gene would show different symptoms of that nervous disease. I studied the structure of that gene and completely disclosed the genetic basis of the illness."

Then why are you playing the fool, with your elegant Latin capital *N*'s, I wanted to ask him.

*Act IV*. 1965. Khrushchev had fallen. Genetics has sprung back to life. But the Armenians hadn't given up. All the key positions at the university, academy, and Ministry of Agriculture were in the hands of Lysenkoists, and they continued to defend their citadel of ignorance. The year 1965 saw the height of rear-guard battles. Lysenkoist bureaucrats firmly counted on retaining the commanding heights—and they were successful.

Martiros Khristoforovich Sarkisyan presented a report in the assembly hall of the Presidium of the Armenian SSR Academy of Sciences. "The time to remember" had passed for him. He was living in the phase of history when it was "time to forget." He laid out his graduate student's work, spoke of mosaicism, and selected types of it that had nothing to do with the matter at hand. He debunked the chromosome theory of heredity. I was sitting in the audience and preparing a classification of types of mosaicism so that I could point out to him precisely which types were responsible for the phenomenon that he had observed.

When he had finished, I went up with my scrap of paper and explained all of this. When the meeting was breaking up, a woman came up to me. Her face, hair, and clothing were the color of dust off the road and she yelled at me irately and loudly, "Never in my life have I been a witness to anything more outrageous or brazen!"

"I think you've seen much worse things, but your evaluation of them was different," I said amicably.

How bad things would have been for Sarkisyan if the cold spring of oblivion had not burst forth onto the dry steppe of the universe!

*Act V*. March 6, 1976. The thirtieth anniversary of the fly *corps de ballet*. It's a tropical spring in Arizona. A plenary meeting of the Conference of American Drosophilists is underway in a marvelous building at the University of Arizona, in Tempe.

In the USA, interest in the study of the Drosophila shows no signs of waning. The place chosen for the conference was as pleasant as one could imagine. Palms, cactuses, and roses were in bloom. Small evergreens with numerous little round white buds and a few flowers that have come out here and there stood by the university. A beautiful strong aroma, like that of the aroma of osmanthus at the Nikita Botanical Gardens, carried for yards. The birds were in a frenzy. Here their calls are more melodic, on the average, than in Siberia, and just as everywhere, they were strikingly insistent and urgent. It was very warm. The dandelions on the lawns had faded. They were already losing sections of their seeds. Sparrows were hopping around in the grass, looking for meat for the baby birds.

Inside, before a full audience of about 400, William Kaplan was giving a presentation. His topic was the neurogenetics of sleep. He spoke about the hereditary diseases of sleeping fruit flies. When put to sleep with ether, the flies move their legs. Their healthy neighbors sleep calmly.

In Drosophila, in contrast to the silkworm butterflies, it is the female who has two sex chromosomes—they are called X-chromosomes—while the male has only one. A female who has the leg-jerking gene in only one of her sex chromosomes sleeps calmly. The disease-inducing gene is suppressed. The loss of one of the two sex chromosomes can be induced in the laboratory during the fly's embryonic stage of development. Then mosaics turn up. Here, for instance, was a female with traits that occur exclusively in males. It was exactly like the silkworms that the two Sarkisyans, teacher and pupil, had shown me. One could select marking traits in such a way that the parts that remained female would be gray, and the male parts, produced as a result of the loss of one of the chromosomes in the cells of the embryo, would be yellow. Some of the legs on such a fly are gray. They are motionless. Only the yellow legs are waved about. Only the male part of the fly jerks its legs. Using mosaicism, Kaplan established exactly which of the nerve ganglia exhibit this comical malady. The mosaics put to sleep by ether jerked only those legs that they should have according to the chromosome theory of heredity. Each leg has its control panel. Certain of the mosaics waved around only one of their six legs in their sleep. Mosaicism allowed one to predict with perfect accuracy precisely which of the legs would come into motion when the fly was asleep. No one was interested in trying to prove or disprove the chromosome theory of heredity. No new proofs were needed.

No other report had such a success as Kaplan's. I asked him whether the sleeping flies reacted to sound impulses. He said they didn't. I told them about Sarkisyan's work conducted under the guidance of Alexander Sergeyevich Serebrovsky and based on his idea. Serebrovsky was well-known. Alas, no one had ever heard of Sarkisyan. Serebrovsky had acquired worldwide renown thanks to his other great discoveries, not to those pioneering investigations into the neurogenetics of sleep.

The reflection of Sarkisyan's worldwide fame, which hadn't come to pass, fell on me.

# *The Truth Is a Lie, and a Lie the Truth*

The *corps de ballet* of sleeping flies has distracted me from the narrative about the events that unfolded at the Department of Darwinism at Leningrad State University. As you may recall, after the chancellor of the university assigned money for the organization of my laboratory, the money was diverted into the channel of Lysenkoism and came under the control of the most pathetic member of the department, the poorly educated Gorobets.

While Gorobets, using money granted by the university chairman to organize the Laboratory of Population Genetics, was making a desperate effort to shatter the heredity of tomatoes through suffering and thus to shatter the very foundations of Mendelism, I was immersing myself in the study of the mutability of flies brought in 1958 from Uman. I attempted to return to the flies the high mutability they had lost. I measured mutability in mutant flies, in flies coming from inbred and crossbred strains. I expected that selection of mutants, inbreeding and crossbreeding, would restore the highly mutable genotype. It was in vain. The loss of high mutability proved irreversible.

A series of new expeditions eventually followed the Uman expeditions. The university chairman never learned that funds granted to me were flowing in another channel. Not knowing that the money for my expeditions was allocated, I was compelled to appeal to the thieves for financing. Out of fear of discovery, and yielding to my urgent pleas, the faculty granted money to organize these expeditions.

There was no limit to the mockery and no end of persecutors. Zavadsky, George, Pinevich, and Chesnokov. Chesnokov resembled a bedbug that had extended its proboscis and was sniffing at its victim. Have you ever seen a bedbug sniffing at its victim? You haven't? I've seen more than enough of them doing that. Chesnokov's sharp nose became even sharper when he spoke. "There's no reason to sniff the tails of bourgeois science."

"Excuse me," I insisted, "but population genetics was born in the Soviet Union. Soviet scientists lead the world. We were the first who—"

"*Were,*" he interrupted me, "and that's all water under the bridge."

"You seem inclined to lay the blame for the destruction of genetics on the geneticists themselves," I said.

"We have nothing to talk about. You won't get any money."

I applied again and again and finally money was given. Results of Uman population studies were summarized, an article was written and submitted to the scientific journal issued by the Biology and Soil Department of the university. It is hard to believe that the criticism it elicited came from a scientist. The edge of the mockery was aimed at the most elementary conditions for conducting any experiment or any observation. Over the course of twenty years, the critic said, I had caught flies in the same places, with the same flycatching apparatuses, had raised their progeny on the same feed, crossed them with flies from the same laboratory strains, and had never changed the magnifying power on the microscope. What was that, if not stagnation?

Trying to fire me, the dean of the biological faculty entered into a skirmish with the chancellor's office. When there are no official reasons, people begin to play on the vanity, dignity, and honor of the person being fired. Direct insults start up. The recommendation that Zavadsky, Agayev, and Khakhina gave me in connection with my groundless desire to receive the degree of Doctor of Sciences, thus leap-frogging over Zavadsky, a Candidate of Sciences and head of the department, was only one of their bullets. They all missed me. Their bullets were like water off a duck's back to me. I had undergone tempering through disgrace in the Stalinist times. The armor forged at that time was impenetrable. Not only did I not think of leaving, but I wasn't even upset. The "recommendation" was a tolerable irritant. Nonetheless, you would not call such a life sweet.

My torments finally ended in 1963.

It was under the most unfavorable circumstances that I received an invitation to organize and head a laboratory of population genetics at the USSR Academy of Sciences (Siberian Branch) Institute of Cytology and Genetics. Yuly Yakovlevich Kerkis, an old, old friend of mine, a co-worker of Vavilov's in the past, and a scientist with a world reputation, passed on to me the invitation from the institute's director, Dmitry Konstantinovich Belayev.

I think that the initiative for inviting me came from Kerkis. He had known me since the time when Muller became the head of the laboratory where the enchanting young Kerkis worked and where I received a workplace in the same room where he worked. He knew and valued my publications.

The conversation took place at my apartment.

We weren't alone. Ervin Zinner, a friend from my school days, was living at my place then. He had only spent one year—the last one—at our school, but everyone came to like him. He was a German, and had become a philologist. He was arrested in 1941. The notation in his passport—"German"—was the sole reason for his arrest. He was sentenced to spend a whole lifetime in a camp. Thanks to Khrushchev's de-Stalinization he only spent fifteen years there.

In the camp he had a few privileges at first, since he was a party member. He was led to party meetings under guard, but later they ceased to be. He was rehabilitated in 1956, worked as an assistant professor at the Pedagogical Institute in Irkutsk, wrote a doctoral dissertation on the influence of Tolstoy and Dostoevsky on German writers, and had come to Leningrad to try to defend the dissertation at the state university. Difficulties arose, as is their wont. He spent a month at a hotel, but the time hadn't yet come when he could leave town. The matter of his dissertation was at a standstill. An administrative law forbids one from living in a hotel longer than a month. And it is also forbidden to move to another hotel after a month has passed. A hotel stamps a passport "lived there from such-and-such a date to such-and-such a date." It was only at that point that Zinner remembered me and I took him in.

Liza, Masha, and I occupied the back room, and Zinner moved into the passageway that served us as a front hall, dining room, living room, and study. The stone women, my neighbors in the communal apartment, had long considered my home a den of iniquity and all of us prostitutes, and they could safely be ignored. The stone women couldn't imagine any relations between a man and a woman other than those requiring a bed. But I was thankful enough that they didn't denounce me to the police for the appearance of a new resident in our apartment—one staying there without a residence permit. Otherwise a fine would have been unavoidable.

We got along marvelously with Zinner. Not even the shade of disagreement in the evaluation of the reality surrounding us arose between us. He had many more reasons for hating the Soviet regime than I did. And he related such horrors about the camps that one's hair stood on end. I cite one of his stories in the preface. The prisoners dug graves for themselves in the summer so that it would be easier to bury them in the winter.

"I'm inviting you, of course," said Kerkis, "but I'd like to know whether you'll be a good comrade."

"I will," I said, taking his words as a pleasant joke, one quite appropriate in view of our very friendly relations, which didn't exclude mutual teasing. "And if you need a live witness, ask Ervin Petrovich. We studied in school together."

"Yes, we went to school together," said Ervin Petrovich, "but I can't act as a witness for the defense in the given instance, because I haven't been witness to anything good." The conversation dried up.

After Kerkis had left, I just looked at Zinner.

"I'm very glad that I played a dirty trick on you," he said. "It's my revenge and you thoroughly deserved it. You knew that I was a party member. How did you treat my convictions? Did you respect my principles? You behaved all the time in a very uncomradely manner, taking advantage of my dependence on you."

"Why didn't you indicate any of this earlier?" I asked. He continued living at my place as though nothing had happened.

I didn't rack my brains over the reason for the completely unexpected attack on Ervin Petrovich's part. Fifteen years in a corrections institution behind barbed wire could turn anyone into a beast. It seemed to me that the reason for the backheel was revenge for the suffering he had endured, the desire to humiliate everyone, to do everyone ill indiscriminately, as soon as an opportunity presented itself. I accidentally turned up close at hand.

Twenty-one years later, in 1984, here in the United States of America, in St. Louis, Missouri, I received written evidence that Ervin Petrovich Zinner really was taking revenge. He punished not humanity for his crippled life, but, rather, me for the fact that thirty-four years prior to the events being described, the year when we were finishing school, I had rejected his love.

The written evidence consisted of a scrap of paper on which was penciled in German, but not in Gothic script, the German phrase: "Die Ewigkeit kann nicht Vergangenheit werden." (Eternity cannot become the past.) I found the narrow little piece of paper in a copy of Hamsun's *Victoria*, which Ervin gave me in 1963. On the title page of Hamsun's book he had quoted Goethe in red ink, and perhaps blood, using Gothic script: "Amor ist immer noch ein Kind." (Love is still a child.) I hadn't even opened the book for twenty-one years. But in 1984, I was invited to West Germany, to the city of Mainz, to Johannes Gutenberg University, to give a series of lectures. In order to brush up my more than fifty-year-old German, I opened the volume of Hamsun.

I was inexact in writing that in 1929 I had rejected Zinner's love. I just hadn't attached any significance to it. And just as thirty-four years earlier I had not understood the shy expressions of his feelings, so in 1963 I understood the bloody inscription only as a reference to the hero and heroine of Hamsun's book. I had not noticed the slightest evidence of any feelings other than those of friendship on Zinner's part.

The inclination of many men to behave abominably toward women who have rejected their feelings is a well-known phenomenon. It does not call forth any admiration in me. As a matter of fact, I have had occasion to encounter quite the opposite emotion.

At any rate, in 1963, first Kerkis, and then Belayev repeated the invitation to head a laboratory. In accordance with the right granted to me, I reserved my Leningrad living quarters, and in September of that year, Liza, Masha, and I left for Novosibirsk's Akademgorodok.

When I told A. D. Alexandrov, the chancellor of the university, my defender and a great friend of genetics, about my leaving, he was filled with indignation. "Go. *Adieu*. See you in Novosibirsk."

When he said that, he hadn't the slightest suspicion that history's crazy wheel of fortune would toss him to Novosibirsk and he would continue to play a role in my changeable fate. We met a year later. His transfer to Novosibirsk ought to have become the start of a giddy new career. Lavrentyev, then president of the Academy of Sciences Siberian Branch, was one of Khrushchev's favorites, and Khrushchev wanted to bring him to Moscow to head up something there. Alexandrov was invited to take Lavrentyev's place. Alexandrov made the move, but he hadn't yet succeeded to the throne when Khrushchev was ousted. Lavrentyev remained in his prior post. Alexandrov instead became the head of one of the departments of the Mathematics Institute of the Academy of

Sciences in Novosibirsk, a very modest position as compared with the post of a chancellor of one of the most prestigious universities of the USSR.

The 1963 expedition to the Nikita Gardens and Uman left Novosibirsk at the beginning of October. And I was continuing to study correlation pleiades as well. I had six copies of my doctoral dissertation with me. The Institute of Cytology and Genetics, my new haven, was most willing to recommend it to the Academy of Sciences Institute of Botany. I would, however, never have seen the degree of Doctor of Sciences had it not been for major political events that had nothing to do with the independence of the sizes of some parts of a plant from the sizes of other parts.

Things on the genetics front were growing ever worse. The very existence of the Institute of Cytology and Genetics was threatened. The institute lacked its own building. The building originally intended for it was transferred to another institute before the roof had even been tiled. The construction of a second building was begun—and the same fate overtook it. When I arrived, the foundation for a third building was being laid. The hope of getting it was minuscule.

There wasn't a single doctor of biological sciences at the institute. Geneticists hadn't been awarded degrees in fifteen years, and some people had even had them taken away. One of these ill-fated scientists was Efroimson, one of the most important geneticists in the Soviet Union. Efroimson had been stripped of the degree of Doctor of Sciences when he was sent to prison in Dzhezkazgan in 1949, soon after genetics as a field of study was rooted out. In order to condemn him to oblivion, the persecutors of genetics looked for and found an excuse to accuse him of a crime. At the end of the war, in 1945, Efroimson—then a lieutenant—and his unit found themselves in Germany. He became a witness to the crimes, thefts, and violence that the victors committed upon a peaceful populace, and he rushed to help the victims. He demanded the government's action to end the lawlessness. For a while he got away with his exposés, but in 1949 Efroimson, a Jew who had stood up for the honor of German women, was accused of slandering the Red Army, stripped of the doctoral degree, and sent to Dzhezkazgan for ten years.

From that time on until the middle of the 1960s, not a single geneticist had become a Doctor of Sciences. When Khrushchev came to power and raised his hand against the Stalin cult, he released from the camps hundreds of thousands of victims of the Stalinist terror. Efroimson was among those freed. The doctoral degree and the badges of honor that he had received for wartime heroism were returned to him. The very existence of the Institute of Cytology and Genetics in the bosom of the Siberian Division of the Academy of Sciences, the miracle of its creation in 1958, were an extremely vivid indicator of the thaw.

But the thaw was replaced by a harsh refreeze resembling the climate of Stalin's times. The institute, the refuge of those people who for fifteen years had cared only about surviving, not about doctoral degrees, became the object of royal wrath. N. P. Dubinin, the institute's first director, was removed at Khrushchev's direct orders. Dmitry Konstantinovich Belayev, a Candidate of Sciences, who deserved the degree of Doctor of Sciences much less than did many other of the institute's employees, was the deputy director at that time. He took skillful advantage of the monarch's wrath, "gave the falling one a slight shove," and replaced him. I hadn't known that when I was on my way there, otherwise I wouldn't have accepted the invitation.

In 1963, when I showed up, the institute was still coming into being. It was being built not even on sand, but in a flood zone. And here I appeared and with me came my dissertation, and the book *Heredity and Hereditary Diseases in Man*.

My first skirmish with Belayev came over a seminar on Marxism-Leninism conducted by him. He insistently suggested that I sit in, so I did. A young man was reporting on whatever chapter of Engels's *Dialectics of Nature* was assigned for that session. After the report I told Belayev, in private, that not only would I not be sitting in on other seminars because of lack of time, but that I wouldn't allow any of my co-workers to waste their time in like manner. If he wanted us to study the methodology of science, then I would organize a permanent seminar on the topic "Control Mechanisms on Various Levels of the Organization of Life." Belayev would do the presentation on mechanisms controlling the life cycle of mammals. He seemed to agree to this. After assigning the roles and confirming the titles of the future presentations with my institute colleagues and the professors and lecturers at Novosibirsk University, I posted the seminars schedule in the vestibule at the Institute. But Belayev tore it down in a rage. My seminar started up anyway, and his ceased to exist.

My dissertation was sharply out of harmony with Belayev's politics. His position was an extremely difficult one. The director of an institute in the Soviet Union is the servant of two masters. The first force that rules over him is the necessity to produce science; the second is the ideological front and the political authorities. A director is raised to the post not because of any scientific or scholarly merits, but because of his ability to save the institute, since every institute needs to be saved. Belayev headed an institute that was not threatened with new personnel, radical reorganization, or a change of leadership—its very existence was threatened.

The Institute of Botany accepted the dissertation, but no date was set for the defense. And it was not only the references to the lackeys of bourgeois science, Schmalhausen chief among them, who were stigmatized by the August Session of the LAAAS, and not only the fact that I completely ignored the world-famous achievements of Michurinist biology that blocked my path and hindered the administration of the Institute of Botany from setting a date for the defense. The real problem was the Chinese emperor Siao Yan. His fate was horrible. I can't think of it without shuddering. He lived in the fifth century and wrote poetry. In China at the time I am talking about, he was anathematized as an emperor, in the Soviet Union as a Chinese. The abstract of my dissertation was adorned with an epigraph, a poem by the unfortunate Siao Yan.

They are so beautiful, these simple Chinese verses of bygone centuries, that I can't resist adorning my memoirs with their gentle sound.

Here is the poem of the emperor:

> *In a land called "South of the River"*
> *The lotus plants have unfurled,*
> *Red glitter has spread over the emerald-green water.*
> *The color of the flowers is identical, the seeds are the same,*
> *The roots are different, and the seeds have no difference.*

Independence of seed dimensions from dimensions of roots, that was exactly the essence of my discoveries.

You will ask me why, at the height of the squabble between two Communist giants, I needed an epigraph from Chinese poetry. We knew perfectly well about the squabble and took a certain malicious pleasure in reading the newspaper descriptions of the brutalities of the Cultural Revolution in China. Much of what was written about China was applicable word for word to the Soviet Union, and the noble wrath of Soviet hacks fueled our glee. Siao Yan's verses, however, were so far removed from politics and so appropriate for my topic that I did not foresee the tempest that burst out apropos of them, and had I foreseen it, I would have thrown out the epigraph immediately. The idiocy of an ideological press exceeds everything imaginable.

Daniil Vladimirovich Lebedev, the science secretary of the Institute of Botany and a great anti-Lysenkoist, was summoned to the District Party Committee. Not the Regional Committee, and not the City Committee, but the District one. The District Party Committee is the organization that manages the entire Leningrad district. And the Leningrad district has a population of many millions. The point had nothing to do with accepting a dissertation written by a Mendelist-Morganist-Weismannist. But not to reject a dissertation that was crowned with a Chinese epigraph would signify an open challenge to Soviet foreign policy, as far as the District Committee was concerned. Lebedev told me that he attempted to defend himself and me, but in vain.

As long as Nikita was on the throne, my dissertation lay motionless. A year passed, Khrushchev fell, genetics lifted its head above the water, and I was immediately called to Leningrad to defend my dissertation. At the time of my defense, I was so fed up with correlation pleiades that I spoke about the contribution of standardization and diversity to nature's beauty, and about the musical rhythms of flowering meadows. I compared standardizing selection with the transformations that Bach introduced into music. One of my three opponents, Alexander Innokentyevich Tolmachev, the one who at first expressed doubts whether a dissertation in which the author expresses gratitude to foreign scientists could be accepted, beautified my defense in an oblique way. By reproaching me for the Chinese epigraph, he gave me an excuse to read the second, the amorous variant of Siao Yan's poem.

The Chinese poet says one thing, but means something altogether different. The spoken and the implied make for two different poems. Every Chinese person familiar with this language of symbols knows what the poem is about. It is about love. A second translation of Siao Yan's poem was now heard at the Scientific Council of the Botanical Institute:

*You and I have entered*
*The time of love.*
*The time has come for the male*
*And female principle*
*To merge in love.*
*Our feelings now are the same,*
*And we have a common future,*

*No matter how our fates may differ*
*Before we met.*

I added that in order to understand the poem it doesn't hurt to know that the roots of the lotus are edible, that lotus flowers change color with time, that the red glitter spreading over the water symbolizes the male principle and the emerald-green color of the water symbolizes the female element. The next day the stenographer gave me the transcription to sign and praised my presentation highly. The main thing, she said, was that it was all understandable.

Efroimson, who had turned up in Leningrad by chance, said that he was absolutely certain that I would be failed. "That's not the way to defend a dissertation," he told me. "You need to go deeply into scientific questions, not entertain the audience with the beauties of nature." But the vote for awarding the degree was unanimous.

The Supreme Attestation Committee, hardly moved by a feeling of justice, but rather by the necessity to trim one's sails to the wind, confirmed the Science Council's decision in a very short time. Celebrations on the occasion of a defense are organized immediately after the meeting of the Science Council. We first celebrated in Leningrad, and then in Novosibirsk. I returned to Akademgorodok with the rank of Doctor of Biological Sciences in order to become the sole possessor of that degree at the institute headed by Dmitry Konstantinovich Belayev.

Breeds of cows are divided into four types: milk cows, meat cows, work cows, and all-purpose cows. For Belayev I was a cow of the all-purpose type—a precious gift to him from fate. If we were ordered to teach genetics, I could, and he couldn't. And if administrative wrath fell on his shoulders because the institute had exceeded the measure of the allowable, then he had a culprit who refused to carry out his demands, which coincided with those of the authorities. And once again I was that culprit.

Reprimands were showered on me constantly. One day Belayev used me as an example to show an administrator's highest virtue—the ability to recognize a potential enemy: vigilance. I received a reprimand for tardiness when I hadn't been tardy. I entered the vestibule of the institute at two minutes before nine. Belayev, wearing a cap, was talking to one of the institute staff members who was not wearing a cap. I noted that detail. I don't know how it is in the West, but in Russia removing one's hat or cap when entering a building or upon conversing with a person expresses one's respect for the place where one is located or for the person to whom one is speaking. Belayev, a priest's son, was perfectly well acquainted with the tradition but did not honor his subordinates with its observance. The university chancellor, Alexandrov, always took off his cap the moment that he entered the vestibule, even if it was empty. I had seen that several times, from a distance.

I was summoned to see the director that day. He asked me to stop being tardy. As I left his office, I applied for vacation time. If I were on leave, I could come to the institute at any time. In that way I assured my usual working conditions for myself. I can work fourteen hours a day without getting tired, and that's true even today, but only if I am completely free from outside compulsion. A leave—two months for full professors and laboratory heads—gave me that freedom. I had neither the opportunity nor the desire to use my leave in any other way. And when the leave ended, I thought, perhaps Belayev's attack of mania to rule would die down a bit and my fellow workers and I would be able to work without his peremptory shouting.

The banquet I had organized to celebrate my having been awarded the degree of Doctor of Sciences fell precisely at the time of my vacation. For those who are not acquainted with the customs of the scientific world in the Soviet Union, let me say that I was following an inviolable tradition and fulfilling a ritual. And not without pleasure, since I enjoy inviting guests and wining and dining them well.

At the banquet Belayev sat near me, of course. We were eating roast duck with apples. Some guests were washing it down with wine, some with mountain ash juice. "I signed the order for your vacation," Belayev said to me, "but I see you every day at the institute."

"The sound of the alarm clock lowers my capacity for work," I said. "Besides which, you organize an awful lot of discussions at the highest level. And I have an experiment under way."

"And what will you do when your vacation runs out?" he asked in a tone that, in accordance with the occasion, was quite friendly.

I, too, kept to the same tone. "Oh, Dmitry Konstantinovich, surely that wouldn't be the first time that either of us has been without work." I was dissembling by linking the two of us. I went for six years without work. But Belayev had never lost his job. He had been the zoological technician for an Institute of Animal Husbandry and remained at that post until Dubinin took him into this institute and made him his deputy.

There was a smell of catastrophe in the air, but no catastrophe resulted. I didn't find myself out of work and I no longer had to adjust my schedule to an alarm clock. Three women who were heads of laboratories, Zoya Sofronyevna Nikoro, Raissa Pavlovna Martynova, and Vera Veniaminovna Khvostova, asked Belayev to close his eyes to my schedule. And he did, making me an exception. But the intractable genius Sherudilo, who was persecuted by him, had to produce at his demand a medical statement that he suffered from a psychic ailment and needed an individualized schedule of working hours.

But it wasn't just the intercession on the part of three women who were influential to differing degrees and in different ways, but the lucky star under which Belayev was born as well that prevented my being fired. He needed me as a major public demonstration of his servility, so until the right moment came—and it did come—he would be patient. The all-purpose cow tilled the soil and she could be milked.

The Institute of Cytology and Genetics was founded to develop science on a world-class level and to instill the knowledge of genetics in all teaching establishments of the Novosibirsk district so that the expanding institute could in the nearest future be ensured a pool of literate personnel. When I turned up, the development of science was running along smoothly, but the preparation of future personnel was just getting underway. Sixteen years had passed since the time of the August Session of the LAAAS, when genetics was forbidden, and school textbooks had been reworked in accordance with the demands of Lysenko and his clique. The teachers themselves had studied from these books and were now teaching from them. Doubt had not impressed itself on their souls. The young teachers believed because they'd never known anything else. The older ones had been forced to believe. But there are different kinds of belief, of faith. One pole is ennobling faith—the feeling of attachment to those who were and those who will be. That is the faith of the fearless. The other pole is faith as a surrender to terror.

Soon after my arrival, I was standing at a podium and giving a lecture on the theory of evolution to Novosibirsk high school teachers who wished to raise their qualifications. During a break between lectures a methodologist spoke and lectured them on the need to raise one's qualifications, urged them to keep up with scientific developments, and to subscribe to popular-scientific and scientific journals and books. With the resoluteness of despair they informed him that the conditions of their lives did not allow such a possibility. The collective farms didn't provide them with food, and there was nothing to buy. There weren't even bread shops. They were allowed to keep a cow. They had to prepare the hay for winter by themselves. The collective farm gave them two cubic meters of firewood and paid for their electricity. But to survive the winter in the Novosibirsk region you needed seven cubic meters of firewood, not two, and the free electricity was for one forty-candlepower light bulb only. It was forbidden to have a second one. If a female teacher marries a collective farmer she loses even those privileges. The hours that the nursery school was open weren't adapted to schoolwork. If a teacher had children, she had no place to send them. The necessity of keeping a cow made the greatest impression on me. Besides the fact that it has to be fed, you have to milk it three times a day, without a day off. It is impossible to raise one's qualifications under such conditions.

I lectured in a strictly anti-Lysenko spirit without entering into any polemics with Michurinist biology. They listened very attentively and didn't ask any questions—if that's what they were to teach, then that's what it had been ordered to believe, and not some other thing. But toward the very end of the course, an elderly teacher asked what the position of Michurinist biology was on the questions that I had illuminated for them. I said that there was no Michurinist biology, that that was a name given to Lysenko's pseudo-teachings, which had nothing in common with science. At that point they protested. "Three hundred kinds of fruit and berry plants derived by Michurin, the most advanced theory, based on the works of Marx, Engels, and Lenin," they said. I took their faith away from them. No one had ever seen a single one of Michurin's plants, let alone 300 of them. And people who had tried the fruit of the offspring created by his followers spat it out. Women who sold apples at the Leningrad market would say, "These are real apples, not any of Michurin's graftings."

But the teachers had been told to believe, needed to believe, that they existed—these fruits of graftings and the remaking of nature. There were portraits of Michurin and Lysenko in their school textbooks, and it was their patriotic duty to follow them. That they themselves had never eaten apples and couldn't tell a pear from an apple, let alone a peach from an apricot—that was the norm. They believed that if not they, their children, then their grandchildren in any case would be eating the fruits of these 300 sorts of exciting plants. And those people, who were the best in the world, were eating them somewhere now.

People reported that when Khrushchev returned from America, he said, "They don't have Michurinist teaching in America, but they do have bread, vegetables, and fruits, while here in Russia we've ruined our Michurinist plants." He knew that Russia had ruined her fruit orchards. The people, of course, didn't know that. But I have had occasion earlier to describe how the apple orchards in Russia were ruined. Until very recently the tax on the harvest from fruit trees growing on private farm plots was so high that it was more profitable to chop a tree down than have to use the income from potatoes to help pay the fruit tax in years of poor harvests.

"But in the genetics course they told us something quite different," said the elderly teacher. That course was taught by Yuly Yakovlevich Kerkis, the same Kerkis at whose initiative I had wound up in Akademgorodok, a student of Yu. A. Filipchenko's and the talented graduate of the first Department of Genetics in Russia. He had worked at the Academy of Sciences Institute of Genetics, first in Leningrad, and then in Moscow, when Vavilov was the director of the institute and in the same division that Muller directed.

In 1937, he used scientific methods to study the vegetative hybridization of tomatoes and showed that grafting did not change the hereditary traits of either of the plants that were united. He certainly knew the uselessness of Michurin's methods for remaking nature. When Vavilov was arrested in 1940 and Lysenko became the director of the institute, Kerkis lost his job immediately. He was the director of a state sheep-breeding farm in Tadzhikistan when his persona and biography attracted the attention of the local authorities in 1948 and he suddenly found himself face to face with death. The vegetative hybridization of tomatoes—his successful disproofs of the Lysenkoist doctrine, his brilliant demonstration that it is impossible to change hereditary traits of tomatoes by grafting—saved him. No one was going to bother to dig down to find out what the results of his research were, and the title of his article didn't give him away. Death stopped for a second at the threshold of his home, looked him in the face, and passed on by. I heard him tell Timofeev-Resovsky about his tenure as director: "I threw myself on my bed and cried like a woman, not with niggardly male tears, but like an old woman." In 1958, when the Institute of Cytology and Genetics was organized in Siberia, Kerkis was invited to head the Laboratory of Radiation Genetics. Belayev hated him violently. If anyone ought to have been the director of the institute, then it was Kerkis, not Belayev.

At the courses for raising teachers' qualifications, Kerkis taught the teachers real genetics, but he referred to Michurinist biology with praise. They told him what I had said, and he grew alarmed—I had clearly exceeded the measure of what was allowable in the spring of 1964. The very existence of the Institute of Cytology and Genetics was threatened. A sword of Damocles hung over it on a thread. He informed the institute's director.

And it was precisely then that it became clear what my longtime friend Yuly Yakovlevich had had in mind when he asked me in the presence of Ervin Zinner whether I would be a good comrade. "Will you dissemble along with the rest of us when bitter necessity forces us to dissemble? Or will you oppose yourself to us, to the collective, prevent us from maneuvering out of the situation, and ruin everything with your fastidiousness?" My answer, that I had always been a good comrade and would always be one in the future, was a lie. Ervin Zinner, telling a lie, had told the truth.

Kerkis considered his complaint a moral victory, though it would have been unthinkable at my German Reformatskaya School. He had placed the interests of the collective above personal friendship.

I was administering examinations at Novosibirsk University when the director's car came to get me. I interrupted the examination and went. There was triumph at the institute. D. K. Belayev, the director, had been elected an associate member of the USSR Academy of Sciences. The election to the academy of a Candidate in Biological Sciences, and in particular the election of a nonparty person such as D. K. Belayev, was a huge victory on the genetics front, a cardinal strategic success. His office was literally swamped with flowers. I had done my bit for his election, having depicted Belayev's modest achievements in a recommendation that the Science Council sent to the Presidium of the USSR Academy of Sciences on the eve of the election.

But it turned out that I had been invited to have my hands slapped for my unbecoming conduct with the teachers, not to participate in the general celebration. "Lysenkoism won't even last another three years," I said to Belayev in response to his paternal rebuke, "and you'll be ashamed to look me in the eyes if you haven't forgotten by then what's going on right now."

Lysenkoism fell along with Khrushchev in October 1964, less than half a year after that conversation. But Belayev remained firmly convinced of his rightness. He was not ashamed. He was always ruled by higher considerations. It was just that the scale of values had been turned upside down by the unavoidable course of events.

Fate intended Belayev to be a leader. If someone is born with whatever it requires to take a position at the helm of power, he will take it no matter how inauspicious the circumstances. And the circumstances for Belayev's being promoted up the ranks were the most negative. He was the son of a priest. His brother, Nikolay Konstantinovich Belayev, a talented co-worker of Chetverikov's, was an enemy of the people who was arrested in 1937 and disappeared from the face of the earth. Belayev was not a party member—not because of any principles, but because of practical considerations. Had he applied to the party, they would have started digging around in his biography. He wouldn't have been accepted into the party then, might lose his job, and could even be arrested. He was too cowardly to become a member of the party. He asked my advice in 1967 about whether he shouldn't join the party. I'm too lazy even to try to imagine why he should have needed my advice on that matter.

Belayev's diplomacy never let him down. It's not the presence of a party membership card, but precisely its absence that can be a trump card in certain circumstances. Lysenko and Dubinin, most ardent in climbing the scale of ranks, were nonparty. (Lysenko never joined the party; Dubinin did it after he was elected to the academy as its full member.) The reasons that one or another public figure should refrain from joining the party are manifold. There are different kinds of nonparty members, just as is the case with party members. There is spontaneous, elemental nonpartyness, and there is purposeful nonpartyness. One may assert with certainty that Lysenko's and Dubinin's nonpartyness was purposeful. Nonetheless, their goals were different. Marks of distinction lose any and all meaning before the face of supreme power. Such power grants marks of distinction without worrying about such trifles as a party card, a university diploma, or a higher degree. The absence of such things only serves to emphasize the might of the rulers in ignoring it. Lysenko didn't need a party card in order to be elected to the academy. If the academy is ordered to elect someone, that's what it will do—moreover, by a secret ballot. Dubinin's nonpartyness, on the other hand, is mimicry. He used it to mask his striving for power. Academicians defend true science in his person. He is the selfless priest of science.

Purposeful nonpartyness is a rarity. Spontaneous, elemental nonpartyness is granted to the millions of those who do not stand at the threshold of paradise, who are not admitted to the helm of power, who are outside the game because of their personality traits, or whose fate is simply such. According to party jargon, they are the nonparty rabble. That was precisely the category to which Belayev had belonged since time immemorial. When Dubinin was driven out of the director's chair at Khrushchev's orders and the chair passed to Belayev, the situation changed radically.

A ceaseless struggle was waged in the academy. Scientists, those on whom the country's military might and the mastering of its energy resources depended, defended their branches of science against the destructive influence of ideology. That was true even under Stalin. Back then they had driven off attacks by philosophers on the theory of relativity and the theory of electromagnetic resonance. Now, under Khrushchev, they broadened their influence, struggling for true science against the forces of ignorance. When it became too dangerous to defend Dubinin, the gazes of the academicians turned toward Belayev. Lysenko's power was constantly increasing, as was opposition to him.

Election to the academy is a many-stepped process. The Science Councils at research institutes propose candidates. Not all the people thus nominated will be recommended for election at a general meeting of the academy, but only those who have been approved by the appropriate academy branch. Biologists are examined by the biology branch. Belayev didn't have the slightest chance of making it through the biology branch. Lysenko had complete control of it. Extraordinary measures were taken to weaken Lysenko's oppression. A new branch of biophysics, biochemistry, and active compounds was created. It was through this branch that Belayev became an associate member in 1964 and a full academician several years later. His nonpartyness acquired a purposeful character.

In order to become a leader, fear alone is insufficient, but you can't do without it either. Belayev knew exactly what one should fear, when, and to what degree. Last night, today's leader was mortally afraid of yesterday's—and he was just waiting for the right moment so that he could make his move. . . .

# *A First-Class Example of Calligraphy*

After defending my dissertation and receiving my doctorate, I fell into the category of the privileged. I became one of the worst among the best, a person having the lowest status among those who are granted elite services. There was a separate food store and luxurious merchandise shop; exclusive, closed-for-"others" hospital and restaurant; special resorts and sanatoria; access to concert tickets and to railway and airplane tickets. . . . The name of the establishment that provides these creature comforts to the elite contains a devilish irony: the Department for Worker Provisions.

Housing is provided according to rank. There is an enormous difference between the cottages of full members of the academy and those of associate members, between their cottages and the apartments of Doctors of Sciences, and between those apartments and the dwellings of Candidates of Sciences. The height of the ceiling drops with the salary. Electric ranges are provided for the elite, and they pay only two kopecks for a kilowatt hour of electricity. The rank and file must purchase their own smaller, less efficient ranges and pay four kopecks per kilowatt hour of energy. While the elite are given a considerable amount of discretion in payment of utility bills, the poor find their services disconnected as soon as a bill becomes overdue.

I was not given privileges partly because I rejected them. I despised the inequality that they represented. I could not believe that such things could be possible in a state pretending to be a socialist one. But partly I did not apply for services granted to the minions of fortune because I knew so little about these grants. Inequality is discreetly hidden, and no one would hurry to expose it. People who have a right to these elite services have to struggle with the enormous unbounded army of bureaucrats belonging to the organization that dispenses the services but reluctantly. As a protest against the deprivations of the poor, I did not register at the special stores. The result of my ignorance about elite grants was that I had to pay large electricity bills. I payed four kopecks instead of two per kilowatt hour.

At times the situations that arose as a result of these inequities were ludicrous, but others were tragic. Let us start with a tragic one. It is tragic when someone from the manager's office of the house comes to turn off the electricity just when your child is ill with fever, suffering an attack of bronchial asthma, and no one else is home. When Masha, suffering from an attack of asthma, was left alone in complete darkness—and the hot plate did not function—I threw a fit, threatened to sue, and demanded the money that I had overpaid for electricity be returned to me. The money was returned. People who received a salary five times less than mine paid exactly as much as I did, but I was privileged. I didn't sue. During my fit I had learned that electricity becomes more expensive the smaller the user's salary is.

On the eves of some holidays, "gifts" were brought to the elite homes. Caviar, canned salmon, Pilsner beer, corn oil. These luxuries had to be paid for, but nevertheless, they were real gifts, presents of fate, not available for "others." Once I received such a gift. I hadn't ordered it. I wasn't on the list of people for whom it was deemed appropriate, although on the basis of my rank I had a right to it. I didn't even know that those "gifts" existed. And suddenly I was brought all sorts of wonderful foodstuffs. I paid. The next day a shop employee asked that the gift be returned because it had been sent by mistake. Two girls came to retrieve it, and I asked them who the gift was supposed to go to. "Doctors of Science and academicians," they told me. "Well, that means I qualify," I said. "But you came to see me and want to take the stuff back because I'm a woman and don't look like a member of the intelligentsia, and so you decided I don't qualify." They laughed, told me that wasn't so, and that I did indeed look like a member of the intelligentsia. They left, but didn't take back the "gift."

And now I'll tell the story of how I learned that I qualified not only for luxury foodstuffs such as Pilsner beer and corn oil, not just cheap electricity, but for all my needs. The topic here will be a refrigerator.

The refrigerator. A requisite of any minimally comfortable kitchen. A symbol of and substitute for the family hearth. It's possible to get along without a refrigerator, however. People did, and even now millions do. As famous a scientist as my father was, he still didn't have a refrigerator. While he was a professor at the first Geographical Institute in Petrograd, which he founded, the windowsill of a northern window served as a refrigerator. And if my father didn't have a refrigerator, then of course I didn't. Later, when children were born, I was too busy to warm myself at the half-ruined family hearth. I was involved with science. And besides, "my husband left me for another," as the poet said. And there was no need to have a refrigerator. Need is akin to habit. If one doesn't exist, neither does the other. If the creators of scientific socialism had thought a bit about what need is, the formula "communism is to each according to his needs" would have been enriched, perhaps, by some sort of epithet.

When the drama unfolded, and a refrigerator became the main prop on the stage, I had achieved the rank of professor, was living in City of Science in the heart of Siberia, and was the head of a laboratory that I had founded. One day I entered the institute's vestibule. An enormous sign announced that anyone desiring to get such-and-such a brand of refrigerator should drop in to such-and-such an office to see so-and-so. Need was born in me. I went.

The young man there, after listening to me, asked with great malicious glee, "And are you in line?"

"No," I said.

"Then you're trying to cut in front of people who have been in line for years? You should be ashamed." And all of that was said spitefully, maliciously, with the pronounced intent to humiliate.

"Why did you hang a sign in the vestibule?" I asked. "I didn't need a refrigerator. I didn't have one and won't have. And I have never yet stood in line with my hand stretched out for a goodie. It would be one thing if it were something needed. But a refrigerator!" And I headed for the exit.

Quite by accident, I assure you, quite by accident I failed to close the door behind me. It slammed shut itself. It sounded as though I had left, slamming the door. But it is not at all like me to slam doors. For as long as I can remember, I've never slammed a door, and I don't expect ever to slam one.

A certain amount of time passed, and I received a note from that young man: "You can get a refrigerator." I got one.

It turned out in the course of events that the refrigerator was an old brand, small, and used a lot of electricity. The people at the front of the line had decided to wait until a better one would be "dropped from above." The sign urged the people at the end of the line to take an item in short supply which had been rejected. But the people at the end of the line didn't deign to do that either.

The refrigerator beautified my kitchen.

The scene changes. A marvelous voice is heard. Maria Callas. It's still City of Science in the heart of Siberia. The marvelous voice was reproduced mechanically—a record was being played in an auditorium of the House of Scientists. The auditorium was full. And the aria was being commented upon by the most enchanting creature ever to visit our vale of tears, Marina Godlevskaya. She was the child Marina who had lived at my dacha in the summer, and who, with great artistry and inspiration, had performed dances that she made up herself. I remembered one evening amid the Komarovo pines. Against the background of the illuminated leaves of the underbrush, streams a fountain constructed from a garden hose; a record plays, and we listen and watch, and go into ecstasy. By "we" I mean my daughters, our guests, and I. Leon Abgarovich Orbeli—the academician, lieutenant-general in the medical service who headed the Military Medical Academy during the war, and great physiologist—wept. I remember his gray head against a snowy-white handkerchief.

Marina had graduated from a theater institute and become an employee at the Theater Museum. At the invitation of the Akademgorodok House of Scientists, she was now giving a lecture on Maria Callas.

Not every Italian girl is a beauty. But if an Italian is a beauty, she is Marina Godlevskaya. Her beauty is dynamic. Besides the harmony of form and color, it is the beauty of gesture, movement, sound. She tells something comic, and her voice smiles, just a bit, but she herself seems not to smile; the smile is somewhere, not in her eyes, not on her lips, but it is there. . . . Wonderful! Without urging, you are invited, an honored guest—not a pupil, not a graduate. . . . It's impossible not to fall in love.

And he did fall in love. None other than Dmitry Konstantinovich Belayev himself. He was a lean, handsome man, who perhaps in his youth had given women a run for their money. Now, however, over fifty, he mostly swaggered, tried to act the young man. From time to time, right before my eyes he had made the wives of his subordinates happy with his directorial attention. To my honorary old maid's practiced eye (and no one knows as much about love as old maids do; and an old maid is any woman who has accumulated a statistically saturated sum of observations of other people's amorous doings; and old maids are divided into old maids *sensu strictu* and honorary old maids, who also have their own experience), it was clear that there was little powder left in the pan. But then suddenly it burst into flame—he asked to be invited to visit.

The scene changes. The guests—the director and Marina Godlevskaya—are having supper in the kitchen. It's impossible to convey the inimitable humor with which Marina later told me about the pantomime of passionate outpourings occurring every time I turned my back.

But now the chores of hospitality are behind me, and I'm sitting with my guests, and the director's gaze falls on the refrigerator. He is astonished. "Where did you get that piece of junk?"

"Why 'junk'?" I asked.

"The store to which you, as a lab head, Doctor of Sciences, and professor are attached doesn't carry refrigerators like that."

Everything became clear. I didn't tell Belayev that I hadn't even suspected the existence of that privileged store for the elite to which it turned out I belonged. I didn't know, but the young institute employee who was rude to me obviously knew. It was hard for him to understand me. A full person can't understand a hungry one, as the saying goes. But a hungry person can't understand a full person either. I was the full person.

The events were transpiring at a time when a curtain was lowering over my privileged satiation, which had suffered damage anyway as a result of my ignorance. My five-year tenure in Akademgorodok was coming to an end. The KGB threw me out of Akademgorodok for my modest dissident activity. But I was not the only one they threw out.

The lean and handsome director was, however, much more zealous than many other directors in serving the KGB in its struggle against dissent. There is a law involved here. The fewer his scientific merits, the more zealous an administrator is in serving the regime. There is also a regular phenomenon that partially justifies the poor director. The closer the discipline studied by the institute is to the line of fire, the greater the toadying demanded of the director.

The director who was burning with passion in my kitchen defended if not the most seditious, then the longest and most cruelly battered science—genetics. It had only just been removed from the category of servant of Wall Street and ceased to be a scapegoat for the failure of agrarian policy. Geneticists had only just become the equal of other citizens with regard to the threat of unemployment, arrest, death in prison or in camp. The fear of a relapse of bloody favoritism did not strike anyone as a mania. When a branch of knowledge being defended barely can keep one nostril above water, a director's slavishness and his police actions are no longer a vice, but a virtue, a selfless defense of scientific truth, of an entire branch of science, and of hundreds of people, from authorities who at any moment can close down both that branch and the institute developing it. A director turns himself inside out to prove his loyalty.

It is difficult to retain the post of director of such an institute. It is a contest of scoundrels. But one so wants to retain the position! Election to the academy beckons on the horizon. And a store with refrigerators of the most perfect construction! It's no joke to lose the chance to buy a refrigerator!

You are dying to find out whether the director who cooperated zealously with the KGB to bring me to the ground was elected to the academy in spite of the complete absence of any scientific merits. He was! And in 1983, at the opening of the Fifteenth International Congress of Geneticists in Delhi, he served in the Presidium, as president of the World Federation of Geneticists, and sat one person away from Indira Gandhi. And after an important person wearing a turban, who was sitting between the prime minister and the president, presented Indira Gandhi with an award, my former director gave her a congratulatory handshake, and there was something gentlemanly, something manfully gallant in his bow.

I received some crumbs from the master's table, but never the most desired objective of every Soviet citizen, the opportunity to travel abroad. Did I want to travel abroad? In Korney Chukovsky's book *From Two to Five*, a little boy asks his grandmother whether she could eat a dead cat sprinkled with sugar. A trip abroad is the sugar, and the process preceding it is the dead cat: the receipt of permission from the party bosses, the ideological examination at the District Party Committee, the long years of ingratiating oneself with the authorities, the impudent competition of fellow citizens who lust to travel abroad. A dead cat sprinkled with sugar is an unacceptable diet for me. I didn't want to go abroad.

I was invited to conferences and to give courses on many occasions, but I never went. Invitations, to the Seventh International Genetics Congress in Edinburgh in 1939, to chair the population genetics section at the Tenth International Genetics Congress in Montreal and to lecture at an American or Canadian university of my choice in 1958 had to be turned down.

In 1963, I was invited to a congress at The Hague and to give a course on evolutionary genetics in Prague. I didn't even appeal for support. A young man named Gnes, who was trying to enlist me as an informer, a personage right out of Kafka's *The Trial*, tried to tempt me with Prague's museums and architecture. He had been much impressed by the museum of alchemy in Prague. I myself was tempted by Prague—Tsvetayeva's Prague, with its stone knight on the bridge, its Gothic architecture . . . But, well, there's no point in reopening old wounds. Gnes hadn't the slightest chance of persuading me.

In 1965, I was in Novosibirsk—a professor, the head of a lab, a Doctor of Sciences. A Mendel anniversary was being celebrated in Czechoslovakia. At least a hundred people received invitations to attend. In addition to the delegates who would be traveling at the government's expense, there would also be a tourist group whose members would pay their own way. I had every chance of going, if not as a delegate, then as a tourist. All the documents were submitted. The abstract of the paper I would present was sent off months before the anniversary, from Turkmenistan, where I was studying correlation pleiades in desert ephemerous plants.

The participants of the international forum were notified by telegram that the time had come for them to report to Moscow. My name was not on the list. Nonetheless, I flew to Moscow, assuming that there had been a mistake in the compilation of the list. I called Dubinin from the Moscow airport. "Your candidacy has been rejected by the Central Committee, and none of my attempts helped anything," Dubinin's passionate tenor voice sang over the receiver. I decided that that was a misunderstanding, too. I headed off for the Presidium of the Academy of Sciences, to speak to V. N. Kirillin, who was the vice-president of the academy and the head of the Central Committee's science branch. He was the one to whom I had written ardent letters asking support for the book that S. N. Davidenkov and I had co-authored.

All the delegates and tourists were gathered together there. And before going to see Kirillin, I took a seat in an armchair upholstered in scarlet velvet, with gilt arms, in the room where all those lucky ones who were to go abroad had gathered in order to be instructed in patriotic behavior. Everyone knew that I wasn't going. "You've yourself to blame. You're being punished for not having sent your abstract via Moscow and the committee that processes abstracts, but using a wild system of your own," I was told by Militsia Alfredovna Arsenyeva, a geneticist of the old school. I remember her tone very well.

We listened in silence to the lecture on how to conduct ourselves abroad. We kept silent; that is all except one person. When it became known that the head of the delegation would be Stoletov, who had formerly been one of Lysenko's vilest apologists, Rapoport raised his voice and said that there was no reason for geneticists to trail along behind Stoletov, that he had seriously compromised himself as an adherent of Lysenko's. No one had any reaction. There couldn't even be any discussion of removing Stoletov. Everyone except Rapoport was willing to follow at Stoletov's heels.

When the meeting ended, I went to Kirillin's office. His secretary was extremely polite, saying that as soon as the representatives of the Society Knowledge came out, I would be able to see the vice-president. Just then I saw Dubinin. "Nikolay Petrovich," I asked, "please go see Kirillin and intercede for me." He agreed willingly. I was distracted for a second, exchanging a greeting with one of the delegates, and then I went back in to go with Dubinin to see Kirillin. The secretary said that Dubinin had given all the necessary information, that my request was denied, and that Kirillin refused to hear me out. There was no trace of Dubinin.

Tatyana Antonovna Detlaf, a student and great admirer of Dmitry Petrovich Filatov, knew everyone and was quite perceptive. "And you believed it?" she said. "Go to the Foreign Department of the academy Presidium and ask at what level your candidacy was overturned. Be sure to do it. A lie concocted by Dubinin will have consequences. It's in your interest to find out the true state of things."

I went. A huge, well-made young man dressed in a bright suit spoke to me from a standing position. "Dubinin turned in the lists," the young demigod told me. "He crossed out your name and put 'no paper' as the reason. Your name was not on the list submitted to the Central Committee."

My paper was read in Prague by Boris Lvovich Astaurov. It was printed in the conference's papers. But in the description of the sectional meetings published in the Soviet journal *Genetica*, where special attention was devoted to the participation of Soviet geneticists, my report was not mentioned.

I received invitations to attend conferences or give courses later as well. In 1967, Humboldt University in Berlin invited me to give a talk at the festivities in connection with the celebration of the Fiftieth Anniversary of the October Revolution. All the papers were sent to Moscow to the Foreign Department and my medical documents were in order. The head of the Foreign Department of the academy Presidium, this time in Novosibirsk, explained to me in great detail that he doubted whether the Foreign Department of the Academy of Sciences, where my documents had been sent, would respond, and if it did—it would probably be too late for me to go. "If the Foreign Department were to begin answering every letter, it would require the entire Presidium . . ." I didn't bother to wait for an answer and left on an expedition—my last out of Novosibirsk.

In 1968, I was included in the delegation to take part in the Twelfth International Genetics Congress. A trip to Japan, the land of my dreams, was in store for me. I was thrown out of the delegation when it became known that in December 1967 I had signed the petition asking that Ginzburg and Galanskov's sentence be reviewed. But I didn't care about the congress. No, I shouldn't have cared about it at all, had I been more sensible. It was only a bureaucratic foul-up that saved me from prison.

That fall of 1968, when I had a pistol at my head, I readily did participate in an international congress, but it was held in Moscow, not abroad. That was the Thirteenth Entomological Congress. I was elected chairperson of the Session on Insect Genetics and Cytology. And it was while I was occupying that post that the punishing hand of the law sought me out, expelled me, and Boris Lvovich Astaurov, the congress's vice-president and a great authority on the silkworm, headed up the session. Nonetheless, I was the chairperson at one of the section meetings. There was only one translator and she only knew English, so when the Canadians addressed questions to the Frenchman Y. David in French, I translated myself, earning the instant reputation as an authority on French, for which I paid dearly. I arranged a hotel reception for all the session participants, and sat at the table next to David's wife, whose question about Russian soups and the difference between *shchi, borshch, okroshka, solyanka,* and *rassolnik* I had great difficulty in answering.

The day and hour arrived when I felt an unquenchable desire to find out whether the fruit fly was the only toy in the hands of the earth's powers, or whether other creatures as well had been seized by flare-ups of mutability. Hereditary information is encoded in all living organisms with the help of the same substances—nucleic acids. The enormous molecules of those acids are not subject to the law that is obligatory in the world of small molecules, namely, the law of constancy in the atoms forming the molecule. These giants are allowed a certain measure of freedom. The disposition of atoms in these molecules varies according to the length of the chain, as well as from chain to chain. It's these geometrical differences, rather than chemical ones, that serve to record the commands that parents put into that little clump of living material out of which the future offspring will develop. The laws for recording information are the same everywhere, whether it be an egg, a stereo record, or a line in a book. Signs go into action in a certain specific sequential order. Their record is preanticipated time. Time is unidimensional and its unidimensionality corresponds to a linear record. The elements of the code for hereditary information are distinguished from each other only geometrically, not chemically, like the signals in the Morse alphabet, or the elements on a tape recording.

The signs for the hereditary code are genes. Once a certain global factor was capable of producing mutations of one and the same genes in fruit fly populations, couldn't its action extend to the genes in other organisms as well . . . to human genes? At one point, the moment arrived when yellow mutants began to appear among flies everywhere. Perhaps some mutation in man had also begun to appear everywhere. It was decided to study the birth dates of patients suffering from hereditary diseases and to add someone with a medical education to the laboratory staff. A graduate position for the young doctor who applied kept appearing and then disappearing, but eventually it was confirmed. The graduate student who was to study the birth dates of patients took her post in the laboratory. But the realization of that plan was as far away as the stars. The administration refused to approve a topic.

It's not the true value of research but rather the ability to satisfy the tastes of those who will be voting that decides the outcome of one's graduate work. There is no room for innovation, risk, or plowing virgin soil. The most important thing is the degree. Without it you won't be able to feed your family; without it you won't even have those crumbs of freedom of research and work that a Candidate of Sciences possesses. I assure you that it's much worse without a degree than with one. A candidate is even granted more living space than is a scientist without the degree. I'll explain. According to the laws, everyone is allowed nine square meters. Let's say you live in a communal apartment, where you have a room of eighteen square meters. You're neither moved out nor forced to take in another occupant. You pay for the extra space out of your tiny salary, and you pay for every extra square meter more than for every legal meter. The increased payment is something like a fine for having them. But then, Oh, joy!—you are granted a degree! Your salary grows, and your rent becomes smaller! You have the right to a room regardless of its measurements. The payment for your legal nine meters and the payment for "excess meters" become equal. The fine is removed.

The most difficult thing is having a dissertation topic approved if it in any way departs from the norm. The Senior Council from the institute or university where the defense will occur will judge the quality of the work, and they of course have specialists in the area in which the pretender to the degree is working. An abstract of his or her thesis will be sent to all the cities and villages in the Soviet Union that have specialists in the necessary area. Their evaluations will be taken into consideration in the awarding of the degree.

The topic is confirmed by the Senior Council at the same institute or university where the research will be conducted. It's possible that the council may not include a single person who can evaluate the likelihood of success. That's exactly how things were at the Siberian Branch of the USSR Academy of Sciences Institute of Cytology and Genetics. And in giving her report, the graduate student assigned to my project tried to emphasize the difficulties and present my idea in the most stupid way imaginable. The threat of liquidating the graduate position had disappeared by then, but the graduate student, unfortunately, turned out to be a product through and through of the era of scientific socialism. The idea of studying the changes in mutation rate in man had been labeled an "unpassable" topic: No one could believe in the possibility of getting material in a country where for many years the very concept of inherited traits was considered seditious.

While the disputes went on, I was searching out sources for materials and funds for paying the doctors who would be providing me with material from their clinics and research institutes.

Success followed success. Nina Alexandrovna Kryshova, a brilliant student and co-worker of Davidenkov's, granted me the use of Davidenkov's archives and assigned me Natalya Gennadyevna Ozeretskovskaya to extract the information I needed from the archives. In order to exclude the possibility of subjectivity in our choice of data, Nina Alexandrovna and I decided to keep Natalya Gennadyevna in the dark. She was not to know that we expected to find disproportionalities in the distribution of patients' dates of birth. The surety of the diagnosis based on a lengthy study of the patient, certainty of the hereditary nature of the illness, and a detailed description of the symptoms served as the criteria for including a patient in our list.

The possibility of paying the doctors could have presented itself only by way of a miracle, and a miracle occurred. In Akademgorodok a new and unprecedented institution arose by fiat of the regional Komsomol committee. The institution was a job placement office for students of Novosibirsk University. This office was called "the Torch." The office took orders from research institutes for the fulfilling of urgent large-scale tasks, such as filling out sociological surveys: The Economics Institute, for example, was studying population migration, employment patterns among young people, and the tastes of newspaper readers. The institutes paid the office, and the office paid the students. The business grew and expanded. As some smart fellow said, private property on earth is capable of turning a stone into a flowering garden.

The office was headed by a Science Council whose volunteer membership was made up of Doctors of Sciences from all areas. "The Torch" flared with a bright flame. In a year, its working capital reached a million rubles. The air began to smell of an approaching thunderstorm. The central newspapers were still praising the regional Komsomol committee's brainchild, and at meetings of the Science Council people were discussing plans for building a factory to produce nonstandardized equipment ordered by laboratories, but people sensed imminent catastrophe. The Torch highlighted the imperfections of paying people by rank rather than for productivity. Miserly workers' salaries resulted in sabotage, a protest passed off as sloth, a secret strike. "I don't care where I work as long as I don't have to work." "They pretend to pay us, and we pretend to work," said the workers. Joke after joke was born. The Torch's flare threatened to be blown out by the cold wind of administrative interference. It would be considered good luck if the rout were limited to administrative measures and none of the Torch employees wound up behind bars. The Torch Council decided to play Maecenas and give money to cover research topics and support a risk. By force of circumstances and chance I became the first object of their beneficence. Since I knew that nothing good would come of this, I rejected payment for my labor out of hand. With unending gratitude and with fear for the fate of my benefactors, I accepted a modest sum for paying the doctors who were providing material for our topic.

At the very instant that any hope for approving the topic at the institute disappeared, our first materials had ripened. Our hypothesis that the birth dates of patients suffering from hereditary diseases would be distributed unevenly in time was confirmed. The peak for such births came in the second half of the 1930s. The topic was approved at the institute. At the very height of our work our good fortune came to an end. The office of Torch, as was to be expected, was closed down.

But I didn't need the Torch's financial assistance or anyone else's for the continuation of my work. The doctors working with me were enthusiasts who treasured the opportunity to publish their works in academic journals. Mental diseases were now at the center of our attention. Three articles written in collaboration with Nina Alexandrovna Kryshova and Natalya Gennadyevna Ozeretskovskaya were published.

Fluctuations of mutation rate in human populations of the USSR were established with high probability. Nine hereditary diseases were observed, mental disorders among them. The distribution curve of more than three thousand birthdates of patients showed two peaks: one in the late 1930s, the other in the late 1950s. But it seemed that the cause of mutability fluctuations in man, just as in *Drosophila*, would remain a mystery forever.

The hereditary nature of high mutability in *Drosophila* has been proven beyond doubt. Specific genes, capable of producing mutations in other genes (the so-called mutator genes), became frequent from time to time, and periods of their high rate were periods of mutability outbursts. But what could cause these drastic changes of mutator gene frequencies? The cause had to be environmental, because of the global nature of mutability fluctuations. It was clear that some global environmental changes triggered natural selection in favor of mutator genes. But what were these environmental forces?

An answer came thanks to God and to a French virologist named Nadine Plus. I met Nadine, my adored Nadine, in 1976, in Belgium, in Louvain-la-Neuve. An all-European *Drosophila* research conference was held there and I reported. My false idea in those days, ascribing mutability outbursts in *Drosophila* to virus epidemics—viruses being a cause of rising mutations—attracted Nadine's attention. We started to work together, and it turned out that viruses had nothing to do with high mutability. The mutator genes ruled over the occurrence of all these yellow, singed, and abnormal-abdomen creatures. And then, suddenly, Nadine decided to apply insecticides to my highly mutable stocks of flies and to stocks preserving their low mutability for decades. The highly mutable flies, descendants of wild captives, proved to be insecticide resistant, while the conservatives were not.

Now the environmental trigger of mutability rise was found and, moreover, it was a trigger acting on a global scale. A natural calamity, use of a new insecticide, wipes out all the *Drosophila* flies except insecticide-resistant mutants. High mutability becomes an adaptive trait; the time of mutator genes has come. When resistant mutants have replaced the extinct unfortunates, the time of mutator genes fades away.

What is true for *Drosophila* could help one to understand the mysterious clustering of birthdates of patients suffering from a hereditary disease. Natural calamities—the war, civil war, hunger, and epidemics—raging through Russia in 1914–21 were a green light for mutator genes. Their owners survived the disaster, and the mutant offspring of these resistant mutants, listed by me as patients, were affected by an hereditary disease. This generation started to be born in the late 1930s. The general collectivization launched by Stalin in the late 1920s and early 1930s, accompanied by hunger and by epidemics in camps of exiled peasants, took millions of lives. It became a cause of the second peak of birthdates of mutant humans in the late 1950s.

I continued to study human genetics even after my expulsion from Paradise.

In 1970, in Riga, we had been peacefully talking with a lecturer at the University of Riga, Raypulis, who was later accused of dissent and expelled from his post. We were sitting in the luxurious lobby of the Hospital for Mental Diseases and waiting for an appointment with the director of the hospital, a woman who was also a Minister of Public Health of Latvia. An elderly man sat down near us and listened to our conversation with obvious interest. Raypulis was talking about an hereditary disease causing mental retardation, which he detected by using biochemical tests during his population genetics studies of Latvian villages. He found that Latvians, carriers of the disease, were normal. This resulted from their protein-deficient diet, since the abundance of particular proteins beneficial for normal persons is harmful for those who carry the pathological trait. Raypulis had not yet finished his discussion when the old man fell asleep and an elderly woman came over to awaken him. Apparently, they were the last ones waiting for an audience with the director. Then the secretary came out to announce that the reception period was over and Raypulis and I would be the last ones received. But I said, "No, we can come later. Our business is not urgent. Let the people behind us see the director." After some bickering, it was announced that all of us would be received at the same time. From a distant corner of the reception hall we noticed that the couple's conversation with the director went well. We also successfully settled our affairs, and after we had left, Raypulis asked me if I knew who the elderly couple were. He told me they were the parents of a young man, Ilya Rips, who had set fire to himself as a protest against the Soviet invasion of Czechoslovakia. The fire was put out and the young man beaten unmercifully before being brought to this hospital for mental diseases.

I lived for five years in Science City. Every fall an expedition was outfitted to study mutability in natural populations of the *Drosophila* fly. Two other expeditions were organized in spring and summer to study the correlation pleiades in plants.

During a fly expedition in 1967, I was lucky enough to meet a remarkable man who has now passed on. He was living on a pension in Bukhara, having retired as head of the Department of Foreign Languages of the Pedagogical Institute there. He spoke the Uzbek language and he had an Uzbek haircut, a short cut with a lock of hair falling onto the forehead, which did not diminish his intelligent appearance in the slightest degree.

Seven people were participating in this last expedition from the City of Science. Three—Katya, Klara, and Lyusya—were members of the previous expeditions: Vitya Koslov was at that time a student at a school for mathematically gifted children (now he is a research worker at the Institute of Cytology and Genetics); Volodya Ivanov was an undergraduate at the university. The sixth member of the expedition was the boldest of the bold—Yuri Nikolayevich Ivanov. During Khrushchev's reign, they badgered and threatened him, imprisoned him, and sent him into exile, but Yuri Nikolayevich refused to be bullied. The seventh member of the expedition was a fellow who joined us only to have a nice trip. Thus there were four men and four women.

We had breakfast in Bukhara in the bazaar, eating baked eggplants and *manti*—a kind of boiled *pelmeni* or Siberian meat dumpling found throughout the Far, Near, and Middle East. Bazaars are museums of folk life, but after breakfast we went to the town museum built on the palace ruins of the last emir of Bukhara. Only heaps of dusty bricks remained of the palace. As we were leaving, Lyusya shouted through the tears running down her cheeks, "Let's go, let's get out of this nasty town right away!" She was shaken by the pictures and colorful statues that depicted with utter cynicism torture and executions, means of oppression used by the rich before the Soviets liberated the suffering poor. We turned to leave behind those first-class models of socialist realism, but did not make it to the door. It became known that the expedition group from the City of Science was in the museum, and an employee came up to us to offer his services as a guide. He told us some trivialities about life in the emir's harem, but after we were enlightened, he asked if we would like to see the minarets of the two Medresse, or Muslim monasteries. "Go there. Sergey Nikolayevich Yurenev lives there. He will show you Bukhara even though he normally refuses to see people." Then the guide turned to me and said, "Sergey Nikolayevich will perhaps make an exception for you, because he is also a Leningrader."

An exiled Leningrader, it went without saying, Sergey Nikolayevich Yurenev did make an exception for us. We found him occupying four cells of a former monastery, two of which were filled with treasures he had uncovered himself at archaeological diggings. He accompanied us to a tearoom and showed us how to hold the Uzbek *piala,* a cup with no handles. Together we celebrated the circumcision of a seven-year-old Uzbek boy. Solemnity was present, but also dancing on stilts, with stones used as castanets. Sergey Nikolayevich instructed us when and to whom money was to be given. We visited the gigantic mosques and cemeteries where the bodies of relatives of Mohammed have lain since long ago.

The streets of Bukhara seem to sneak among the clay walls. When the walls are new, the straw contained in the dark, not-yet-dry clay glitters in the sun like gold.

At night we drank tea in Yurenev's upper cell, which served as the library, and ate real cane sugar. N. I. Vavilov, describing his travels to Afghanistan, mentions the rows of shops in Kabul that sell hard clumps of cane sugar having a caramel taste and weighing 150 to 250 grams. The dark, opaque clumps we were now eating were cut in small pieces, and sugar was the only food Yurenev served to us. Our gifts to him were firmly refused.

"Sergey Nikolayevich," I said, "I am a member of the Science Palace Art Council in the City of Science. I invite you to come and talk to us about the history of Bukhara—the council will pay your travel expenses."

"I'm not a scholar. I have nothing to speak about."

"I doubt that. Did you publish the results of your excavations? Would you please show us your articles?"

"I did not publish any articles, but I will show you alone something I wrote. First the young people must leave."

The expedition members went out. Sergey Nikolayevich pointed to a small curtain covering a part of the wall. The scarlet velvet was edged with gilded cord, and along the bottom was a golden fringe. Thick gilded cords hanging from the ceiling supported the curtain and Sergey Nikolayevich told me to "ring up." The open curtain revealed a simple but elegant gilded frame, containing a first-class example of calligraphy. In black letters against a white background were the words *to shit.*

Three cats lived in the Medresse. Their names were Kissochka, Koshechka, and Kotyashechka, the diminutives of the Russian word *Koshka* or cat. Upon parting I said to Sergey Nikolayevich: "An essay which I wrote, entitled 'What is the Difference Between Cat and Dog,' will soon be published by the magazine *Knowledge Is Strength.* Would you like to have a reprint?"

"Yes, send it to me, please. But I don't write letters, so permit me to thank you in advance for the beautiful essay which gave me so much pleasure."

I wrote to him asking to send me a map of Bukhara, indicating the location of storks' nests. The storks of Bukhara! Each couple build their nest year after year in the very same place. They use the old nest as a foundation for the new one. This foundation grows and forms a pedestal-like column that reaches in some nests a height of 150 centimeters. The soil of a town rises by 1 centimeter a year, and archaeologists call this a "cultural layer." I did not care too much about the location and the cultural layers of the storks' nests on the minarets and on the mulberry trees decorating the main streets of Bukhara. I passionately wanted to have a letter. I did not receive a single one. Several years later, at my request, Mishenka Meilakh and Nikolay Nikolayevich Vorontsov went to visit Sergey Nikolayevich, and from them I learned that he had died.

He was the last golden knight I met, a Russian intellectual of the era of the building of communism.

# *The Relay Baton of Terror*

I had the good fortune to read Pasternak's novel *Doctor Zhivago* when I was still in Leningrad, before my move to Novosibirsk. It is practically impossible to get this book in the Soviet Union. Its publication by an Italian Communist press and the awarding to Pasternak of the Nobel Prize served the Central Committee as an excuse for demonstrating the committee's might on the ideological front. Meetings, condemnations, breaking up student demonstrations in defense of Pasternak, a hideous campaign of persecution of the poet in the press, his expulsion from the Union of Writers, threats to expel him from the Soviet Union, "workings over" by his fellow writers, searches at the apartments of his admirers with the goal of seizing the seditious book—which, by the way, contained nothing particularly seditious—arrests and prison terms for reading the book and distributing it are among the many crimes of the Khrushchev era. The Khrushchev thaw had no intention of melting the icebergs of the Stalin era. The persecution of Pasternak, the conviction of the poet Brodsky on the basis of Khrushchev's decree about parasitism, the glorification of Lysenko, the trial of Sinyavsky and Daniel, the expulsion of Solzhenitsyn, the use of psychiatric hospitals for the suppression of dissent—all of that on the ideological front, and, finally, the suppression of the Hungarian Revolt and the execution of its leader Imry Nagy (1895–1958) are all icebergs created by the post-Stalin era.

An impudent lie aimed at the ignorant indicated a campaign against the intelligentsia. The pages of newspapers were given over to the censure of Pasternak and his books by people who had never heard of Pasternak and who had no need to read a book that carried with it the risk of a prison term. The wrathful words of the workers in defense of the inviolability of the prison rules operating on the ideological front were printed in *Pravda*. I remember that one worker was very exercised by Pasternak's attack on Mayakovsky. No one had glorified Mayakovsky as much as Pasternak had. The worker was told to lie, and he did. Pasternak's real crime in regard to the regime was not the content of his book but the fact that it got around the censor and was published abroad. Writers—good boys and girls who had never in their whole lives published a single line without permission—fell greedily on the chance to demonstrate their loyalty. Fierce envy played a not insignificant role. I thought that I'd never manage to read *Doctor Zhivago*.

It was given to me by friends—a marvelous, exotic married couple. He is a theoretical physicist and a very handsome man with salt-and-pepper hair, sparkling eyes, a magnificent profile, and a beautiful timbre to his masculine voice. His face glowed with intelligent goodwill and self-oblivious interest in the fate, opinions, and thoughts of his partner in conversation. She is a teacher of Marxism-Leninism, the head of the Ideology Department of one of Leningrad's countless technical VUZes (schools of higher learning), a piquant snub-nosed beauty, youthful looking, merry, who's a superb hostess.

One night at their home I decided to make fun of the bourgeois ways of the Soviet intelligentsia, responding to the question of what I would like for supper (the choice was among salmon, caviar, roast beef, and similar delicacies) by asking for pearl barley gruel. Barley gruel, millet gruel, lentils, *vobla* (dried fish)—all those things that assuaged hunger in the hungry years—have always remained unsurpassed models of gourmandizing for me. But my well-off hosts, my coevals, had traversed the same path as I. Rarities among extreme rarities—briquette of barley gruel and dried *vobla* were removed from the refrigerator. It turned out that my tastes were not exceptional.

It was from them that I received *Doctor Zhivago*. Both of them were secret dissidents in equal measure, but the Marxist-Leninist expressed her dissidence more freely than did the theoretical physicist. She offered me a book, ignoring his not very insistent protests. The package was wrapped up in newspaper. I unwrapped it at home. It had seemed to me at first that in her agitation, the insanely brave Marxist had given me something other than what she had claimed the boldness to give me. It was a volume of Balzac. But in the event of a search prudent members of the intelligentsia had put Pasternak's book inside the cover of one of the volumes of a multivolume Russian-language edition of Balzac.

When I arrived in Novosibirsk I was offered a gift—*Doctor Zhivago. Samizdat* was flowing toward me like a river. The ball comes to the player. In 1963, Zhores Medvedev's book, typewritten, wherein history itself spoke through the lips of an authority, condemning Lysenko, was given to me by none other than the institute director—Dmitry Konstantinovich Belayev.

The windows of my luxurious three-room apartment had sills that were covered with iron from the outside. The iron covering could be lifted a bit, and it was under there that I kept *Doctor Zhivago*. The volume propped up the ceiling of its place of concealment and starlings made a nest next to it and started a family. The baby birds' chirping guaranteed the reliability of the hiding place. Hardly anyone read that copy that was kept by the starlings.

But then there was a young mathematics teacher whom I'd asked to prepare my daughter Masha for the entrance examination at Novosibirsk University. I secured their promise that they would work diligently, and flew off to Kamchatka. It turned out, however, that I had let a goat into the garden.

My father and stepmother had not spent a single farthing on my education. One of the main goals of my life was to give my daughters a good education. I firmly counted on Masha's going to Novosibirsk University at Akademgorodok immediately after high school. But we had very bad luck with her high school education. In the school that Masha attended, the woman who taught sociology, the Armenian Nelli Rubenovna, made Masha the target of her anti-Semitic attacks. Masha was an excellent student. Nelli Rubenovna so terrorized her that on those days when there was to be a sociology class, Masha would have an attack of bronchial asthma. I went to the school and tried to correct the situation. In vain. I was told—Nelli didn't speak, she shouted—that she would make certain Masha left school with a black mark in her passport and would not be able to enroll in a single institution of higher learning. I had to take Masha out of the school and enter her in an Evening School of Working Youth. The curriculum at that school gave no chance for passing the entrance examinations at the university. And that's when I invited a mathematics teacher, a student at the Physics Department of Novosibirsk University, the charming Sasha Shenderov, to prepare Masha for the examinations. The nice young man not only agreed to tutor Masha but even brought her home from school on the bus in the evening.

It is appropriate here to talk about anti-Semitism in the City of Science. The monarchical command to build a town of science in the heart of Siberia was issued in the middle of the fifties by Khrushchev. In 1958 the town with its fifteen institutes, constituents of the Siberian Branch of the Academy of Sciences, came into being. The very idea of creating the city was born by the "thaw era." An atmosphere of freedom wafted over the early years of its existence. When governmental anti-Semitism raged in the Ukraine and the doors of the Kiev, Kharkov, and Odessa universities were closed to Jewish young people, the youngsters headed for Novosibirsk University. There Jews were accepted on a par with Russians. Yet when I arrived in Akademgorodok in 1963, anti-Semitism at the university and in the schools of the City of Science had assumed monstrous proportions. In Leningrad I'd never witnessed anything to approach the excesses that occurred in this city. Students organized anti-Israeli demonstrations during one of which a skeleton depicting Golda Meir was burned. Things went as far as pogroms. During a fiercely cold period in the winter of 1965, Russian students patrolled a dormitory. Their goal was not to let Jewish students into the building. The temperature was −46°C. The Jews were bound to freeze to death. On the doors at the entrance hung signs reading A CHICKEN ISN'T A BIRD, AND A JEW ISN'T A HUMAN BEING.

A battle finally broke out. Russian students who had not participated in organizing the pogrom took the Jews' side. The anti-Semites were defeated. A group of Jews came to see me for advice. I told them that the actions by the anti-Semites were covered by the Criminal Code. They should appeal to a court of law. They did so. Only one of the criminals was tried—he had stolen a watch and was tried for that. Nonetheless, a community trial of the organizers of the pogrom was held at the university. Alexander Danilovich Alexandrov served as the prosecutor. The court condemned the pogrom, but none of its organizers were expelled from the university.

I was ready to wail and bash my head against the wall when I received a telegram at Petropavlovsk-on-Kamchatka: "I won't apply to university. Left with Sasha to visit his parents. Don't know when be back. Masha and her Sasha." I arrived and found the apartment empty, clean, the flowers watered. My faithful friend, the young artist Slava Voronin, was looking after them.

Masha and her Sasha returned when it was too late to take the exams. Sasha moved in with us. Not long afterwards, he told me the following story. When he was leaving, Sasha asked Slava to look after the plants and gave him the keys. Sasha had promised *Doctor Zhivago* to his friend

Tolik. Slava, knowing that I wouldn't have objected, organized a chess club in my apartment, and Tolik came by one evening to get *Doctor Zhivago*. Tanechka, Senechka Kutataladze's young wife, met him at the door.

I already knew that Senechka had been assigned to me as an informer, and I had had occasion to show him the door. But here I must interrupt my narration of Sasha's agitated tale for a moment and tell a bit about Senechka. Senechka was preparing to go to the United States to study "on an exchange," and that involved him in certain obligations. I remember our last conversation perfectly: I said that one shouldn't condemn a salesman who trades under the counter, who engages in some exchange with the goal of extracting the greatest profit, since there was no other way to make a living. Senechka insisted that such practices should not only be condemned, but punished. The path to a secure existence lay through education, and that was available to everyone. I objected, saying that along the path to receiving a diploma, and especially an advanced degree, lay obstacles that were insurmountable for an honest person. One barrier was the examination on ideological subjects, where you were asked for a confession of faith, not for knowledge. "Can you imagine a creative person, Brodsky, for instance, at an examination in Marxism-Leninism?" I asked. "The road to the university is closed for Brodsky." Senechka lost control of himself. "We demand so little from them, and they refuse to carry out even that," he sputtered. "Senechka," I said to him, "I am ashamed to have acquaintances like you. Leave here and never come back." He retreated and begged me not to drive him away, but our friendship had come to an end.

Sasha told me that Senechka's wife had given *Doctor Zhivago* to Tolik. Tolik read it and gave it to a female acquaintance of his to read. That girl was a *tabula rasa*—something untouched, free of all influences. She didn't even suspect the existence of *samizdat* and had no idea that Yura Mekler had received four years in prison camps for reading and passing around that same volume which she was now reading with exultation. A party lady asked her to do some service work—to go help with the institute's wall newspaper. "I can't! I have such a wonderful book at home, such a wonderful book! I'm reading *Doctor Zhivago*!"

She was summoned immediately to the Special Department. (Every establishment in the Soviet Union, whether a university or an automat, has its Special Department, a group of people who earn their salary by being on the watch for deviations from the conduct prescribed by the Communist Party. Praiseworthy conduct involves no meetings with foreigners, no reading manuscripts passed from hand to mand—manuscripts by Soviet citizens that have evaded the vigilance of the censorship hierarchy and circulate under the ironical term *samizdat* [meaning self-publishing]—and, above anything else, no reading of books, journals, or newspapers published abroad. These latter items, zealously read by Soviet citizens eager to enjoy the forbidden fruit, have a most capacious collective name, *tamizdat*, something published *there*, a word signifying that the very mention of the West or other foreign countries is taboo. *Doctor Zhivago* was a first-class specimen of *tamizdat*. Ironically, Pasternak's novel was published by a Communist press in Italy, heralding a now-faded phenomenon of Eurocommunism. Anyhow, the Special Department had to enter upon its duties.) "They say you're reading *Doctor Zhivago*. Is that true?"

"Yes, it is."

"And who gave you the book?"

She named Tolik.

"And wouldn't you let us read it too?"

"All right, but first I'll have to ask permission," and she named Tolik.

She didn't have time to ask him permission. Tolik was called in to the Special Department. "They say you're reading *Doctor Zhivago*. Is that true?"

Tolik was a graduate student who knew what was what. He was being sent to a conference in Bulgaria. "What are you talking about?" he said, "I've never laid eyes on it. Maybe you'd let me have a look at it?"

"We'll arrange a double interrogation—then things will be worse for you."

"But whom do you want to arrange a double interrogation with? I've never even seen that book."

At that they let him go. Tolik ran to see Sasha, and Sasha ran to see me. Sasha concluded his tale with the proposal that *Doctor Zhivago* be delivered to the police station, since the inspectors expressed such interest in reading the renowned novel.

"No," I said, "send Tolik to see me." Tolik turned up. "You said that you never saw *Doctor Zhivago*. Stick to that story, and send the girl to see me." The girl came around, more dead than alive. "You said that Tolik gave you *Doctor Zhivago*. Now go to the Special Department and say that Tolik never laid eyes on the book, that it was given to you by Raissa Lvovna Berg and that she asked you to tell them that." The girl left.

I wasn't at all being a Don Quixote, nor did the hypnotic power that forces a gerbil to leap into the maw of a boa constrictor play the slightest role here. I had no need to leap—I was already in the boa constrictor's maw. "They" knew perfectly well where the girl had gotten *Doctor Zhivago*. Tanechka, Senechka Kutataladze's wife, had taken it to Tolik. You don't get sent on "an exchange" to America gratis. My action prevented the "supervisors" from recruiting Tolik, Sasha, and the girl as informers.

Two days passed. The girl came to see me. It turned out that she and Tolik had gone to the Special Department together. There she said: "Well, I told you that he gave me the book. That's not true. He has never even laid eyes on the book. The book was given to me by a certain person and that person ordered me to come here and name her name. And now you can cut me into little pieces if you want, but I won't say who that person was." She told me that and wept. Apparently she had wept at the Special Department too. They told her that no one was planning to cut her into little pieces. The whole affair ended right there. *Doctor Zhivago* took its place next to the abandoned nest. The roof over its hiding place became the lid of a coffin. Then it was covered over by snow. Tolik left for Bulgaria. Sasha remained with Masha and me. I realized only much later that Sasha had already been recruited and assigned to me as an informer.

A short time later I received a request to present myself at the Special Department of the Automation Institute, under whose supervision our institute was as well. I had a considerable number of sins in my recent past, and I was trembling. It turned out, however, that I was being called in to participate in the marathon of fear as an intermediate link rather than an ultimate one. The head of the Special Department asked me to sign a statement to the effect that I had read a certain important document. All the heads of institute laboratories were supposed to put their signatures on the back side of the last page of the document. Several signatures were already there. This document—not really a letter, more like a circular, but I don't really know what to call it—contained a condemnation of Agambegian, the director of the Institute of Economics. Taking advantage of his high rank as an associate member of the Academy of Sciences, he had allowed himself indecent attacks against the country that had raised him and given him all the conditions for unfettered scientific research. And it wasn't just the audiences at his many lectures, but foreign correspondents as well who extracted from him false information about the economic condition of the USSR. The essence of Agambegian's slanderous attacks was not expounded, of course, in order to avoid reorienting the target. The signature of M. V.

Keldysh, the academy president, was there on the circular. My first and very strong inclination was to write "I don't sign just any garbage" and to sign that statement. But they were only asking me to state that I had read this vileness, and I really had read it. I signed.

I had heard one of Agambegian's lectures. He spoke at our institute and was very mild and moderately critical. He spoke about slipshod planning, the criminally slow construction tempos, frozen capital investments that produced nothing but losses; he spoke about the dislocation among offices and institutions that ought to be located near each other and about the catastrophic situation with transportation. The truths that he offered were obvious ones, and the examples cited were in the nature of jokes that corresponded strictly to reality. He didn't speak of the system of labor compensation as the reason for the decline in worker productivity and production quality. He didn't mention the fact that computers, many years after their introduction into the planning process in the festering West, were considered an ideological diversion aimed against the bases of dialectical materialism.

"Mstislav Vsevolodovich [the academy president] is mistaken, and I'll write him about that," I told the head of the Special Department.

"What a good idea," he said, "and how I envy you scientists. You approach everything from scientific positions. I would write him, too, but I don't have any education."

"I don't think it's a matter of education, but rather human decency and a knowledge of the constitution," I said.

"Here's some paper. Why don't you sit down and write," he offered, "and I'll send it for you myself."

I wrote that I considered the condemnation of Agambegian a mistake: "As it is, people are frightened half to death, afraid to open their mouths even for the most moderate, constructive criticism, and now you, too, from your post as academy president have included yourself among those who through clumsy management are putting the brakes on the economic development of our country. Agambegian's lectures don't disgrace the merits of an academy member, but rather elevate them. There can be no question even of his having violated constitutional rights. The circular must be canceled."

"I'll have to get the directorship of your institute to approve sending this letter," the head of the Special Department told me. "I'll see your director tomorrow at two."

The next day at 2:00 P.M. I was sitting in the director's waiting room. The head of the Special Department, a file under his arm, went in to see the director. With a worried look, without exchanging greetings with anyone, Salganik, the deputy director, who apparently had been summoned by phone, went in, too. I was not asked to participate in the discussion.

The head of the Special Department left the director's office without noticing me. That's all that I know about my letter. I had no answer. Top officials do not answer letters.

But when forty-six employees of the institutes of the City of Science, including myself, wrote a letter defending political prisoners, among us were some economists from the institute headed by Agambegian. Agambegian had "to take measures"—that means to root out dissent from his institute, to fire and to help the KGB persecute those who have raised their voice against lawlessness. It was said that he was less cruel than other directors. Rumors about my letter defending him probably reached him, because I did not keep silent about my futile attempt.

It is embarrassing to admit one's own stupidity, but what happened, happened. In Chekhov's story "Vanya," the little boy describes in a letter to his grandfather how his master's ferocious wife beat him in the face with a herring because he had done a clumsy job of cleaning it. He then sends the letter to the nonspecific address: "To Grandfather, in the village." But at least he was smart enough not to show the letter to his master's wife.

# Judgment and Punishment

I was one of the last to sign the "Letter of the Forty-Six." This was in December 1967. That year and the one following it were marked by mass protests against governmental violations of the law. The intelligentsia, with Andrey Dmitriyevich Sakharov among its ranks, attempted to enter into a constructive dialogue with the government, appealing to it to observe the laws registered in the law codes of the country and guaranteed by its constitution. In a letter brought to me for my signature, institute fellows and university teachers protested against the absence of public trials for political crimes and demanded the reexamination in an open courtroom of a particular case against Ginzburg, Galanskov, Dobrovolsky, and Lashkova. Had I been the author, I would have expressed the protest in a more diplomatic form, but the signatures were already in place. I added my own. The text of the letter follows:

To General Procurator of the USSR, Rudenko
To the Supreme Court of the RSFSR
Copies: To Chairman of the Presidium of the Supreme Soviet Podgorny
To General Secretary of the Central Committee of the CPSU, Brezhnev
To the Lawyers Zolotukhin and Kaminskaya
To the Editorial Board of the Newspapers *Komsomolskaya Pravda* and *Izvestiya*

The absence of any coherent and full information in our newspapers about the existence and course of the trial of A. Ginzburg, Yu. Galanskov, A. Dobrovolsky, and V. Lashkova, convicted according to Article 70 of the Legal Code of the RSFSR, has disturbed us and forced us to seek information in other sources—in foreign Communist newspapers.

What we have managed to learn has caused us to doubt that this political trial was conducted in accordance with all the measures required by law, for instance, the principle of open proceedings. That alarms us.

A sense of civic duty forces us to declare in the most decisive way that we consider the holding of de facto closed political trials intolerable. We are alarmed by the fact that behind closed courtroom doors illegal acts may be carried out, and unfounded verdicts delivered on the basis of unproven accusations.

We cannot allow the mechanism of justice in our country to again escape public control and again plunge our country into an atmosphere of judicial arbitrariness and lawlessness. Therefore we insist that the Moscow City Court rescind the verdict in the case of Ginzburg, Galanskov, Dobrovolsky, and Lashkova, and we demand the reexamination of this case in conditions of full publicity and the scrupulous observance of all legal rules with the obligatory publication of all materials.

We likewise demand that the parties guilty of violating the requirement of publicity and the rules of legal procedure guaranteed by law be called to account.

*Izvestiya* and *Komsomolskaya Pravda* were the only newspapers that published information about the trial, and that information was scanty and obviously distorted. Zolotukhin and Kaminskaya were attorneys of the defendants in court.

We made demands, but at the same time we were extremely modest, because we did not protest against political reprisals, but rather about their being carried out behind closed doors. I told the authors of the letter, just as I later told Andrey Dmitriyevich Sakharov when I joined him in signing a protest against the death penalty and a request for amnesty for political prisoners, that I was adding my signature because it is better at least to do something than to do nothing. I went on to say that our action contained an element of falsehood. We were pretending that we believed that there was someone to appeal to for help, when we actually knew full well that we were complaining about one evil spirit to another.

The defense of Ginzburg and Galanskov contained another falsehood as well. We knew the true reason for their prosecution. Ginzburg and

Galanskov had betrayed the secrecy around the trial of Sinyavsky and Daniel, writers who had dared to speak the truth. Sinyavsky and Daniel did not only make use of Russia's return to the pre-Gutenberg era, to the time of minstrels and bards. Their stories and novels appeared not only in *samizdat*, but abroad as well, over there, in *tamizdat*, in the rotting, decadent West. *Tam* means "there" and allowed wags in the Soviet Union to call publishing abroad *tamizdat*—"publishing over there." The result was a perfect parallel with the half-ironic term *samizdat*—"self-publishing."

The transcript of Sinyavsky and Daniel's trial—*The White Book*—had been published in England. Their trial was open; the publication of the transcript done by Ginzburg and Galanskov could not become a corpus delicti; it could not be considered libel or the dissemination of libelous information. The fabrication of accusations against Ginzburg and Galanskov in connection with subversive Western organizations demanded time. The transcript was not even mentioned at the trial. Besides that the four accused did not form any sort of group united by a common goal. They were tried together, as though it were a single case. The inquisitors needed aggravating circumstances for delivering the verdict.

This all-powerful means of aggravating a criminal's guilt—the group trial—was invented at the time of the French Revolution. Political criminals and common criminals were prosecuted as accomplices at such trials. History or the cynicism of the falsifiers themselves—I do not know which—gave this procedure the name "amalgam." Stalin was a great tinsmith. Khrushchev did not need amalgamation; he changed the legal code in accordance with circumstances and made his new decrees retroactive. The "collective leadership" headed by Brezhnev returned to Stalinist norms.

We knew all of this and pretended that we did not. We lied in our letter, fearing that we would wind up behind bars along with those whom we were defending.

We knew the truth from many sources. One of my enlighteners was a young artist. He came to see me after reading a newspaper article that condemned us. "Do you know Ginzburg and Galanskov? You don't. Well, I do. They have nothing to do with Dobrovolsky. They were put together in order to stick them with foreign currency speculation. And have you read *The White Book?*" I had not heard a thing about *The White Book*. "It's the transcript of Sinyavsky and Daniel's trial that Ginzburg made and which has been published in England. If you have a search, they'll be sure to find it in your possession, just as they found a copy machine at Galanskov's that he didn't in fact have." If you listen, you learn.

My other enlightener was Vadim Delauney.

Before telling you about him, however, I will tell you about Alexander Galich. Neither Delauney nor Galich is now alive. I learned of the existence of the great poet I consider Galich to be from Elena Sergeyevna Ventsel, otherwise known as I. Grekova, a professor in mathematics and a writer. Grekova brought a tape recording to my Novosibirsk apartment in 1965. She asked Liza and Masha to leave the room. The girls left us, and the voice of an unknown bard whose name was supposed to remain a secret from me sang out from the tape recorder. "Mama, that's Galich!" Liza shouted from the other room.

The summer of the same year, near Moscow, on the Moscow Sea not far from Borodino, amid the wondrous Russian countryside, the Komsomol committee of the City of Moscow organized a summer school and turned over its sports camp to the many students and teachers. We, the faculty, lived in a house, and the young people lived in tents. In the depths of the night, having closed all the windows tightly, we would listen to a tape recording of Galich. In the morning, in front of the shelter where breakfast was served, a tape recorder blared. The Komsomol members along with faculty members listened to Galich. The first in line was Timofeev-Resovsky, and it was all he could do to keep from yelping from pleasure, and maybe he did yelp—I could be wrong.

Galich appeared in Akademgorodok in 1968, when our luck ran out. The Letter of the Forty-Six had been sent off, but the storm had not yet broken. Nowadays, it is difficult to believe that the Festival of Song actually took place. Bards were invited from all over the country, and twenty-seven people arrived. I bought tickets in advance for the recital at the House of Scientists, but I did not buy one for the opening. The performance by the twenty-seven singers did not entice me. But when the time came for the opening of the festival, I could not hold out. I went. I was standing in front of the glass door outside the lobby of the concert hall. Beyond the door, Galich, already with his coat off, was walking along with my friend Golenpolsky, a professor of English. Golenpolsky introduced me to Galich and bought me a ticket. "I've heard about you from Elena Ventsel. Please tell me what songs you think I could sing," Galich said to me.

"To your great honor I have to say that you couldn't sing any of the songs of yours that I know. Sing 'On the Death of Pasternak' and 'Near Narva.' "

He sang them and also "The Ballad about Surplus Value." Three-quarters of the audience gave him a standing ovation. He performed at a recital at the House of Scientists and at a banquet arranged in honor of the bards at the House of Scientists. There Delauney said that Galich had given back to poetry the quality of bread. I should have kept quiet, but they asked me to speak. I said that I was there with mixed feelings. We had been administered the sacraments of freedom. Millions of people outside the walls of the palace where we were savoring freedom were deprived of even the crumbs. We should think about them. And I also said that our feast was a feast during a plague.

When people had drunk quite a bit, an angry voice was raised. A woman with white eyes, as he later called her, yelled at a tipsy Galich that he praised Pasternak, Zoshchenko, Tsvetayeva, and Mandelshtam, but that he did not praise those people who had defended the motherland on the field of battle and had upheld socialism, and that he wrote lampoons. I entered the fray. "You are violating the rules of an honest fight," I told the woman with the white eyes. "You know your opponent's name, but he doesn't know yours, nor does anyone else here." Her rage fell on me, but it was like water off a duck's back.

At midnight we saw Galich to his hotel, and I headed home. I slept through what could have become the pearl of my biography. At two o'clock in the morning people came to get Galich, the Movie Theater was opened, students from the Akademgorodok Mathematics School filled it, and Galich sang for them.

The next day his final solo recital took place. I was standing in the lobby before the start of the recital. Galich came up to me. An anonymous caller had warned him that two busloads of workers from the agricultural implements plant in Novosibirsk had been sent to his performance in Akademgorodok. They were supplied with heavy spill-proof inkwells and were ready to throw them at him, because everyone of them considered him an "anti-Soviet kike."

"I'm not afraid," he said. "I came on my own two feet and I'll leave on them, but what will happen to the people who invited me and organized the festival?"

"I'm an old woman," I told him, "but if anyone in my presence allows himself a vocal attack, not to mention throwing an inkwell at you, I'll slug him in the face without the slightest hesitation. And ninety percent of the people in the audience will be like me, only they'll be young and strong. Those champions of the Soviet regime do their business in the boss's waiting room or during searches and arrests, with weapons in their hands. None of them will so much as make a peep." And that was exactly how it turned out.

Galich sang at my apartment. He told me that he had been called in to the Special Section of the academy Presidium, where they had conducted an educational session with him, reproaching him for anti-Soviet propaganda. He defended himself by saying that he was a satirist, a poet. The festival was the end of Galich's career. The publication of a book of his songs was canceled. He was expelled from the Union of Writers and from the Union of Film Workers. He did not even manage to remain a member of Litfund (*Literaturny fond*—The Literary Fund) as Pasternak had been lucky enough to do. Galich had sung of Pasternak: "And he's not a stepchild of fate; he's a member of Litfund, a deceased mortal." Galich was making fun of the newspaper announcement of Pasternak's death, in which Pasternak had been called neither a poet, nor a member of the Union of Writers, but a member of Litfund—a feed trough for the meekest of the meek which had continued to offer its services to him after his expulsion from the union.

Just before the "stepchild" Galich was forced to emigrate, I saw him in Leningrad and in Moscow. He told me about the meeting where they "worked him over" with the aim of expelling him from the Union of Writers. His disappointment in his fellow writers was boundless. I told him that every cloud has its silver lining: They were acting out a magnificent play for him. One of them, outraged, had said: "It would have been all right if he had sung his ravings for himself at home. But no! One public performance after another! His own difference of opinion with the Soviet regime isn't enough for him. He has to bring others over to his anti-Soviet views. And it's precisely anti-Soviet propaganda that our collective is condemning." Another person, a latecomer, said, "If you have an opinion of your own, if you disagree with your native country's system of government, you should say so openly and honestly. But Galich disseminated the fruits of his creativity secretly, hiding in corners and in dark alleys. And for that he deserves our condemnation as well as that of the whole of society." When the sick and exhausted Galich told me about that meeting, he was in no mood for laughter.

Vadim Delauney came from a remarkable family. His great-grandfather's great-grandfather was the commandant of the Bastille. On the day it was stormed, he was thrown down onto the insurgents' bayonets. His son, a doctor in Napoleon's army, wound up a prisoner of war in Russia, married a noblewoman, and founded a glorious dynasty. His great-grandsons, brothers, were the geometrist Boris Nikolayevich Delauney, a member of the academy, and the famous cytogeneticist Lev Nikolayevich Delauney. His great-granddaughter, their cousin, was Elizaveta Yuriyevna Pilenko, a gifted artist and poet who emigrated to France and became a nun. The German occupation caught her in Paris. She—Mother Maria—and her son Yury saved Jews by obtaining baptismal documents for them. They were arrested in 1945 and perished in the death camps. During the last selection Mother Maria was not among the doomed, but she exchanged places with a woman who was being sent to be slaughtered. The mathematician Boris Nikolayevich Delauney, A. D. Alexandrov's teacher, was Vadim's grandfather.

Before I met Vadim, he had spent a year in Lefortovo Prison after being seized during the demonstration against the arrest of Ginzburg. Alexandrov, who by then had already lost his position as rector of Leningrad State University, took Delauney to his lovely Novosibirsk cottage and arranged for him to study at Novosibirsk University. The stay at the university was a brief respite in Delauney's prison life. The fall of that same year, 1968, Delauney was a participant in the protest demonstration against the invasion of Czechslovakia and was sentenced to three years in the camps.

After leaving camp, he happily married Ira Belogorodskaya, just as courageous a defender of human rights as he himself was. When she was arrested, KGB agents made emigration a condition of her release from prison and threatened to send both of them to a camp in Siberia if they didn't emigrate. Vadim and Ira wound up in Paris. We saw each other many times. Vadim was in despair over the fact that he had been forced to emigrate. His longing for his homeland and for his friends did not abate. On June 13, 1983, he died suddenly, in his sleep. He was thirty-six. His memoirs and a book of his poetry were published posthumously. Delauney entitled his memoirs *Portraits in a Barbed Wire Frame* (London, 1984). I would give the volume another name as well: "A Book Not About Myself." On the pages of this sadly short book, Delauney tells not only about the fate of unfortunate fellow prisoners, but about the fate of all of Russia, which as a result of a tragic convergence of circumstances wound up behind barbed wire. One hears the voices not only of prisoners but also of guards, those people to whom falls the hard lot of apprentice executioners. Kindness and nobility themselves guided the brush of the master who pictured these portraits on the canvas. To be behind the barbed wire is painful and unbearable for everyone, but there's not even a hint of despondency in the book. The contrast between the frank swearing, the camp jargon, and the refined language splashing up to the surface of a sea of all-round savagery, the spray of world culture, is extraordinarily exhilarating. Chivalric valor can be heard in every passage of the saga of hopelessness. Comic elements are increased by the interspersed examples of high-flown official banality. That's how things really are, and that's how one should write about them. The essence has been captured, and, moreover, it is funny. The main characteristic of the author of this striking book is an aristocratic quality—the presumption of innocence of the people with whom he had occasion to come into contact, the projection of his own unconscious noble qualities onto everyone, regardless of rank and station. Contact with Delauney forced people to rise above their own interests and take the side of the insulted and injured. That was how things were in Akademgorodok in Siberia, and that's how things were in Vadim's depiction of camp, again in Siberia. The book *Portraits in a Barbed Wire Frame* has been published in French. The French edition is entitled *Pour Cinq Minutes de Liberté*, referring to Vadim's five minutes on Red Square, protesting against the invasion of Czechoslovakia by Soviet forces. The words "for five minutes of freedom one doesn't mind paying with years of camp" were spoken by Vadim at his trial.

While living in Akademgorodok, he did not lose his connections with the people together with whom he had protested against Ginzburg's arrest, and was well informed about freedom-loving actions among young and old. Were it not for him, the Letter of the Forty-Six might never have been written at all.

On March 27, 1968, the letter was read over the Voice of America. The names of all those who signed, along with their positions and institutes, were listed. According to rumors, the *New York Times* published the letter. The next day we found out that we would be "worked over." We had written that we feared a return to the total lawlessness of 1937. Now we were going to learn just how well-founded our alarm was.

All seven copies of the letter had been sent by registered mail, with return notification of receipt. The return address on all of them was mine, and I received all seven notifications of receipt. A young man, one of my co-signers, told me: "Mozhin, the secretary of the Regional Committee, claims that the letter was not sent anywhere except to the CIA. In his opinion, the addresses listed in the letter are a camouflage for a libelous counterfeit." I had never seen the young man before, nor had I seen the majority of those who had signed. It was decided to make photocopies of the notifications of receipt and deliver them to Mozhin and to the people who were in for a "working over." The public "executions" were to take place simultaneously at all the institutes represented by signatures on our letter. Some institute directors were away on vacations, business trips, or conferences, and people had to wait for them. Some directors were in no hurry to punish. But Belayev was in a rush. His golden opportunity had arrived.

In answer to his announcement that I was about to be purged, I told Belayev, "I won't come to the meeting unless the announcement and agenda are put up in the institute lobby." The next day I received an agenda with an invitation to a closed meeting of the Science Council. The duties of my position required me to be there. Entry into the institute was permitted only by passes. The watchman checking passes in the lobby was instructed not to admit university students on the day the Science Council was to have its closed meeting. The reason? A reason can always be found. It had been prepared ahead of time, one and the same for stores, restaurants, research institutes—a cleaning day. It sounded ambiguous. Belayev was afraid of revolt.

I had once consoled Galich that in expelling him from the Union of Writers, his fellow writers had performed a magnificent play for him. Now another wonderful play—that very Science Council meeting—was being devoted to my humble person. The play was perfect! I recorded a word-for-word transcript performance. It is included as an appendix to this book, so that you can appreciate all its subtlety and drama.

I see my play on the stage of a theater. There are two levels. Up above are those who run things; below, the Science Council is in session. Strings run from those who are above down to the marionettes. The rulers pull the strings, signaling that it's time for a performance. Belayev has a double. He chairs the meeting, and he is also there, up above. Only two characters lack strings. The defendant is obstinate, and with mad daring Zoya Sofronyevna Nikoro stands up for the right to have one's own opinion. She is the only one to do so. In all other respects everything is running along smoothly. Raushenbakh, an irreproachable model of Soviet man, performs his role irreproachably. Minor details do not count. He lies too coarsely, but that is a minor detail. Antipova speaks to no effect about some secret information that she has received during her stay in Germany in recent years. Is she hinting at the bloody incidents? Does she mean the workers' uprising in East Germany of June 17, 1953, when thousands were arrested for attempting to cast off the yoke of Soviet communism? (Unlike the events in Hungary, few people knew about that uprising. I did not learn of it until I landed here, in the West.) Are not Khristolyubova's remarks, her mentioning *The White Book*, a miscalculation? The people attending the meeting have to pretend not to know anything about the trial of Sinyavsky and Daniel, nor about the transcript of the trial, nor about the people who informed the world about the trial. The newspaper articles had not attracted their attention. They learned of Ginzburg's and Galanskov's existence for the first time from Belayev's and from secretary of the institute party committee Monastyrsky's introductory remarks. And the deputy director of the institute, Salganik, gives himself away completely by sharing his secret turmoil with those present: Unheard-of audacity! He not only listens to the Voice of America but even dares openly to admit doing so. These are all minor details that are incapable of destroying the performance's divine harmony. Only one remark introduced dissonance. Nikolay Nikolayevich Vorontsov said that the letter could have landed in the West via the editorial office of *Komsomolskaya Pravda*. Vorontsov had his own reasons for casting a shadow on *Komsomolskaya Pravda*. Some years before, this newspaper had launched an ideological fight against basic research. Vorontsov's brilliant work on the evolution of the digestive system of rodents was ridiculed. The moment Vorontsov had chosen to avenge his humiliation was a most inappropriate one.

Let us, however, turn our attention to the most dramatic moment in the show. The marionettes get out of control. Not all of them, but the monotonous string pulling was immediately replaced by feverish activity. They refused to vote. In their humiliating position they were clutching at the right to silence, the right to their own individual opinion, even if it diverged only a bit from the one dictated from above. They did not want to vote "aye," but were afraid to vote "nay." Would it not be better to limit themselves to a vote of censure? Belayev did not lose his wits. He put his cards on the table. He spoke of a crime punishable by law and carrying a sentence of seven years in the camps, with subsequent disfranchisement. The subtext of his harangue was the following: An underground organization has been discovered. Its aim was the transmittal of libelous information to subversive forces in the West. Civic duty obliges us to hand over the criminal to the organs of punishment so the criminal may be dealt with according to the law. I move that we vote for a resolution condemning the irresponsibility that expressed itself in the signing of the letter. I am doing this with the sole aim of saving the criminal. Everyone voted. Everyone, except Nikoro, was for censure.

A vote in which the freest, best-informed, most virtuous people participate has just as many defects as it has virtues, if not more. The agreement of the majority is not a criterion of truth. In the ideal, the resolution of every question is a matter for authorities, the result of qualified expertise. Bring into balance an analysis of everyone's needs, regardless of number, and the suggestions of experts who know how to get things done, and you will have an ideal democracy. A vote by compulsion with a predetermined outcome, however, is a profanation of democracy.

Both the text and subtext of Belayev's statement, with the help of which he forced the obstinate members to vote, contained nothing but falsehood. The play had been run through earlier by the people on the upper level. At the conclusion of the play I was to go behind bars. Belayev had handed me over to the organs of punishment so that they could deal with me.

Events followed one another at a head-spinning tempo. It became known that the Presidium of the Siberian Division of the academy passed a resolution to transfer the Humanities Division of Novosibirsk University to Krasnoyarsk, along with all the teachers and students. The Humanities Division was considered the main breeding ground of sedition. The resolution—which everyone knew had absolutely no force—certified that the disease had passed from a subclinical phase to a clinical one. Hearing of the catastrophe that threatened, students at the university gathered signatures on a petition asking that the teachers be exiled but that the students be left to study in Akademgorodok. The regional committee of the Komsomol cut short the gathering of signatures. An assistant professor of philosophy at the university, after having been "worked over" for irresponsibility expressing itself in such-and-such or something-or-other, was about to be fired. The rector of the university rushed to leave before the end of the meeting, asked that his support of the most severe punitive measures be taken into account, opened the door part way, closed it again, came back, and persuaded the members of the council not to take any action against Alexeyev and to apply the breaks to the case. Outside the door were students ready to defend their favorite teacher to the hilt.

The party leadership of Akademogordok began firing teachers selectively, and not all in a row. They were especially severe in dealing with the teachers of the Physics and Mathematics School. But the Presidium of the Siberian Branch of the academy rejected its own plans for transferring the Humanities Division to Krasnoyarsk.

The slander against us on the part of the District Party Committee differed from Belayev's only in certain details. Belayev had asserted that an underground organization had been uncovered. The secretary of the District Committee explained to his audience that we had not needed either to organize ourselves into an underground group or to write the letter. The CIA had assembled morally lax residents of Akademgorodok and proposed that they sign a blank sheet of paper. "And whatever libel I write is my business," the CIA agent had told us, or so went the version described by the Regional Party Committee. The secretary described it precisely in that way—the CIA agent himself called his future scribblings libel, although don't you think, dear reader, that it would be more likely for him to have called it the truth? And after all, the story of our fall was being recounted to full members of the Academy of Sciences, not to illiterate peasants. According to the Regional Committee secretary's version, we signed that blank sheet with servile alacrity. All we cared about was libeling, but what sort of libel would best suit the CIA was for the CIA to decide.

In order to keep open the possibility of at least somehow defending the wretched signatories, Alexander Danilovich Alexandrov, the former rector of Leningrad State University and a member of the CPSU, my friend and protector, a student of Delauney's grandfather and the guardian of Delauney's grandson, published an article condemning our letter. That article obliged him to be silent when at a meeting of the Presidium the secretary of the Regional Committee laid out the details of the preparation of the counterfeit document that allegedly was printed in the *New York Times* and later was heard on the airwaves of the Voice of America. But apparently his heart won out. Alexandrov interrupted the secretary in midsentence. "You know very well," he said, "that Raissa Lvovna Berg received notification of delivery of all seven letters. Photocopies of the notifications have been transmitted to the Regional Committee."

My days in Akademgorodok were numbered. The five-year term for my temporary residence permit in Akademgorodok had run out. In such cases an institute petitions the regional police to extend the permit. Deputy director Privalov, a member of the CPSU whom the institute maintained as a torturer, and at a very high salary, informed me that the institute would not take any measures to have my permit extended. That meant that Belayev had been ordered to hand me over to the Security Police for punishment, and in order to do that all he had to do was have Privalov deny me an extension of my permit and to deny it temporarily, and then he, Belayev, could not be held to blame at all.

I left for a conference on medical genetics in Leningrad. Quite by accident I met Kalinin, the same Oleg Mikhaylovich Kalinin who had participated in my Kamchatka expedition. Our friendship had not been interrupted even for a moment. When he learned of my misadventures, he ordered me not to lose heart. He worked at the Agrophysics Institute and would immediately inform the director that a specialist in population genetics was planning to move from Novosibirsk to Leningrad. The Laboratory of Population Genetics of the institute was interested in an experimentalist. The director called me the next day and invited me to come see him. I presented myself to him and his deputy and agreed to organize the experimental work in the laboratory headed by the deputy director. Everything was shaping up nicely. My dissidence, however, an underwater rock, could sink the ship. "Aren't you afraid to hire such a seditious person as I?" I asked. "But we're all seditious characters," the director exclaimed. Later I realized that we were talking about different things. The director had genetics in mind. His institute was under the jurisdiction of Lobanov, the ignominious chair of that same August 1948 Session at which genetics in the USSR was brought to a nationwide end. As far as Lobanov was concerned, even twenty years after the session, the director, his deputy, and I were all seditious characters. They did not know the extent of my dissidence, and I did not know that they did not know. The invitation was confirmed.

Back in Akademgorodok, no sooner had I crossed the threshold of my apartment than Masha and Sasha handed me an unsealed envelope. They had taken it out of the mailbox. There was no address on it, just initials and the last name. It was obvious from the look on their faces that they knew what was in the envelope and that it was bad news.

An official document in the kingdom of bureaucracy has a strictly prescribed format. It is printed on a special form, has a seal, a signature or several signatures, and is dated. If the organization does not have typographically reproduced forms, the organization's stamp is used. The document that I received had nothing on it but the signature of the chief of police. No stamp, no seal, no date. Underneath the typed text stood the signature—Likholyetov—written with an oblique flourish. There is a Russian saying "God marks a rogue." By a whim of fate, the last name of the chief of police fits into the folk formula laughably well. *Likholetye* means "hard times."

Here is the full text of the document:

Copy: To Roza Lvovna Berg

To the Chief of Police of the City of Leningrad
You are requested to cancel the authorization for the living space occupied by Roza Lvovna Berg, since she has a permanent residency permit in Novosibirsk.

I was referred to as Roza, and my Leningrad address was not indicated. And thank heaven for that. While they were tracking down the unfortunate Roza in order to throw her into prison, I would have time to cancel my permit from Novosibirsk, return to Leningrad, and move into my apartment. Then no one would be able to take my residency permit away from me and I would not be put in prison.

"But what does prison have to do with anything?" the bewildered reader will exclaim. It was a bad business. In taking away my residency permit in Novosibirsk, the Novosibirsk police had decided to prevent my return to Leningrad. The possession of two permanent residency permits is a crime covered if not by the Criminal Code then by the even more threatening administrative regulations that turn the violator over to the hands of the KGB and the police without benefit of trial or investigation. The lack of a residency permit is no less serious a crime than a double residency permit, and is punishable, again, according to administrative regulations, by prison. The goal of the action was clear. The work was clumsy, however.

Please note that the document indicated that I had a permanent residency permit in Novosibirsk. I had never had a permanent permit. I had a temporary one, which was compatible with a permanent permit in Leningrad. Now I did not even have a temporary one. The score for a concerto for three instruments—the chief of police, Belayev, and myself—had been written by the KGB. Neither rehearsals nor a conductor were needed. One must know how to prepare scores! There was a conductor, of course, off in the wings. The overture consisted of two parts—first, the refusal to petition for an extension of my permit and second, the document I had just received. No even slightly sensible person would give up a Leningrad permit of his or her own will, even if in return he or she received piles of gold outside its city limits. Moscow and Kiev can entice, but Akademgorodok, in the vicinity of Novosibirsk, not at all. According to their plan, once I received the document, I would head for Leningrad. But the document would already have done its work. Three days would go by, and I would be living without a permit. That would mean a fine. Another three days, and still no permit. Another fine. Yet another three days, and then the finale: a jail term for violation of Administrative rules.

That is what would have happened had the composer been a VIP of Soviet culture. But the composer was a genius. And thanks to collusion with the KGB, the conductor was Belayev himself. In driving me away in the direction of prison, Belayev was playing up to the authorities in the most effective way imaginable. The likelihood of the party bosses supporting his candidacy for election as a full member of the academy was increasing. But he needed to reckon as well with the fact of a plenary meeting of academicians where there would be a vote, moreover, by secret ballot, to decide whether an animal breeder holding the Candidate of Sciences degree who had never made the slightest contribution to the storehouse of knowledge was to be a member of the academy or not. The brilliant diplomat realized that in sending to prison the daughter of a famous scientist, a member of two academy branches, the renowned and respected Berg, a woman who had managed to acquire a certain recognition in scientific circles, he was risking losing the favor of the people on whom to a certain extent the fulfillment of his ambitious plans depended. He needed to have me put in prison by other people's hands—and not by firing me from the institute, not for political reasons, but for the violation of administrative regulations about residency permits. Don't forget that there are no political criminals in the USSR. Everyone is a potential common criminal. The time had come for me to become one.

I received an invitation to report to the office of Deputy Director Privalov. The time had come to rescind the earlier decision not to grant me a petition for a temporary residency permit in Akademgorodok. In combination with the document sent to the chief of police of the city of Leningrad, the institute's petition did not change matters. They could claim they wanted to keep me; that they had not fired me, and that they had taken measures in my behalf. I had resigned on my own, left for Leningrad, and there I was arrested, it is difficult to say for what, but most likely for links with subversive Western organizations. At that point Belayev could say he had absolutely nothing to do with it.

With the document signed by Likholyetov and with the petition, I set off for the police headquarters. I went to a secretary and showed her the document. "Could you please tell me," I asked, "when this document was sent?"

"But that's not a document. It doesn't have a seal or a date or a stamp."

"It's a copy, and I would imagine that there were a stamp and seal on the original."

"A copy of a document is the same as the document," the secretary explained to me. "Likholyetov is busy. Go talk to the deputy chief."

"Excuse me," I said to the deputy in a meek voice, "a mistake has been made. The institute is petitioning for a temporary permit, not a permanent one. I would like to ask you to issue me a temporary permit and to have the chief of police of Leningrad send off a document with a notification of the error."

The greatest amazement was portrayed on the deputy's face. "They issued a petition," he whispered. He lifted the telephone: "Roza Lvovna Berg is in my office. She is outraged! I'm waiting for instructions." He did not even have time to finish speaking before I said, in the same meek voice, "Not only am I not enraged, but I'm even flattered. I'm only asking for a temporary permit, and you're guaranteeing me a permanent one." He did not have time to answer. Likholyetov turned up, accompanied by the procurator. It was almost as if one-and-a-half people entered the room. No, there were two of them. Anything that one lacked the other had. The procurator was extremely thin, while Likholyetov was extremely heavy. A black shirt with a strap emphasized the procurator's gauntness, while a police uniform made the chief even more massive. In addition to his resemblance to half a person, the procurator turned out to be one-armed. One sleeve, empty from the shoulder down, was tucked under his belt. All Likholyetov needed was a third arm in order to pass for the person and a half in this phantasmagorical pair.

I repeated my requests. As far as I remember, the person and a half did not say a word. The half a person yelled: "You have already broken the law! You're living without a permit. These learned people imagine that laws are written for other people. You have been living without a permit for longer than the prescribed seventy-two hours. We'll teach you to respect the law!"

"In the situation that has been created the person who should be upset is me, not you," I said. "And the situation has been created through the institute administration's fault. There was a delay in issuing me the petition." And I repeated my requests.

The procurator suddenly calmed down. Apparently he had been yelling in order to make me lose control of myself, and then he could have added a charge of insulting a representative of the Soviet legal process. "We are denying you a permit. And we'll recall the document that was sent to Leningrad, but only on one condition—that you give a written promise to leave Novosibirsk within a week."

"But how can I leave Novosibirsk within a week? I'm giving a course at the university. I'm a professor."

"You're no professor. Resign and leave." There was no need to argue. "Here, on the back of this document, write: 'I promise to leave Novosibirsk within a week.' " The document was the same copy that had Likholyetov's flourish on it. I gave in. Likholyetov took the paper, pounded his hand against the wall, a door covered over by wallpaper opened up, a camouflaged little window was revealed, and the paper flew off to a room that seemed to be separated from the chief's office by a thick wall. I watched with great sadness as that paper, a superb model of bureaucratic concoction, a pearl in a historiographer's archive, disappeared into the depths of the bureaucratic cesspool. If only I could have kept it! But that was not to be. The score had been written by a genius, but the performers were not geniuses, since the homeless Roza was suffering in Leningrad instead of me, but they had made no blunder here.

When I turned in my resignation at the institute, Belayev acted out unfeigned surprise. Without theatrical artistry Belayev could be neither the director nor a full member of the academy.

I do not think that anyone even considered recalling the document signed by Likholyetov. In Leningrad the police looked for me and found me. I was told about that by the apartment management passport clerk later, when I was already living in my Leningrad apartment and working at the Agrophysics Institute. They did not take my permit away. The Novosibirsk permit had been annulled. There was a stamp to that effect in my passport, and that was the end of the matter.

In Novosibirsk, meanwhile, steps were taken. I received an administrative penalty for living without a residency permit. I was fined. I was supposed to report to police headquarters, where an administrative committee would conduct an educational session and assess the fine. I told the woman who came to inform me of the date of this new trial that I would not go to see the administrative committee. "I have given all my explanations to the chief of police and the procurator. The delay was not my fault. I am leaving by order of the police. There is nothing to fine me for. Drunkards and prostitutes are fined, but I'm a professor, I am not guilty of disorderly conduct. I have two daughters and a granddaughter. Let me write a note that the misunderstanding has been cleared up. Otherwise you'll get into trouble. They'll accuse you of not informing me."

The woman started crying and said that she also had children and that she had to work and carry out all sorts of horrible assignments, and that I should not be angry at her, because she had nothing to do with it. The committee met in my absence, and the fine was deducted from my final pay at the institute Accounting Department. The bookkeeper, comrade Monakhova, a bribe taker and flunky, eventually gave in to my insistent demand that she issue me a document stating that I had been fined for "residence" without a permit. She resisted with the resoluteness of despair, as well she should have. After all, I had really been fined for "residence" in Novosibirsk without a permit by the same police who informed Leningrad that I had a permit for Novosibirsk, and moreover, a permanent one. The document issued by the bookkeeper annulled that other wise, sacramental one!

While the police were turning themselves inside out trying to find the unfortunate Roza Lvovna in order to strip her of her Leningrad residency permit, my life in Novosibirsk was going along normally. I finished my course on the theory of evolution.

There were seismic shocks, however. Once I arrived to give one of my regular lectures, and instead of the usual forty or fifty students there were only five or six in the lecture hall. "What does this mean?" I asked those present. The class was taking an English exam, I was told. They would be coming. Twenty minutes passed. The number of students in the hall had reached seven or eight. I went up to the rostrum. "My dear friends," I said to those seven or eight students, "I have no grudge against you. You are here. I have no intention of hurling thunder and lightning at those who are present because of those who are not. I ask you to tell the people who did not come to the lecture that I regard their absence as a demonstration of solidarity with the authors of the petition to expel the faculty defenders of human rights. I will not be giving any more lectures." I left the lecture hall. One of the students chased after me. "This is a misunderstanding. They didn't even know about the existence of the petition. They would never ever boycott your lectures," he told me.

"We'll see," I said. "I lecture for the students in the evening division and will continue to do so, and every one of them will come to my lectures. If they wish, students from the day division may join those from the evening division." The student promised to pass the word along.

I liked the students in the evening division. It was from them that I heard the unforgettable words of reproach addressed to scientists:

"Why aren't geneticists fighting for freedom of enquiry? After all, writers fought. Some are even in prison. That means that it's possible to fight." My heart belonged entirely to the people from whom I heard those words. The students at an evening division are exhausted toilers, the progeny of people just as hopelessly poor as themselves. The children of the party elite do not enroll in evening classes.

When I arrived at the tiny lecture hall to teach my evening students, I found an announcement on the door: The lecture had been moved to room such-and-such. That huge hall was full of people, more than there were room for.

Around this time I was informed in writing that I had been excluded from the delegation to the Twelfth International Genetics Conference in Tokyo. But the International Entomological Congress in Moscow was coming up, and I was preparing for it. In order to improve my English, I went on a cruise along the river Ob from Novosibirsk to Surgut and back. The trip was organized by Golenpolsky. We studied English by the immersion method. There were thirty-six students—thirty-five from Akademgorodok, and an engineer from the Novosibirsk agricultural implements plant. This engineer told me that a condemnation of the people who had transmitted libelous information abroad was organized. At first they wanted to involve workers in the resolution of condemnation. But then they decided to limit it to engineers, two hundred people who could not at all understand what precisely had taken place, and besides, were not much interested. They did not participate in the vote. They did not abstain; they simply ignored it. The record of the meeting, however, stated that the resolution condemning the subversive activity of an underground group of members of the intelligentsia was accepted unanimously. And people reacted to that lie indifferently. Indeed, the engineer told the story listlessly. He never suspected that he was talking to one of the subversives.

The cruise ship on which our immersion in English took place under the wise leadership of Golenpolsky was called *The Mikhail Kalinin*, and Golenpolsky was called "Galyunpolsky." The nickname was extremely malicious and terribly appropriate for the occasion. *Galyun* is the nautical term for a toilet, and *polsky* means "Polish," and both are included in the name Golenpolsky. He no doubt made that mockery up himself. It was very like him to make up such a thing about himself.

At Kolpashevo I decided to take a stroll around town—it was a regional center, after all. The ship would be at the dock for an hour and a half. I walked along the wooden planking past little houses and looked around. Women wearing tarpaulin boots, scarves tied under their chins in the ancient Russian manner, and extremely modish Italian all-weather coats were hauling milk cans on their shoulders on yokes. The women had just gotten out of an enormous rowboat. They had gone to the other side of the river Ob to milk cows.

I stopped for a moment at a large window. A kangaroo with a baby kangaroo in her pouch looked out from the window. A wildly flourishing pink geranium pressed its clusters against the window, providing a beautiful decoration for the toy. A young, well-dressed woman came out of the house. "Please come in. We have a guest who knows you." I found myself in a room where at a table set for a feast sat a single woman—an old lady. On the wall was an advertisement for a Georgian song and dance troupe. In it, a huge Georgian, nearly life-sized, or so it seemed to me (the old woman was small) was making a devil-may-care leap, soaring in the air. "We decided to trick you. No one knows you, but we'd like to treat you. What will you have—vodka, beer, home brew?" The young woman, the old woman's niece, had come from Novosibirsk to visit her aunt. For that reason they were having a feast. The radishes with sour cream to which they treated me could not have been better. They made me promise to look in on them on the return trip. I did. Nothing remained of the merriment except for the single Georgian dancing his unending *lezginka* in the sky. The old woman was sitting at the table in the same spot where she had been before. Before her lay an unopened letter. The old woman displayed no joy upon my arrival.

We opened the letter right away, and I read it to her. Her daughter was writing from a prison hospital. Her feet were swollen, and she had neither shoes nor money with which to buy shoes. "Mama, help!" Her mother told me that her daughter had been a stewardess, and that wicked people, her drinking companions, had corrupted her. The daughter's son was in an orphanage and he visited his grandmother, but she had nothing to give him. Fortunately, I had money with me. I took down a letter that the woman dictated to me and then hurried to the wharf. There was no trace of *The Mikhail Kalinin*. It was getting dark. I went to the rescue station. In a motorboat we could catch *The Mikhail Kalinin* in no time at all. But the station chief was absolutely drunk. There was no gasoline. One needed a key to untether the boat from the mooring post, but another employee had it. We went to find the employee. He had the key, and gasoline could be found. The employee, however, by his own admission, was also so drunk we decided it would be better not to set out. They walked me back to the wharf, and we found out that the next morning there would be a hydrofoil that I could take to Magochin or Molchanov, where I could wait for my ship. That meant I had in store a night on the wharf waiting for the hydrofoil. The plan was reliable. A hydrofoil does seventy-five kilometers an hour; *The Mikhail Kalinin*, going against the current, could manage fifteen at best. It took no great effort to calculate where the hydrofoil would overtake it and what station I needed a ticket for.

The wharf was swarming with people. There was a huge couch in the station but it was forbidden to sit on it. A woman tried to put her twin boys to bed on it. A guard chased her away. I did not witness that scene, but the mother told me about it. While I listened to the story the twins turned away, covered their eyes with their hands, and started crying. They acted identically, as if obeying an order. Gypsies had put their feather beds on the station floor and had settled down in comfort. There were hosts of them. A short, stocky red-haired young man of healthy appearance turned up beside me. "Go to the hotel," he told me, "there'll be a place for you." I went and was given a place: a bed in a room where another bed stood empty. I carried the mattress and pillows out into the hall so that the twins could sleep on the floor there. No one offered any interference.

In the morning I went to the ticket office and asked for a ticket for Molchanov. Nothing of the sort. I knew exactly what was going on. Everyone knew I desperately needed a ticket, so I was supposed to give money equal to the cost of the ticket to the ticket clerk or the station director. He would put the money in his pocket and "arrange" a seat for me on the hydrofoil, and I would not need any ticket. But I was afraid that they would take the money and then not let me onto the boat without a ticket. I asked them to give me a ticket and told them that I did not need a seat. I could stand on the deck. At that point the stocky red-haired fellow materialized again. He brought the station director. They sold me a ticket, exactly according to the rules, and I joined the flood of hydrofoil passengers.

I did not try to claim my rightful seat. I stood right at the side and intently watched the foam and spray that the engine heaved up toward the morning sky and gulls. Beside me I found the same stocky, red-haired fellow. "Your seat is empty," he told me. "Let's go." I left the raging foam with regret and took a comfortable seat in the glass-enclosed lounge. The red-haired fellow turned out to be my neighbor. And at that point I heard a sacramental sentence that thoroughly explained the stranger's conduct: "I told them who I was." But he did not tell *me* who he was. That came out between the lines, bursting to the surface like a confession, like repentance, like a farewell to his past. He said that he had graduated from a sports institute and had worked as an instructor for some sports organization. Now he was fed up with his present job and wanted to enroll in the law school at Leningrad State University and become a lawyer.

It was not difficult to understand that he had been assigned to me as a spy and was compelled to endure all my scrapes and misfortunes with me. What it was precisely that caused him to strip off his disguise and confess to me that he was sick of his position interested me very little.

"Did you really not notice me on *The Mikhail Kalinin*?" he asked.

"No," I said, "I didn't."

"I'll take this hydrofoil back to Novosibirsk," he said, "and when *The Kalinin* docks at Novosibirsk, I'll meet you." I did not see him at the wharf in Novosibirsk, but he came to visit me both in Novosibirsk and in Leningrad. He did not manage to become a lawyer—he was none too bright. He married a Leningrad woman, received the right to live in Leningrad, and worked as a waiter at a restaurant.

The question arises as to why the people at the top needed this noisy and excessive campaign against those who had signed the petition. We did not hide ourselves. The letter was read on the Voice of America. It would have been simple to have forty-six investigators summon the forty-six people immediately and simultaneously interrogate them and collate the evidence. The initiators, intentions, and connections would have been brought to light without a mistake. But none of that was done. The persecution of members of the intelligentsia is a goal in and of itself. The government needs to bait the intelligentsia with workers. Boris Lvovich Astaurov predicted that I would become a victim. But neither I nor any other of the victims had occasion to experience the wrath of the workers. The workers at that time were under suspicion together with the intelligentsia.

The explanation for the hysterical actions of which we became victims was the "Prague Spring," a spring in Moscow, Leningrad, Akademgorodok. "Tanks on alien soil" (That is a line from one of Galich's poems:

"Citizens, the Fatherland is in danger. Our tanks are on alien soil!") and "workings over," persecutions, false accusations, and fictitious trials on one's native soil are phenomena of the same order.

The Czechoslovakian events broke out while I was still in Akademgorodok. I was on my way out to order shipping crates. There was a knock at the door. It was a young man whom I hardly knew and who was beside himself. "Please hide me as quickly as possible! They have the dogs out after me!"

"What happened?" I asked.

"Just take a look out your window, and you'll see what's happened. We've covered your building and many others with slogans that read 'Freedom for Socialist Czechoslovakia!,' 'Invaders, Hands Off Czechoslovakia!,' 'Freedom for the People of Fraternal Czechoslovakia!' Soldiers are painting over the signs now, and the police are tracking us down with dogs. They'll come for me any minute."

"Undress," I said. "I'll put you to bed. If they come, I'll tell them you came here late in the afternoon, that you complained of a headache, and that now you're sleeping. You had a fever. Now your temperature seems to have gone down." I wrapped up his clothes and put them in the closet where I kept paint. I put a thermometer and some pills next to him. No one came. "I need to go to Novosibirsk to order shipping crates. Get up," I said, "get dressed, and let's go." I gave him some of Sasha's clothes. We left the building. The yard was swarming with soldiers and police. There were no dogs. "Don't look to the sides. Smile, and don't hurry." A bus was pulling up to the stop across from the building. It did not go from Akademgorodok to Novosibirsk, but only a few stops to the end of the line at Pearl Street. "Now we run," I said. "We run and get on the bus." At the last stop I told him, "Go around Akademgorodok through the woods, get on a bus outside of town, and go to Novosibirsk." Neither he nor any other of the people who had painted up Novosibirsk were caught. It was only later that they tracked down every last one of them. Someone had denounced them.

While I was hiding the young man's clothing in the place where I kept paint, I did not notice that the paint had disappeared. The young freedom-lovers had stolen it from me and painted up the walls of the building where I lived. Hiding the clothes in a closet that reeked of paint, which seemed to me the height of ingenuity, was sheer stupidity. They let the dogs sniff the slogans and then they went looking for the paint. The dogs led a policeman to the apartment of Slava Voronin, an artist who had the same sort of paint that I did. He had absolutely nothing to do with the wall writing and, thank God, was above suspicion. If a policeman with a dog had shown up at my apartment, the sham sick person's clothes would have been discovered immediately, with all the consequences that would result therefrom. We had set a trap for ourselves and baited it, but luck was on our side.

At the institutes, meetings were held to show solidarity with the people responsible for our tanks being on alien soil. Attendance was mandatory. It was not enough to attend; everyone had to express solidarity.

Silence turned out to be under prohibition. I did not attend a single meeting. I was fired.

# *"On Venus, Oh, on Venus . . ."*

And in Leningrad, oh, in Leningrad life immediately went back to normal. The "oh" came to me from a poem by Nikolay Gumilyov, one of the people who saved my expedition back in 1938, whose verses I recited to myself at night. "On Venus, oh, on Venus . . ."

I awoke amid the hideous devastation of my Leningrad apartment, full of concentrated bliss. I felt a tightness in my chest. What could the cause be? Ah—I was in Leningrad. My surroundings would have been suitable for the sets for Gorky's play *The Lower Depths*. In my absence my rooms had been inhabited by my school friend Zhenka and his new wife. Zhenka had been the husband of another school friend, Lidia Mikhaylovna. He had wanted to add more residents to his two-room apartment—to move into one room with his new wife, and offer his former wife and their son the other. Lidia Mikhaylovna, a woman of great emotional civility, had not ceased loving her husband for an instant, but she did not agree to that idea. I offered my rooms, and Zhenka's new wife had no trouble at all blending into the landscape of my communal apartment. They lived in one of my rooms, the back one, and moved my furniture into the hallway. By the time of my arrival, Zhenka and his wild woman had moved into their own cooperative apartment, leaving behind hideous devastation. The abomination of desolation was additionally aggravated by the steam heat. In my absence rusty radiators and pipes, bits of metal, dust, plaster, and broken glass had replaced fireplaces.

But none of this disturbed me. I felt blissful. It was Ukhtomsky's principle of a dominant excitation center in action. If a center of stable excitation is formed in the brain, the organism ceases to react to anything else. Other influences switch over to the dominant center and intensify its excitation. The dominant sensation—happiness at returning to Leningrad—was intensified by everything, even the most revolting things. The residence permit that I had almost lost and the Novosibirsk harassment now behind me played no role at all in this triumphant exhilaration. It was Leningrad itself that produced the exaltation. It itself was success, my love. Ultimately, I lived in Leningrad for six years after my expulsion from the puppet city. All the blessings of the socialist system were at my complete disposal: the improvement of my living conditions, free medical treatment, free education. I fought for them, and my tenacity in pursuing my goals knew no bounds.

My most urgent desire was to get Masha a police permit to live in Leningrad.

It was not at all disdain for the decrepit institution of marriage that guided me when I ordered Masha and Sasha not to register their marriage. Masha was pregnant. There was a month and a half left before the birth. I was acting on the basis of my knowledge of Soviet reality. She would need to return to Leningrad with the baby and with a cohabitant (that is the official designation for a lover in Soviet jurisprudence)—that was allowed. Then she'd need to arrange for her permanent residence permit and one for her child. That done, she'd have to make the marriage official and set about getting a permanent residence permit for her husband. Masha and Sasha did not obey me. They registered their marriage. Masha lost the chance to return to Leningrad with me. The Geographical Society took up the cause of a deceased president's granddaughter in vain. "He'll hear us out and then, without saying a word, he'll press the button to invite the next person to come in," Masha said to me in the waiting room of the Leningrad chief of police, where we had been standing for three hours. That's exactly what happened. Masha left for Akademgorodok with the firm intention of divorcing her Sasha. And it was high time. He was a weak person. The role of an informer was corrupting the charming youth, and he kept growing bolder and bolder, all the while losing his charm. The mental illness of people recruited as informers is paranoia. Sasha suffered from megalomania. He lied in a grandiose, self-oblivious, refined way. I had believed him. I won't recount his fairy tales or tell about his vices. They paint my credulity, unconcern, and stupidity in a too-embarrassing manner. Masha realized before I did that she was involved with a traitor.

Liza and I lived in three luxurious rooms. The divorce took about a year. Finally Masha and little Marina moved into apartment no. 6 on Maklin Prospect. Masha acquired the right to live in the apartment where she had lived three-quarters of her life.

At first the improvement of my living conditions boiled down to getting my own telephone. When in my absence the telephone had become a communal one, I had not protested. I could not lay claim to a private telephone in a communal apartment that already had a phone. Instead, I petitioned for an extension, so that I could speak from my room instead of from the hall. The institutes, publishing houses, and psychiatric hospitals where I worked and taught, where my works were published and my co-authors—doctors—worked, provided me with documents. I amassed about nine of them. I had things with which to bombard the citadels of the offices. In those documents I was called what I was—a doctor of sciences, professor, laboratory director, author, privileged person. The object of conquest was terribly modest. I had more than enough material and ammunition.

Just for a moment, imagine the place of an official of the telephone administration at any level up or down the ladder: He (or she) couldn't care less about my privileges. He realizes clearly that once I live in a communal apartment, my privileges won't work. A communal apartment is the clothing that determines how I will be met and seen out, or rather, sent packing. (The old Russian saying that you're met according to your clothing and seen off according to your intelligence has long since faded into oblivion. It is a child of the ferment of minds; it sprang from a sense of social inequality; it is a protest against privileges. There was no room for it here, where the chain reaction of privileges begins with toadyism, treachery, or in the best case, with a bribe.) There is no question of a bribe. I traversed the entire pyramid of the telephone administration and received a refusal. Men with square faces, exact copies of the ones who filled the rows of those invited to the community trial of the parasite Joseph Brodsky, without a moment's thought pressed the button to summon the next petitioner. In order to appeal the rejection to the next higher office, I had to wait for a written rejection. When I had received one from the highest level of the telephone administration, I appealed to the District Executive Council. The council then demanded that I submit all the petitions. The fruit of great efforts, the only trumps in my almost hopeless hand, they were not appended to the written rejection. I made an appointment with the top official for the sole purpose of getting the documents back. "You won't get any documents," he said. "You seem to be an educated woman, yet you don't know that no one submits originals. You need to have copies made. Go see the District Council's jurist." I could hear unconcealed glee in his voice. The ammunition was lost. All that was left was to admit defeat.

Next I appealed to the City Council to give me a separate apartment in any district in exchange for my three rooms. I described my fireman neighbor's anti-Semitic pranks and wrote that the building across the street bore a memorial plaque dedicated to my father, and that I considered the anti-Semitic attacks against me an insult to my father's memory. I was willing to accept a smaller living area, but a rejection came instantly. The City Council did not take an interest in the fireman's conduct.

And what on earth was I counting on when I submitted my application? While cooling my heels at the District Executive Council office, hadn't I seen on the door of its housing division a sign informing citizens of the rules for granting them living space? People who have fewer than four and a half meters per person are put in the line for improved living conditions. The sign states that only families that have been living in reduced per-person meterage for at least a year are put in line. The sign did not notify citizens that they would have to wait long years: seven, eight, or sometimes even twelve. It was also silent about the fact that even if they had the most sacred right to receive the blessings for which they have stood in line, they wouldn't receive them without a bribe, and the longer the wait, the larger the bribe. The curve depicting the relationship between the length of the waiting period and the size of the bribe originates at the point of intersection of the coordinates. The fortunate privileged don't stand in line and don't give bribes. A zero on the absciss axis, a zero on the ordinate axis.

I know that you can't make sense of this, dear reader, if you haven't lived under the Soviet regime for years and if you don't possess the potent imagination of a Dante or at least of a Swift. Pondering the points in the sign, you are confused. "Let's say," you reason, "that a husband and wife live in a room with an area of three meters by three. They do not have the right to be in line for improved living conditions. But then they have a baby. Does that mean that they have to wait a year before they can get in line?" you ask. The answer is yes, if they don't have connections and can't give a bribe (because in order to give a bribe and take one you have to have connections—a guarantee that you won't be denounced). They won't see their rightful twenty-seven square meters before the child starts school. "Why do they have to wait that year, too, to get into line?" you ask, confused. That point has been introduced by governmental wisdom in order to forestall speculation in housing. In your imagination draw the following picture: A family is living in a separate apartment; they have plenty of space, and any moment now they'll have their space reduced. Isn't it simpler to exchange the apartment for a room so that they deliberately have less than four-and-a-half square meters per person, make a huge killing on the black market, and immediately get in line with enough money to give a bribe? Well, it was precisely so as to forestall such outrages that a year of waiting became necessary. As I spent endless hours waiting to see officials, I had a good time solving housing riddles and penetrating the secrets of government wisdom.

An exchange was the only recourse. The psychiatrists advised me to appeal to the father of one of their patients, a beautiful Jewish girl who suffered from a manic-depressive psychosis. Her father came from Byelorussia, and had almost the same name as a famous artist. His name was Max Chagall. He was no less a genius than the artist, they said. He organized apartment exchanges. His wheeler-dealer skills surpassed anything conceivable and related to the sphere of the miraculous. An exchange with ten participants was child's play for him. He had once moved eighteen families, and to everyone's satisfaction.

He turned out to be a goggle-eyed runt who was not yet old. He had lost his right arm in the war, and he spoke with a monstrous Byelorussian-Jewish accent. He lived with his daughter in a communal apartment in terrible poverty, although he was not only an invalid of the Great Fatherland War, as World War II is known in the USSR, but a bearer of every possible decoration. His medals added up to the title Hero of the Fatherland War. In addition to his pension he had the right to free use of any form of public transportation in the city. He had lost his wife. Tears rolled down his face as he told me about her illness. He mentioned the precise cost of her medicines. I had already visited him several times and become one of his clients when the rumor reached me that an article had appeared in the newspaper exposing his illegal activity. I set off to see him to express my sympathy. I met him in the courtyard of his building in the company of new clients. He was on his way to examine their housing. He knew about the article but was not in the least upset. As it turned out, the author of the article had come to see him in the guise of a client, had become persuaded of how well and how selflessly Max Chagall worked, and had written the article in order to embarrass the government exchange offices, which do nothing except to post announcements of exchanges. A tax inspection was supposed to take measures against Chagall irrespective of the quality of his work and the size of his "underground" earnings, but everything blew over.

I spent many hours in the room where Chagall collected his clients—correction, where his clients desirous of an exchange streamed in in masses: "My mother-in-law is just about to die. The authorities will move someone into her room in the two-room apartment, and they'll wind up in a communal apartment. We need to exchange the two-room apartment for a one-room apartment right away, and then my mother-in-law's death won't have any fatal side effects."

I learned that dog breeders do not sell purebred dogs to residents of communal apartments. Present a document showing that you live in a separate apartment and then get your puppy. A mongrel dog may live in a communal apartment with the permission of all the residents, but if a mongrel creature moves in without their permission, the owner receives an administrative penalty and has to pay a fine.

And I also learned that if you exchange a lesser area for a greater one, you have to pay great sums for each additional meter. The black market has its own rates. The quality of housing is taken into account. I had heard about extra payment before. Now the treasurehouse of my knowledge was enriched with a diamond of the purest quality. The size of the payment is based on the difference between the number of inhabitants in the apartment being vacated and the number in the new residence.

Our apartment had on its rolls a nonexistent tenant, a mythical Ivanov, who'd saved the rat king from having to pay for excess living area. So, if I had had occasion to conclude an exchange, would this least conspicuous of all the residents, Ivanov, have been taken into consideration in calculating payment? The answer is unequivocal: I would have had to pay for him. Without a document from the house management about the number of residents in our apartment, without a document issued in accordance with the unshakable rule of the impeccable bureaucracy, a deal about extra payment could not have come about. The inhabitant of the other world had a Leningrad residence permit. He was included among the inhabitants of our vale of sufferings. And if he was included, then pay up.

The reader is ready to ask the ironic question: "But what if the new residents of apartment no. 6 had to pay for the difference in number of residents, and not you?" I'll answer that. Ivanov would lessen the difference and the payment. His nonexistence would not prevent him from lowering the value of my housing. Because of him my goods became cheaper on the black market in housing exchanges.

Max Chagall broke a leg, wound up in the hospital, used a crutch to defend justice, was transferred to a psychiatric hospital, and died there of pneumonia. His daughter's doctor didn't have time to move him to his hospital, and I didn't have time to visit him.

He had visited apartment no. 6 in building no. 1 on Maklin Prospect and had said that it was a hopeless cause. We would have to break up and exchange the rooms for two in one apartment building, a communal one, of course, and for one room in another, also, of course, a communal one. Masha, Marina, and I decided we would move into the two rooms and separate from Liza. But that too was a hopeless cause. Lots of people want to move apart, but no one wants to move in together. No one wanted to move into our luxurious rooms. The huge balcony with its antique railing and a flowering chestnut did not entice anyone. Max Chagall's efforts and his genius led to nothing. Negotiations never once reached the stage of a deal. The Gypsy-like fireman and informer; the highly intellectual Maria Ivanovna—a jurist with higher education; the rat king, Anastasia Sergeyevna with an undisguised form of tuberculosis; and the insane Elena Kirillovna's coarse swearing were and remained my daily inexorable fate until my departure.

One of the very brightest impressions of my predeparture life is linked to the improvement of living conditions of Soviet citizens. I was waiting to see the head of the housing management in order to get from her, in exchange for an amber necklace, documents that were essential for my departure. The woman next to me in line must have been boiling over with frustration if with tears in her voice she turned to me and said, "But we have bedbugs." The expressive force of her story was worthy of the Gospel. The tears that I heard in her voice will never stop ringing in my ears. She had had the place fumigated. "He came, but Mama wasn't there. She's blind, Mama is, and we had to take her to the neighbors' during the fumigation. And so he made the recommendation that our conditions were not eligible for improvement." It was clear even without words who "he" was—the inspector for the District Executive Council. He had come to check to see whether the mother and daughter really needed improved living conditions. Apparently it was their turn. But at that point, as luck would have it, the bedbugs turned up, and Mama was not in residence. They had the right to better housing. Their room had no windows. In my position as an émigré I could only be just as little and just as much use as the horse in Chekhov's story "Heartache," where the horse patiently listens to the driver who has lost his son. "Go to the City Executive Council and make a complaint," I told her, drawing in my imagination a picture of how a square-faced inspector would shout at her.

I didn't regret either the time or energy spent in trying to get out of the apartment where I had lived more than a quarter-century. The extreme circumstances fed my stubbornness. Here is an episode connected with the improvement of my living conditions. We had bedbugs, too. Neither our own actions nor governmental offices have prevented bedbugs from multiplying and populating the earth. We appealed to an underground fumigator, a parallel to Max Chagall. He guaranteed success, but his magic compound stank. I was in the kitchen when Maria Ivanovna attacked me with coarse swearing. I tried to save myself by fleeing. She chased after me. I didn't have time to turn the key. She jerked the door and spat in my face. I needed to change apartments no matter what. My titanic zeal did not lead to anything, however.

My next mission was to obtain free instruction for Masha; and I have described how Masha miraculously ended up a student at Leningrad State University. And finally, we very much needed free medical treatment. It was because of Sim.

Sim made it to the clinic on his own feet and was sent to the hospital with the diagnosis "acute appendicitis." During the operation they discovered a malignant tumor. He died after the operation. He had just turned fifty-nine. He died just as a heavenly angel should. He didn't complain about anything and only when I had run out of money and could no longer pay for "free" medical treatment did he say: "Why have they all abandoned me?" The tank of oxygen lasted until his death. First there was no one to deliver it, then someone was found after I gave the doctor on duty five rubles. You have to pay for everything—for every injection, for changing the bottles for intravenous feeding, for a doctor's visit. Sim didn't worry about payment, because he didn't have a kopeck. His savings and salary went to the bank in his wife Olya's name. And Olya belongs to the species that the brilliant Nadezhda Yakovlevna Mandelshtam described in her *Hope Abandoned* as "unintimidated." Olya pretended that everything was going along just as it should, and I didn't dare even so much as mention that she should cough up some money. That was how she had managed to behave around me all her life. Still and all, I underestimated her unflappability. Otherwise, giving her up as a lost cause, I would have raised some money and continued spending my own.

Sim died right before my eyes. I don't know whether one can speak of faith in God in describing the pantheism that undoubtedly formed the basis of my father's lofty world view and toward which my soul is inclined, but Sim shunned anything lofty his whole life long. He needed neither rewards nor consolation. He rejected life, not to mention immortality. My father spent his whole life creating. Sim had not lifted a finger for the sake of immortality. During his final hours he slept and seemed to be getting a little better. Suddenly he opened his eyes wide, his face expressed ecstasy, and he said, "God has come." He fell asleep and never woke again.

To this day I doubt that Sim would have died when he did or even that he died of cancer and not because of ghastly postoperative care. And I also doubt that he had cancer rather than appendicitis.

When Galich nearly died in a hospital in Leningrad, I was with him. I took his temperature with my own thermometer. It was 104°F. I showed the thermometer to the doctor on duty. She said that the temperature curve indicated blood poisoning and that they would take every possible measure, but that they didn't have the medicine that they needed to save Galich. She told me the name of the medicine—metamicin. If I could get some, fine, if not, it was her job to warn me that the danger was mortal. I was not alone at Galich's bedside. The main role in saving him seemed to belong to Yura Mekler. There were also doctor friends there. They got in touch with the Institute of Antibiotics, undertook a search, and the lifesaving antibiotic was found. One of the doctor acquaintances explained to me why the doctor on duty had given me the name metamicin. It was known for certain that that medicine was not manufactured in the USSR at all. When Sizov, the secretary of the Leningrad District Party Committee, had been ill, a special plane had been sent to London to get metamicin. By warning me and foisting off the search for the foreign medicine on me, the doctor was relieving herself of all responsibility.

The hospital personnel and patients thought very well of both Sim and Galich. Like Galich, Sim possessed the magical gift of conquering hearts. They had much in common, even in their looks. Galich was saved by his admirers. I was unable to save Sim.

# *Work*

My work with flies did not cease for a moment, despite these events. The blessed times when mutations had arisen with remarkable frequency had long since passed. A long, infinitely long period of low mutability set in. Then, in 1969, my torments ended. Mutations began arising everywhere in great quantities. Earlier, up until 1941, the mutants in most cases had been yellow. In 1969, the mutants were not different in body color from the normal ones. They had disfigured abdomens. The gaudy black stripes on the abdomen of the female were distorted, the symmetry hideously destroyed, and pieces of the chitinous covering were missing. The brilliant black surface of the tip of the male's abdomen was missing. The brilliant black surface of the tip of the male's abdomen was mutilated, slashed to pieces, serrated, faded.

During the first period of high mutability in 1937–46 the golden knights arose among the bronze ones with high frequency; they did not, however, have the slightest chance of leaving any progeny. The mutational process produced them, and natural selection swept them out of the population. The probability of encountering a yellow male among the regular males corresponded exactly to the frequency of mutation of the gene. But flies with disfigured abdomens apparently extracted some sort of profit from their malady. Natural selection turned out to be on their side. From year to year the probability of encountering a fly with a disfigured abdomen kept rising. The mutational process and selection were working hand in hand. Before 1969, flies with an abnormal abdomen turned up with a frequency of 1 or 2 per 1,000. In 1969, they could be found in 10 to 20 per 1,000. In 1972 and 1973, more than half of the flies had abnormal abdomens. The hereditary disease seemed to be saving the flies from some other ailment even worse than it was. The outbreak of mutability in the complex of genes controlling the chitinous covering of the abdomen had not had time to die down before another complex of genes, in 1973, arrived at a labile state. Flies with singed bristles, carriers of the mutation "singed," began to show up with a frequency unheard of until then. The mutation "singed bristles" provided no advantage in the struggle for life. The pattern of its population distribution was very reminiscent of the yellow mutation and differed from the pattern distribution of the abnormal abdomen mutation.

Persistent Raisska—mulish Raisska as Nurse had called me—did not lose heart. In the fall of 1968, an expedition financed by the Agrophysics Institute set out from Leningrad. In the fall of 1969, I was invited to teach an introductory course on medical genetics in the Kirghiz Republic at the Frunze City Medical Institute. Thanks to this invitation, I could add to populations studied since 1937 in the Ukraine, the Crimea, in the northern Caucasus, Transcaucasia, and in Kazakhstan, the populations of the fruit distilleries of Kirghizia. Part of the expenses for my expedition were borne by the medical institute and part by the Agrophysics Institute.

In 1970, I was fired from the Agrophysics Institute. This event, one of the most dramatic in my life, occurred by order of Lobanov, the president of the LAAAS at that time, the almighty sovereign of the Agrophysics Institute. Word had reached the Presidium of the Academy of Agricultural Sciences as to whom the institute director was sheltering. There were very many potential informants, but only one real one was needed.

My last days at the Agrophysics Institute were full of threatening foretokens of my impending expulsion. One of them occurred in 1969. I had every reason to enjoy life. I had a job, and the flies were mutating. From Frunze we headed to Burakhan, in Armenia. Big and Little Ararat were on the horizon. One of the members of the expedition, Tamara Zagornaya, was the daughter of a member of the Party Regional Committee. She had used her father's pull to get herself a job as a junior research assistant. There is more than one kind of daughter, and I did not sense any catastrophe in the making, although bright people had forewarned me. Also on the expedition was "the Centaur," her ardent admirer, a graduate student and senior research assistant at one and the same time, a fantastic chimera of secret dissidence and careerism, passive dissidence and active careerism, and a successful candidate for the laboratory that I had established. In addition, there were two Galyas—Galya Ioffe and another, who belonged to the Centaur, who was in love with him hopelessly, intensely, and secretly. She was going to work for me, but as soon as she saw the Centaur, she decided to devote her lonely life to him.

We lived at the Astronomy Institute of the Armenian Academy of Sciences. The institute's driver nursed tender feelings for Tamara Zagornaya. It requires very little to cut short an Armenian's attentions. He begins with reconnaissance. He announces his unambiguous intentions very energetically and impudently. The lady demonstrates her inaccessibility. The swain immediately retreats and adopts a pose of wounded innocence. What a filthy imagination one must have to interpret his respectful obeisance to beauty as importunity! The lady's reaction decides the case. The driver was a man. Tamara Zagornaya was no little girl. I don't know what stage their romance had reached when he came to lavish his compliments.

The institute and observatory were headed by Viktor Amazaspovich Ambartsumyan, a full member of the USSR Academy of Science and the president of the Armenian SSR Academy of Sciences. The luxuriance of the buildings, the adjoining park, and the interior decoration of the institute are astounding. We had set up our laboratory in one of the institute buildings. It was here that our driver came to pay homage to beauty.

We were staying at the observatory hotels. I was in one hotel, and everyone else in another. Hot water is a rarity even in quarters intended for foreigners. Perhaps it is heated specially for them, but that year there were no foreigners. The driver's visit and hot water in my hotel coincided. We had decided in the morning that we would take turns bathing. I interrupted the driver's outpourings and reminded Tamara that it was time for her to come to my hotel room to bathe. It turned out that she didn't have to go, because Galya Ioffe turned up with the news that there was no more water.

In the evening a huge scandal transpired. The day before, Tamara had asked me for a stamp—she wanted to mail a letter to her mother. I remember the stamp that I took to Tamara that night as though I were seeing it right now—a lynx with her kittens, a lovely symbol of maternal love. It turned out that the graduate student–chimera, his Galya, and above all, Zagornaya herself, had all been mortally offended by me. They believed I had invited the driver to come to my room to rendezvous with Zagornaya. They knew everything, not only about my being a procuress, but also about my mercenary intentions: I needed the institute's bus in order to make trips around Armenia, and consequently, in their version, I also needed a driver. Zagornaya was my advance to secure the driver's future services. The chimerical graduate student had ferreted out my intentions. He was obviously judging me on the basis of himself. He ascribed both moral and physical unscrupulousness to me. The offended daughter of the Regional Committee left my expedition the next day. Soon both graduate students left it too—the leader and his Galya. The game at the Agrophysics Institute was obviously coming to an end.

When I returned from an expedition in the fall of 1969, the Centaur's claims to head the laboratory created by me were based on a carefully developed strategy. He decided to incline me to leave the institute of my own volition. "You're a target for firing practice. They're demanding that you be dismissed. Because of you the entire laboratory could be liquidated. Wouldn't it be better for you to resign before that happens, so as to draw fire away from us?" the Centaur asked me.

I told Boris Solomonovich Meilakh about the Centaur's proposal.

"Write an article about your institute for the Leningrad newspaper, and make it read so that your laboratory's work looks like an integral part of the institute's functions. Then it will be difficult to fire you," Boris Solomonovich advised me.

An article bearing two signatures—Deputy Director Poluektov's and mine—bearing the title "Biology and Cybernetics," appeared in the *Leningrad Pravda*. We wrote about nature's mechanisms of control and about the necessity of knowing them in order not to destroy nature through clumsy interference. In response to clumsy interference, mechanisms of control either return nature to its original state or, if man's interference is too high-handed, the controlling mechanisms break down and the whole system perishes. We wrote about the mechanisms controlling diversity within species and about diversity as a condition of support for the stability of a species. There were also some remarks about the role of mathematic models in the projection of economic activity. We spoke about the analysis of correlations and about correlation pleiades, since all of those things were being studied at our laboratory.

Two phone calls followed immediately after the article's publication. The newspaper editorial office asked us to go into the question of correlations in detail, because TASS had requested information—the United States was interested in the article. The second call was from the film writer Zilberberg. He had written an application to do a film script. Basing himself on our article, he would tell a story about an urgent need to protect nature. He needed scientists to critique his project.

Only two years had passed since I had had the chance to intercede for Lake Baikal, the largest freshwater lake in the world. A joint meeting of the Academy of Sciences Presidium and the Presidium of the Siberian Division of the academy was devoted to its protection. The entire Science Council of the Siberian Division, of which I was a member, was present. The discussion continued even after the meeting, at a banquet in Belayev's office, which was superbly equipped for those sorts of reception. I was present at the banquet not in my capacity as a member of the Science Council, but as a woman. They needed women for the sake of a festive atmosphere.

The danger threatening Baikal was discussed. A huge paper plant was being built on its banks, and the waste products would of necessity drain into its crystal-clear waters. When one of the very high-ranking officials in attendance asked a direct question about the possible harm, Boris Evseyevich Bykhovsky, the secretary of the Academy of Sciences Biology Division, fumbled around and said that they should undertake experimental work at the academy's Zoology Institute, of which he was the director, and that in a year or two they would be able to make a prognosis.

"But you've had a hydrobiology laboratory for ages and ages," I said. "Georgy Georgiyevich Vinberg heads it, and he's a world-class authority. Can it really be that they haven't accumulated sufficient knowledge for making a prognosis?" Instead of playing the silently beautiful woman, this woman was playing the role of Andersen's little boy. The king was naked.

Zilberberg came to my apartment with his application to write a film script. "You have a double—someone with the same name," he said. "An essay by him—'What's the Difference Between a Cat and a Dog?'—was published in *Knowledge Is Strength*. Have you read it? He's a brilliant writer! And it's obvious that he loves nature."

"I not only read it, but wrote it," I said. "The editors of *Knowledge Is Strength* are very proud of the fact that they 'got me out on the street,' as their own definition of journal work puts it."

"Then I'm your client!" Zilberberg exclaimed.

His application was accepted. I was appointed the science adviser for his film. The film script was written; I received a fee; and that was the end of it.

The film was killed in some high office. Zilberberg had documented the destruction of natural ecosystems. Special attention was devoted to the felling of precious cedar forests surrounding Lake Baikal for the manufacture of paper at the newly built pulp plant. Yellow smoke from the plant damaged forests and the sewage system spilled contaminated water into the lake. Zilberberg had shown the pollution of other reservoirs, the undermining of chances for saltwater fish to reproduce because of the total destruction of their spawning grounds, and many, many other things. Zilberberg's film script was proclaimed slanderous. There was no pollution in the USSR and none was foreseen. Zilberberg suffered a serious heart attack.

The article I had written according to Boris Solomonovich Meilakh's wise advice did not prevent my being fired. In 1970, I was removed from the staff. But by that time I had a contractual topic—the analysis of correlation pleiades—financed by the Department of Applied Mathematics at Leningrad State University. An expedition to the Kurile Islands was outfitted to study correlations between separate behavioral acts in fur seals. The director of the Agrophysics Institute personally gave his patronage to the research and the expedition, the more so as they did not crimp his budget. The Kurile Islands are a border zone. One needs an entry permit from the KGB. The permit had been received, and preparations for the expedition were in full swing. But someone squealed, the director was summoned by telegram to the authorities in Moscow, and he was ordered to call off this research and return the money to the university. I never got the chance to create a model of a husband, this time by using material quite close to man—the harems of fur seals—rather than with the help of flowers, as in my earlier efforts.

The Centaur became the head of the laboratory, and now both Galyas were completely at his disposal. I had lost my salary and my laboratory. Life as an unemployed, vegetating in a communal apartment, began again.

To organize *Drosophila* expeditions as an unemployed turned out to be more difficult but not impossible. The Laboratory of Population Genetics I had headed in Novosibirsk became part of the laboratory led by Zoya Sofronyevna Nikoro. She took charge of my young co-workers, funded their expeditions, and permitted them to include me on their team. The Frunze Medical Institute did not withdraw its invitation to lecture.

In 1970, I received a second invitation to lecture there, but I decided to study the Uman population before going to Frunze. Leaving Leningrad for Uman turned out to be not so simple. Cholera was raging in the Ukraine. I was afraid to travel without travel credentials. I could have been detained. The Pedagogical Institute, where I gave a course without being on the official faculty, refused point blank to give me travel credentials, even with the designation "without remuneration." I went to see Bykhovsky, the director of the Zoology Institute. "I won't give you credentials from the Zoology Institute," he said, "but you can get them from the Presidium of the Academy of Sciences. Go there, and I'll order a set of travel papers to be ready for your arrival." I received a travel document with the designation "without remuneration," signed by the Academic Secretary, i.e., by the head of the Biology Division of the Academy of Sciences, Bykhovsky. For the next two years I traveled without official papers indicating assignment and remuneration. Frunze and Akademgorodok financed the expeditions. Belayev closed his eyes to what was going on.

In 1972, his connivance came to an end. I was giving a report on the results of my research on human genetics at the Institute of Cytology and Genetics. The institute's modest auditorium, accommodating about 250 people, was full. I had just arrived at Akademgorodok as a member of an expedition and was still processing my expedition material. In the report I spoke about the distribution of birth dates of patients with a hereditary disease. Belayev felt nervous. He kept interrupting me. He wanted to channel my account in the direction of official ideology, for which I had absolutely no use. I described the family trees of patients. It was important for us to know whether the patient had inherited his disease from his parents or from none of the members of his families as far as could be traced. I spoke about manic-depressive psychosis, and its connections to suicide. I said: "In and of itself suicide is not an indicator of psychic disorder. The question of honor can in fact drive even super-normal people to suicide. Only in combination with other symptoms—"

Belayev interrupted me with: "There is no suicide without mental illness."

"Perhaps there is no honor without mental illness, either," I said, the audience quiet as the grave.

Chairman Belayev's final remarks had exposed his ignorance. He lost control of himself.

In private conversations I had condemned Belayev's actions with regard to Goldgefter, a young scientific worker and a Jew, who had been mercilessly "worked over" at the institute, accused of crimes that he had not committed, and expelled when they learned that he was planning to emigrate to Israel. Belayev was told of my remarks. He was thoroughly in control of himself when he came to the laboratory and told me that the institute lacked sufficient funds to pay for my participation in expeditions. I said that the Frunze City Medical Institute paid for my travel, and the institute only used laboratory funds to pay me field expenses, a grand total of eighty rubles. "Yes, it's not a large amount of money," Belayev said ironically, "but in the future you won't have even that."

In 1973, I did not go on an expedition. The trusty Galya Ioffe, a former fellow employee of the Agrophysics Institute, brought me flies from the Nikita Botanical Gardens, Yerevan, and Dilizhan. It was among them and their progeny that I discovered male carriers of the singed bristles mutation. In 1974, Galya brought me flies again, and I worked with them until my very last day in Leningrad.

I stopped going on expeditions not only because of the lack of money. I was now fed up with their difficulties. An expedition is an isolator, paradoxical as that may sound. The unhealthy phenomena typical of any isolated group of humans are characteristic of an expedition as well. The sense of confinement, protest against constraint—more often than not imaginary—and the mutual dissatisfaction of members of the expedition grow like an avalanche. People begin taking refuge in illness and seeking excuses for quarrels. Centrifugal forces threaten to gain the upper hand over reason at any moment. I would set out on expeditions knowing beforehand the outcome of the inevitable tempest in a teapot. But finally I just got disgusted. Moreover, my dissidence did not help matters. Dissidence can be the litmus test of human relations. I saw many manifestations of nobility, kindness, and self-sacrificing help silently offered as a sign of solidarity. But evil was aggravated, too. The unforgettable protest of the majority of my Burakhan expedition against my ways as a brothel manager was by no means an exceptional event in the everyday life of an expedition.

A place to continue population studies was granted immediately after Novosibirsk was lost. I even had some choice. My former students offered me places at the laboratories they led. I did not stop teaching.

I had taught courses at the Herzen Pedagogical Institute and the Psychoneurological Institute in Leningrad, at the University of Riga, and the Medical Institute in Frunze. Invitations to organize and head laboratories of population genetics followed one after another.

Nikolay Nikolayevich Vorontsov dreamed his whole life of becoming the director of an institute, a corresponding member, and finally a full member of the Academy of Sciences. In Novosibirsk he found himself in a strange and ambiguous position. Belayev, his antipode, stood in his way. As a scientist, Belayev is an absolute zero, but his abilities as a mesmerist border on the unimaginable. Vorontsov is a real scientist, a charming person, but his charm comes out in personal contact. He has no hypnotic gifts. In Novosibirsk he took over the position of secretary of the Biology Division of the Academy of Sciences Presidium. As a Presidium bureaucrat he turned out to be Belayev's boss. But at the institute of Cytology and Genetics Belayev was the one to decide whether or not to give Vorontsov a laboratory, and he didn't give him one. He made him the head of a group and merged the group with my laboratory. Vorontsov turned out to have two bosses at the institute—Belayev and me. Belayev thought that by uniting Vorontsov and me into a single administrative unit, he would create all the necessary conditions for conflict. He was mistaken. No conflict resulted. We existed as two independent, self-governing laboratories. Relations between us were superb.

By the time that I was thrown out of the Agrophysics Institute, Vorontsov had gained the longed-for position of director. When he found out that I did not have a job, he invited me to his institute in Vladivostok. I was to go there for preliminary negotiations. I needed to send on ahead a notarized copy of my work record. The reason for my dismissal could have served as an insurmountable obstacle to my being hired. They could have written all sorts of things in it. Dismissed according to article such-and-such. The person dismissed has no idea what that means, but the secretary of the party organization, the head of the Personnel Department, and the head of the Special Department know full well what it means. In addition to a copy of my work record, Vorontsov asked for a document from the notary's office with a deciphering of the reason for dismissal. Article number such-and-such, in accordance with which I was dismissed, contained nothing pernicious—the research topic had been eliminated and that was all.

After sending off the pile of documents, I headed for Vladivostok too. I had to make two flight changes: in Moscow and in Khabarovsk. At Khabarovsk it turned out that planes could not land at Vladivostok. The airport was snowed in. I decided to take the train. I was walking down the snow-covered track in Vladivostok when I heard over the public address system: "Professor Raissa Lvovna Berg, please come to the information desk." I was treated like royalty. But I could tell that something was wrong. That night in my hotel room everything became clear. The Presidium of the Far Eastern Division of the Academy of Sciences had refused point blank to offer me a full-time position. It was a blow not only to me but to the new director's prestige as well.

Two actors—Vorontsov and I—had to act out a private scene. The scene required a change of equipment. Silently, without letting me in on his mysterious actions, Vorontsov got down to business. He twirled the phone dial and locked it in place with a broken match. Then he put pillows on top of the phone. We began talking. The mysterious actions had as their goal preventing the KGB from listening in on our conversation. The telephone with the jammed dial could not be either an instrument of communication or an apparatus for secret tailing. It seemed to me that the expression of horror on Vorontsov's face and the muffled ringing of the telephone were not links in a single chain of cause and effect: The expression of horror did not lag behind by so much as a second. While piling on the pillows, Vorontsov had dislodged the broken match, and the dial had returned to the position required for its carrying out its dismal ordinary duty, but another of its piquant missions was re-established as well. Someone had dialed a wrong number. I pulled the phone out from under the pillows. By then the need for the broken match had passed.

Left alone, and out of sheer curiosity, I called the president of the Far Eastern Division of the Academy of Sciences, the geographer Kapitsa. The president and I were descendants of obstinate academicians. He, my late father, and I were all geographers. I was dying to find out what the president's reaction to my simple-hearted call to him would be. I was planning to play the part of someone who was not frightened. In order to protect highly placed men from petitioners, there are secretaries at the place of work and wives at home. I was unable to get through to the president.

The next day the director's car delivered me to the airport. The only person seeing me off was Klara—the Klara we all called Katya—a wonderful companion on many of my expeditions.

For a long time after the match episode Nikolay Nikolayevich didn't show his face to me, and when we saw each other two years later, he was no longer the director. He had been replaced. He did not become either a corresponding member or a full member of the academy, but perhaps he still will. His antipode—Belayev—has long since been elected a full member of the Academy of Sciences.

During the first two years of my unemployment—1971 and 1972—invitations for titled positions followed one after the other. An invitation to create and head a laboratory of population genetics at the Institute of Medical Genetics in Moscow was couched in extraordinary pomp. The Academy of Sciences Institute of Ethnography also rendered fabulous honors to me by inviting me to go there for negotiations. I knew exactly how the affair would end, and gloomy ennui overcame me as I set off for the negotiations. The position of laboratory director is an elite one. Candidates have to be approved by the Central Committee's Science Division. When rejecting a candidate, the Central Committee does not bother with explanations. The job of inventing fictitious reasons for rejection is given over the institute director.

The royal honors surrounding the invitations are a function of fear. When the director of an academy institute gives you a personal long-distance call from Moscow, that indicates that he wants to do without a secretary and is avoiding witnesses, documentation on paper, and publicity. Rejection does not need these precautions. I was rejected via secretaries. The director would make an appointment with me, but when I showed up, the secretary would say: "The director is busy. He can't see you. He has Germans (or Frenchmen, or Yugoslavs) with him." That was the end of the conversation, and the deal fell through. When Baroyan, the director of the Academy of Medical Sciences Institute of Microbiology and Virusology, was seeing Frenchmen under the above-mentioned circumstances, I said to his secretary: "Pasteur comes to apply for a job but is told, 'I'm sorry, but the director is busy. He is seeing some Frenchmen.'" The secretary didn't understand. She was not obliged to know who Pasteur was.

I had not planned to apply to Baroyan for a job: a genuine well-wisher had sent me to him. He thought that Baroyan, who had a direct line to the Kremlin, could help. When I went to see my well-wisher to say good-bye, *he* was seeing Germans.

During one of my visits to Moscow, Alexander Alexandrovich Malinovsky put me in a taxi and carried me off in a direction unknown to me. It was thanks to him that I made the acquaintance of Lev Nikolayevich Gumilyov, the son of Akhmatova and Gumilyov. Lev Nikolayevich's wife lived in Moscow, and in the summer he lived in her room in a communal apartment, while in the winter he lived in his own room in Leningrad, which was also in a communal apartment. He worked at the Geography Department of Leningrad State University and taught a course in ethnology. Alexander Alexandrovich thought that Gumilyov and I ought to share our ideas. Gumilyov felt that the pulsations in human history, the interchange of short periods of cultural activity and periods of stagnation, were perhaps biological in nature. Eruptions of activity were linked to the mass appearance in the world of people with increased energy. Gumilyov calls them "passionaries." They are not geniuses or necessarily even positive characters, and they do not lead the masses in their wake, but they give a nudge to the processes of reshaping or expansion, strengthen them, and bring them to the condition of historical events. The conquests of Genghis Khan and Alexander the Great, the liberation of Spain from the Arab yoke, the emergence toward the end of the tenth century of a mighty Russian power with Kiev as its capital, and the American Revolution were all the work not just of leaders, but of passionaries as well. Without concentrations of births of such people within specific time periods, history would have taken a very different course.

Malinovsky felt that the eruptions of mutability that I had discovered ought to interest Gumilyov. Malinovsky saw in them a possible mechanism for the mass appearance of passionaries. The link seemed all the more likely since eruptions of cultural activity and outbreaks of mutability had a global character. Similar phenomena occurred in regions that were deliberately isolated from one another.

My relations with Gumilyov took shape in a highly asymmetrical way. It seemed to me that the irregularity of the historical process and the phenomena that I had observed could have something in common. Gumilyov, influenced by hazy Lamarckian ideas, rejected not just eruptions of mutability, but genetics in its entirety. In his view, there was something astrological, which I found completely unacceptable. Our difference of opinion did not hinder our friendship, and he did not refuse to compare the results of his research with mine. Three offspring of giants of Russian culture—Gumilyov, Malinovsky, and I—got along splendidly; persecuted, tormented offspring of persecuted, tormented forebears. The trip that Alexander Alexandrovich and I made to see Gumilyov quite brightened my unsuccessful wanderings from office to office.

In Leningrad Gumilyov and I gave talks at a meeting organized by the Department of Applied Mathematics at Leningrad State University. There were hordes of people in the audience. There was plenty of criticism, quite sharp and witty, but the polemic was not conducted from Marxist positions. It was as though we and the crowd of young people had found ourself at an oasis in a desert of discredited science.

As it happened, I did not even respond to the most prestigious invitation, which had been issued in 1968, immediately after the International Entomology Congress. I was invited then to head the Department of Genetics at the University in Gorky. Until 1948, it had been headed by Sergey Sergeyevich Chetverikov. How it happened that the position had gone empty, I do not know, but I feared its surroundings. The school's bad reputation was awe-inspiring. Melnichenko, the head of the Department of Darwinism, was enough to instill dread all by himself. The people who had expelled Chetverikov still controlled the fate of the school. I was not about to apply for the post.

I had already retired when in 1972, I was invited to Gorky, not to become a professor at its university, but to participate in a conference in memory of Chetverikov. Chetverikov's pupils—Astaurov, Rokitsky, Nikoro, Efroimson, and Kirpichnikov—had timed the unveiling of a monument on Sergey Sergeyevich's grave to coincide with the conference. All the departmental chairs were there. And rather than be silent, they all spoke. Efroimson gritted his teeth as he listened to former persecutors praising their victim's merits and achievements.

A quarter-century earlier these same people had fed themselves on Chetverikov's destruction, and now—on his fame. In the university's largest lecture hall, standing before a sea of listeners, Boris Lvovich Astaurov spoke about the tragic fate of the founder of a new branch of science—experimental population genetics. Efroimson gave a superb talk. He devoted his lecture to the population genetics of mental illnesses.

I spoke about the change in frequency of mutations in fruit fly and human populations. I already knew that both for the flies and for humans the frequencies of mutation occurred in waves, that the rises and falls of these waves had a global character, and that at the crest of the wave for each period the mutations were one and the same everywhere. Each rise in mutability is individual. The crests of the waves bring to the surface of the world's ocean of time new mutations in each instance. What meaning do the rises and falls in the appearance of new mutations have for the transformation of a species in geological time? They are the reason for the pulsations of the evolutionary process. Thanks to them evolution is unpredictable. It is unpredictable and guided by laws. Not only the inventiveness of the engineer and technologist type dictate the laws of evolution. The very process of hereditary mutability has its own laws. Evolution occurs on the basis of laws and in that sense is nomogenesis, but at the same time it is neogenesis—the birth of something new.

The future of humanity is unpredictable. Our concern for the well-being of future generations should be limited to the protection of nature. The most important thing to protect is the nature of man himself. Just take a look at what remains fixedly beautiful for the ages, at what enters the treasure house of world culture. Creators of the beautiful created for their contemporaries, and all future generations enjoy their creations. Rome, Leningrad, Riga, Florence. "You geniuses sitting in this hall," I appealed to those present, "don't impose your ethical and esthetic rules on future generations. Create for us, your contemporaries, and both we and future generations will be grateful to you. The world is wonderful in its diversity. The future is unpredictable." I spoke about the irregularity of the historical process, about the periods of acceleration and deceleration, and about the independent global eruptions of human activity of which Teilhard de Chardin and Gumilyov spoke.

From the point of view of Marxism-Leninism, this was the greatest sedition imaginable. According to Marx-Engels-Lenin and today's rulers of the country, the members of the audience were not at all supposed to create for the good of their contemporaries. Having maximally limited their needs, the rulers forced them to build communism—the bright future for coming generations. They were to eradicate the vestiges of capitalism, with its flawed bourgeois ideology, from their own consciousness as well as from that of their fellow workers and their neighbors in communal apartments. And they were to educate themselves and the people around them in the spirit of Marxist-Leninist Communist morality, firm in the knowledge that consciousness, morality, and ethical norms have a class nature, are determined by social conditions and by membership in a specific class, with its class interests. They were supposed to know full well that the laws of biology in general and the laws of genetics in particular were not applicable to human beings. My genetic bases of the rhythm of the historical process fundamentally contradicted official ideology, and it was with all the greater pleasure that I offered them to my listeners—to "geniuses," as I called them.

It was an anniversary meeting, and polemics were considered out of place. I did not hear it myself, but I was told that the poker-faced department chairs talked about my lecture with indignation. "If she came our way, we'd tame her"—such was the scientific polemic that went on behind my back. I didn't come their way. I knew perfectly well what would have awaited me in Gorky and did not put my hand in the flame.

# Consequences

In 1972, after I joined Andrey Dmitriyevich Sakharov in signing a petition for the abolition of the death penalty and for an amnesty for political prisoners, my chances of receiving a regular full-time position were completely eliminated. Without a regular position, the possibilities of publishing, teaching, and working declined as well.

I am often asked whether there are many dissidents in the Soviet Union. There are few overt ones, but a multitude of secret ones. The activity of secret dissidents is limited to the avid reading of *samizdat* works and to helping people who are already in hot water. The slender threads of personal sympathy and secret dissidence on which my fate hung were getting more and more frazzled.

The termination of my existence "on the street," i.e., of my journalistic activity, was marked by the refusal of the editorial board of *Knowledge Is Strength* to publish my essay "Are Traits Acquired in the Process of Individual Development Inheritable?" The editorial board itself had commissioned the essay. I did not enter into polemics with Lysenko, but of course the essay was anti-Lysenko.

How can one talk about the inheritability of acquired traits when traits cannot be inherited at all? It is not traits that are passed from generation to generation, but genes. A gene may not even manifest itself, may not give the slightest sign of its existence, if its action is suppressed by a more active partner. The transmission of a gene to the next generation does not depend on whether the gene has realized its action. Both genes, the suppressed and the suppressing, are passed on to the next generation with equal probability. The molecular level of the repository of hereditary information—no, it would be better to say the informational principle of transmission of traits from one generation to another—excludes the inheritance of traits acquired in individual development. A wise Greek, I don't remember who, once said that it was absurd to believe in the inheritance of anything that in itself had not been inherited from one's parents. Anyone knows that if you carve a bench out of a newly felled tree and a twig sprouts from it, it will grow into a tree and not a bench. The channel of direct communication goes from the gene to the trait. Over the course of an individual's life no signals from a trait are sent to a gene. No matter how much you milk a cow, its daughters, in spite of Shaumyan's opinion, will not become record-breaking milk producers because of that. But if the genes know how to manage their business so that the organism survives and leaves viable progeny, then they manage to do so because the genes that did not know how to perished. The loss of inefficient genes and the predominant reproduction of efficient ones perfects genes, and the mutational process makes genes diverse so that there is competition and something to select from. Does that mean that a feedback signal is nonetheless sent from the trait to the gene? Yes, it is sent, but the transmission of information is accomplished via a feedback channel in the process of generation replacement through the selection of the best genes. The possessors of "stupid" genes die, and the possessors of "smart" ones survive and pass them on to their children. It is precisely this loss of certain genes and the reproduction of others that signals the environmental imperatives to the genes. Signals reach the population of offspring, and the offspring tend to have "smarter" genes than did their predecessors. The selection stands guard over the correspondence among the gene, the trait, and the environment.

It is important not only for people who work with animals, plants, and microorganisms to know the truth in this matter, but for doctors and patients as well. A doctor and patient should know that having cured a hereditary illness (and more and more congenital diseases are yielding to treatment), the doctor does not prevent the transmission of the disease to the next generation. An acquired trait—health—is the property of the cured person and is not passed on through inheritance. If children inherit the pathogenic inclination, they will require the same treatment. Is that important? Yes, it is. The idea of inheriting traits acquired in individual development is pure alchemy. I was especially proud of the fact that I had discovered why alchemists' ideas have long since receded into the past, while their parallel in biology had hung on to the twentieth century.

From time to time, the voice of a biologist proclaims that he has succeeded in reproducing in an experiment the inheritance of an acquired trait. But the alchemists' conviction that common metals pass through a series of distinct compounds and turn into precious ones has sunk into oblivion. The difference between the biologists and the chemists, who have completely rejected the notion of certain substances being transformed into others, lies in the degree of stability of molecules and genes. A molecule of oxygen, no matter what compound it is shoved into, whether that be water, ozone, ferric oxide, or glucose, if it manages to leave the compound, will leave it unaltered. Genes, like molecules, and being molecules themselves, are not influenced by the organism carrying them. But being a giant molecule, the gene is not subject to the law of constancy of atomic composition to which small molecules are subject. The atomic composition changes, but the gene remains itself, without losing its principal function of self-reproduction. A change in the atomic composition is a mutation. An altered trait corresponds to an altered gene.

Some mutations that the researcher sorts out imitate the inheritance of traits acquired in individual development. The great defender of alchemist doctrine, Kammerer, took fiery salamanders, with their bright orange pattern on a slate-black background, and placed some of them on an orange background and others on a black one. The coloration of many animals changes depending on the background on which they live. The fiery salamanders reacted accordingly—they turned light against a light background and dark against a dark one. They transmitted their distinctions to their children. But that depended not at all on the inheritance of an acquired trait, but rather on the unconscious selection by the researcher of the lightest salamanders against a light background and the darkest ones against a dark background. In the intensity of their coloration the inherited distinctions imitated the inheritance of traits acquired under the influence of external conditions.

Take a white hen that has been covered by a white rooster, hang it upside down, every night at midnight show it a portrait of a black rooster, carefully measuring out the amount of time you display the portrait, and in the offspring of the white hen you'll get a black chick along with white ones. The more mystical the nature of your actions, the more likely that you will attribute the birth of the black chick to them. Black chicks are born from white hens without any outside influence, and it is the investigator who is hypnotized by his or her manipulations, not the chicken. The mutational process and unconscious selection are the reason that the idea of the inheritance of acquired traits, alchemistic in its essence, has survived into the twentieth century. That is what I wrote in that article that was never published.

The editors of the journal *Knowledge Is Strength* were pleased with my text. It was not the contents of the article that turned out to be the reason for the rejection, but rather the author's biography. My whole life long I had kept one nostril above water and had not drowned, thanks to the disorder, the colossal entropy, the lack of coordination between individual links in the system. Now orderliness was on the rise.

The termination of my teaching activity, as well as of my work with flies, occurred as the result of the mental illness that struck one of the students at the Herzen Pedagogical Institute.

The laboratory where I was offered the opportunity to count flies and the department of advanced instruction for biology teachers where I gave courses were headed by Pyotr Yakovlevich Shvartsman, "my pupil," as he called himself, although he had only attended my classes. It is impossible to imagine a creature more charming than Petya Shvartsman. His features are not impressive; he resembles a chimpanzee. But his wonderful, manly voice, his intelligent, understanding smile, his radiant kindness, the entire makeup of his speech and actions win one's heart. The necessity of paying tribute through toadying in exchange for the opportunity to head a laboratory caused him terrible torment.

He was in a black mood when he told me about the meeting at which Efim Grigoriyevich Etkind, the institute's pride and joy, a professor adored by students, was expelled and stripped of all his degrees and titles. I told Efim Grigoriyevich about Petya's despair. "He voted for expulsion," said Efim Grigoriyevich.

"How do you know?" I asked.

"The motion to dismiss me and to petition to have all my degrees and titles rescinded was approved unanimously."

Petya Shvartsman nearly cried when he asked me not to come to the laboratory anymore. He had received an order not to admit me. He would give me a microscope, test tubes with food for the flies would be brought to my house, but he could not admit me to the laboratory. "I'm a bastard," he said, "but I'm not thinking only of myself. If they close the laboratory, so many people will suffer."

The basis for the order to expel me had been provided by a phone call that the insane Sasha made. Sasha became infected with a persecution complex. He told his parents and his brother that he considered their concern about his health KGB intrigues and that he would see a psychiatrist only if I thought it necessary. I found all of this out from his mother. At my request, psychiatrist friends of mine put him in the Bekhterev Institute Clinic, and it was the doctor on duty there who called me to ask me to persuade Sasha to take some medicine. In response to my persuasion, Sasha babbled things quite inappropriate for a phone call but so obviously pathological that it should have been clear to any idiot that the person speaking was ill. The people who eavesdropped on—excuse me—who listened to the phone calls at the institute bearing the name of the freedom-loving Herzen had no grasp of the subtleties of psychiatry. There were more than enough sins held against me even without Sasha. Now Sasha's ravings drew the authorities' attention to me.

The dacha that I had inherited from my father, the dacha wafted by the marvelous Komarovo air that my father so appreciated, that dacha ceased to be. How many different people tried to take it away from me! In fact, they began taking it away from me even before I received it. Immediately after Father's death in January 1951 I was informed—orally, of course—that the decree about the right of members of the academy to bequeath their dachas had been rescinded and that the dacha would be taken over before I came into the right of inheritance. This misinformation failed to produce the desired effect. I persisted in my right to have the dacha. Threats began.

Without raising his voice, but putting a lot of metal in it, the square-faced chief of the academy's Housing Division told me: "We'll ruin you. We'll force you to make repairs on the dacha. You'll bankrupt yourself on legal fees."

"I'd like to introduce you to my lawyer," I said politely and introduced him to dear Margolin. I knew the laws very well myself but used Margolin as a scarecrow to ward off vultures. It worked superbly.

I redeemed two-thirds of the dacha from my stepmother and Sim and came into the right of inheritance. I was living in Novosibirsk when a fire occurred at my dacha. It was wintertime. The fire was the fault of Yasha Vinkovetsky, an artist, and the only person living at the dacha, who was smoking around buckets of paint which, as he put it, ignited somewhat less well than gunpowder, but somewhat better than gasoline. "Let's be glad that you weren't transformed from an artist into what artists draw with," I wrote from Novosibirsk, meaning charcoal, lampblack, and charred bone. A spark from that fire ignited the flame of enmity in the family that my stepmother was constantly trying to destroy.

When he received the insurance money, Yasha found out that at the Housing Division of the Regional Executive Committee three inheritors were listed for the dacha. Statements were demanded from Sim and my stepmother, and the bureaucratic mistake provided my stepmother with a reason for contesting my right to the dacha. Sim defended me. The matter went to the courts. My stepmother was not petitioning on her own behalf. There was another relative, about whom I have preferred to remain silent in my story, and whom I will introduce now only in connection with the dacha saga.

With a shaking hand, my stepmother, over eighty years old, wrote a declaration to the court stating that my father had disinherited me. The relative wrote directly to the Central Committee, saying that the dacha should be taken away from me because of my anti-Soviet activity. Neither my stepmother nor the relative achieved anything. The relative's lawyer could not contain himself. "I cannot participate in this villainy," he said, shook my hand, and walked out of the courtroom. It required extraordinary circumstances for the elderly Gornshtein, a famous lawyer, to display such lack of restraint. My stepmother and the relative appealed the court's decision. I had to visit both the Procurator's Office of the Republic (RSFSR) and the All-Union Procurator's Office.

Certain personages from the reception office at the latter establishment have fixed themselves in my memory forever. A middle-aged Estonian, Latvian, or Finn, to judge by the blueness of his eyes and the millet blondness of his hair, wearing very clean blue work overalls, stood there motionless, like a statue. His face expressed unshakable decision. An irony of fate had made him a superb model for a statue of a Latvian marksman. An enormous old peasant woman, dressed poorly and not at all in city style, apparently knew only what her grievance was, but had not the slightest idea of how to go about her complaint. She seemed uncertain of whether to stay or go. She had not addressed anyone, and no one had offered her help. On her back she had a bag, not a backpack, but a simple bag made of sacking. The contents of the bag were hemorrhaging, and a spot of fresh blood was increasing in size. She obviously had gotten some meat at a store, and the frozen meat, not wrapped up in wax paper, had thawed and soaked the bag. The old woman's face expressed infinite submissiveness to fate.

At the Procurator's Office of the USSR I found out about my relative's appeal to the Central Committee. The complaint about my subversive actions had been sent from the Central Committee to the Procurator's Office, and the Procurator's Office directed it to the regional court where the case was tried. Liza and I made a copy of the complaint and enjoyed doing so. The case dragged on for years. I had given Sim one-third of the dacha's price and he knew that Marmikha had also gotten a third. He testified at court that I was the sole owner of the building. Neither Sim nor my stepmother lived to see the denouement. First Sim died, and then my stepmother.

The main threat to my keeping the dacha came from the academy and its members. They aimed their sights at heirs from the very first day of the heirs' coming into the right of inheritance and kept those sights aimed there. There were a number of snipers: the academy's Housing Division; the Directorate of the Academy Village, whose members are academy members elected at a general meeting by fellow academy dacha owners; and Vasily Filatovich. The most dangerous sniper was Vasily Filatovich, "Filatych." He was the head of the management office and was in charge of looking after the plumbing, wiring, roads, and rent collection. He seemed the embodiment of modesty, and he dressed correspondingly—poorly. According to rumors, he was fantastically wealthy. Bribes ran to him in rivers. He was all-powerful: He could help anyone with repairs and he could refuse anyone. But repairs were the least of it! He could declare any of us a violator of Soviet laws and administrative rules and call us to account through the whole huge apparatus of compulsion: the police, tax inspection, the pyramid of executive committees—that very apparatus of compulsion which, according to Lenin's promise, was supposed to die off as socialism was constructed, but has kept on growing right before our eyes and becoming stronger. Who leads whom by this leash is not at all clear. The directorate made up of academy members danced obediently to his tune. They were terribly afraid of him. "He's a KGB agent," I was told in a whisper by an academy member at his own dacha.

Heirs are divided into four categories of vulnerability. The least vulnerable are widows, and after them come in ascending order direct offspring, distant relatives, and people having no blood link to the member of the academy. Academician German, my dacha neighbor, willed his dacha to his housekeeper. Prince Shcherbatsky, an Orientalist and academy member, married his housekeeper. Andrey Petrovich Semyonov-Tyan-Shansky is the son of Senator Petr Petrovich Semynov-Tyan-Shansky. The Senator was a famous politician, a geographer and historian, a traveler and writer. He presided over the Geographical Society for nearly half a century. Andrey Petrovich was himself a famous scholar and poet, the translator of Horace. He took his cook Natalya to the altar. Academician German did not get married, but willed his dacha to his housekeeper, because the law did not prevent that.

The goal of sanctions against heirs is to force them to sell the dacha to the academy, and according to statutes the elite dachas where academy members enjoy tranquillity can be sold only to the academy. It is then that a dacha, purchased for a price established by the state, becomes the property of an academy member who craves it. A bonus—*podachka*—in the form of a dacha—a *dachka*, as a poet would say—was a one-act comedy played out by Stalin in strict accordance with the classical rule of unity of time and place. It was granted once, and that was that. Newly elected members of the academy foamed at the mouth in anticipation.

In the feudal structure of Soviet society a former housekeeper who owns an academy dacha is a highly vulnerable creature. The sanctions against heirs are economic. Heirs are strictly forbidden to rent the dacha, although other citizens are not deprived of the right to rent out living space. In order to raise extra money, academy members can rent out space and not pay the tax ordinary citizens have to pay. If an heir is not an academy member, maintaining a dacha is too expensive for him or her. The following document can serve as a model of sanctions: "The Directorate of the Academy Village proposes that you repair your living quarters, garage, and fence in short order. Your dacha does not serve as an adornment for the Academy Village. The Directorate proposes that you remove the added rooms and walls that you have built with the goal of rental speculation. Failure to adhere to resolutions is punishable by law." I received such documents many times, although I had added no rooms and walls and kept the dacha in good shape. They once forbade me to water the garden—there might not be enough water for academy members. They constantly threatened to disconnect the light and water. What does the unfortunate heir do under such circumstances? He or she takes in renters under the guise of relatives and pays a bribe to Filatych and his secretary and passport clerk, dog-faced Nadka (that's behind her back, of course—to her face, it's Nadezhda Mikhaylovna). If he gives too little, the directorate of academicians—not Filatych—will demand proof that the residents are blood relatives and not second cousins twice-removed. In exchange for a bribe one can obtain that sort of document, too, but after all, in order to pay, you have to have money! A vicious circle. Economic sanctions were being used more and more energetically, and more and more money was being pumped out of heirs. Not only an American reader but a Soviet one may well ask: Why were the participants in the drama, the bureaucratic hierarchy guarding the carnal interests of academicians, so zealous in persecuting the heirs of academicians? The answer is the bribe. Do you think that the academician having a claim to buy a dacha from the academy—and there was a long line of such aspirants—would get a dacha without giving a bribe to every member of the nearly endless hierarchy of those charged with watching over the interests of every academician? He would be obliged to express his gratitude, starting with the chairman of the economic department of the academy and ending with Filatych. The directorate of academicians has the final say, and its members are puppets in the hands of those who distribute the creature comforts. But there were many academicians, neighbors of mine, who were prevented by human dignity from joining the directorate. The strategy aimed at forcing the heirs to sell the dacha to the academy and to pass the newly obtained real estate to a newly elected immortal, failed. Nobody wanted to leave the inherited home.

And then suddenly—in 1972 or 1973—we were called to a general meeting of homeowners. Until then only academy members attended the general meetings, and heirs did not have the right to vote. Academician Nalivkin and Vasily Filatovich presided at the meeting. Nalivkin informed us of a bit of good news: German's heiress had sold the dacha to the academy. That was the result of the wise measures that the directorate had taken in order to safeguard academicians' tranquillity. It is very strictly forbidden to disturb the silence in the academicians' village. German's heiress had offered her dacha for rent and those who moved in had not had respect for the rules of the place they illegally entered. My neighbors showed literally no signs of life. "Severe measures will be taken in the future as well," Nalivkin said. "Wouldn't it be better to part with one's dacha before disaster strikes?" The pleasant bit of news stunned heirs. They received it with extreme hostility. The atmosphere began to smell of exposé. Filatych's crystal-clear blue eyes grew dim. His nose turned indecently and alcoholically red and swelled up. The perfect square of his face became stretched and distorted. The nephew of Polkanov's late widow said that German's heiress apparently did not have enough money to maintain her dacha, but he did, had moved from the provinces to Leningrad, had lawfully come into the right of inheritance, knew the rules about renting out extra space, did not plan to violate them, had no other living space, and saw no reason to part with his dacha. The grandson of Academician Poray-Koshits spoke as a government figure. He said that the privileges that the government so generously extends to academicians did not give them the right to break laws and arrogantly put themselves in a more privileged position than the one granted them legally by the authorities. Academy members were compromising themselves by abandoning less affluent people to the tyranny of the head of the management office. He spoke calmly, as though he were not a victim of oppression. A lawyer in a court of law could have spoken that way. But the son of academician Barannikov yelled, with a pale, distorted face and trembling lips: "Academicians ought to consider the fact that they are mortal too. No one can escape the law of natural selection. Me today, you tomorrow. Today academy members and this character"—he pointed at Vasily Filatovich—"are persecuting us. Tomorrow their heirs will be persecuted in exactly the same way. If you don't want to think of us, then at least think of yourselves and your loved ones." I was especially surprised by the appeal to natural selection. Barannikov had identified it with death.

I was next to take the floor and said that the protection of our rights was in the interests of academy members, as Barannikov had correctly stated. The academy members who wanted our dachas would be better off joining together and petitioning the Council of Ministers to allot them land where they could build dachas for themselves. But they should leave us alone. And I also spoke about the water that was not the least in short supply on the Karelian Isthmus. I asked whether it wouldn't be simpler to take the money that we paid as rent and use it to dig artesian wells than to forbid certain people to water their flowers so that there would be enough water for others. And I also spoke about the nonexistent walls. I proposed that a representative of the heirs be appointed to the Village directorate and named Poray-Koshits. I was very tempted to name Barannikov but resisted. My proposals were not accepted. The harassment only increased.

I decided to emigrate, and the question of money arose. I had no way of getting any. I decided to sell the dacha. The decision coincided with the decree requiring those who leave the country to pay, retroactively, the cost of their education. We had not even dreamed that in response to the demands of world public opinion and as a result of the extreme necessity of buying wheat abroad, the decree would be rescinded. What good fortune that the Soviet Union needed American grain in order to feed Italian Communists with it! The scandal around the sale of food grain to the Italians had penetrated even the Soviet press at that point. By the time I was preparing to leave, the decree had either been rescinded or they had ceased applying it. Had it been in force, I would have had no chance of leaving. Retired people, I among them, were not exempt from paying for their education, although you would have thought that the debt to the state had already been repaid. Just as they now cut off the pensions of retirees who are emigrating, so they did then. No amount of money raised by selling the dacha would have been enough to pay for my education—university, graduate school, doctoral study—the state had spent money on me for thirteen years. I would have had to pay back thousands of rubles.

I could not hurry with my departure. Leaving without my expedition journals would be the equivalent of suicide for me. I spent two years summing up results and preparing microfilm. Meanwhile, two events occurred: A buyer for my dacha was found and the decree about repayment of costs of education ceased to be enforced.

The buyer was academician Kantorovich, a legitimate candidate for an elite dacha. An unbelievable set of proceedings ensued. The more legitimate the demands, the less are the chances. There was no smell of bribery. In the summer of 1973, the deal fell through. In the winter, another fire occurred at the dacha. It broke out at night when none of us was there.

My ordeals during the investigation of the case were rich in humorous details and coincided with Akaky Akakiyevich's ordeals following the theft of his long-suffering overcoat, as related by Gogol. Immediately after the fire the Village directorate and the Housing Division, by way of a great favor, granted Kantorovich permission to buy the dacha. I reduced the price and we went to the Regional Executive Committee to get permission for the deal. It turned out that the Regional Executive Committee already had an inspector's report according to which there was no dacha, since it had been destroyed by fire. My saying that I had received an insignificant insurance payment was of no use. Neither was Kantorovich's wife's showing photographs of the dacha taken from all four sides. Nor did it help when I spoke up again and said that there were two buildings on the lot. Besides the dacha itself there were a garage and a two-room guardhouse with plumbing, and they had not suffered in the least. The resolution of the Regional Executive Committee directed Vasily Filatovich to confiscate the dacha from me, disconnect the electricity and water, so that I could not use the guardhouse, and directed me to turn over the lot, "with the trash removed," to the

Regional Executive Committee—to raze the buildings and remove from the lot the dust and ashes left of my homeowning.

Two secretaries kept a record of the meeting. Their silence suggested some sort of bewilderment, and in them there appeared that same sense of sham that the old librarian had when upon orders from above she blacked out Vavilov's name with India ink.

In order to tear down the two buildings I needed a sum that greatly exceeded what the insurance office had paid me. I mentioned that. They yelled at me. That did not concern them. I appealed the decision at the City Executive Committee. They yelled at me there, too, the same way they did at all the other visitors. I appealed the decision of both the Executive Committees—the Regional and City—at the Council of Ministers, so that I would not be forced to raze the buildings and clear away the rubble. I can imagine what efforts clearing away the mess would have required! I would have needed a decree from the Regional Executive Committee allotting a plot of land for all the rubble. And the trucks, drivers, loaders? I slipped away without waiting for the decision by the Council of Ministers and without cleaning up my lot.

The police did not even think of troubling themselves to determine how a fire broke out at night in the empty house. But a friend of mine found out. The offspring of academy members—not their children, but their grandchildren—were amusing themselves with petty thefts and gibes at the petite bourgeoisie ways of homeowners. They did not so much steal as destroy despised comfort—they broke mirrors and defecated on the table. The raid on my dacha was their second one. The first time, not having found any regalia of wealth, the romping grandchildren first wanted to steal my statuette of Don Quixote but then threw it into the snow, where it was discovered in my absence by a police dog. The police called me and told me that they had Don Quixote and that I could come to pick him up. I said that there was nothing to steal at my dacha and that I was donating Don Quixote to the police.

The second time, the little thieves stole several valuable books—Andrey Platonov, Camus in French—and apparently they accidentally dropped a match in a storeroom where the light did not work. It is absolutely certain who was responsible. Finding out the truth did not change matters. The dacha burned down.

# *Moirai*

The Moirai are the Greek goddesses of fate. According to one version, they were born of Themis, the goddess of order, and according to another, of Ananke, the goddess of inevitability. The three Moirai of my fate could with equal right have been the daughters of either. Three elderly women, short, stout, with their hair drawn back and gathered into tight little buns at the back of the head for orderliness. In appearance they are indistinguishable from each other, except that the crimson-yellow streaks of a bruise around the eye of one of them set her apart from the doleful rhythm. The Moirai spun, stretched, and severed the threads of life.

The first of the Moirai, my Atropa (in Greek the name means "inevitable"), severed one of the threads. I was at one end of it, and Yura Valter was at the other. Yura Valter and I had studied in parallel classes at the same school, he in the Russian section, and I in the German. It was his topographical maps that bore my signature and hung at the displays of the best student works. I did not write compositions for him, however. His classmate, Milusha Denissova, who wanted to make friends with the girls in my class, introduced Yura to me at his request. Yura was impressed by the idea of playing the role of a knight to a girl who, if not a princess, was the daughter of a famous scholar and traveler. Of the two of us, however, he was the noble. He grew up in the marvelous family of a Norwegian forestry scientist. I have no idea what brought his father, Karl Voldemar Valter, to Russia. His father lived with his beautiful wife Lyutsia Gustavovna and their son and three daughters in the apartment on what was Officer Street before the Revolution and is now called Decembrist Street, in the same building on the corner of the Pryazhka Embankment where Blok's celebrated Carmen lived. Out the huge windows of the huge rooms, beyond the linden trees, on the other side of Officer Street, one can see the building in which Blok lived and died. The apartment where the Valter family lived—there were also a large number of aunts living there—seemed an enchanted realm to me. I remember its smell and the voices of Lyutsia Gustavovna and all three sisters. I quailed before Yura's father's gray sideburns, saw him only a handful of times, and my memory has preserved nothing of him except his attractive appearance. Lyutsia Gustavovna and the aunts spoke with a slight accent, but Yura and his sisters had absolutely none. The beautiful Venetian chandeliers, the dinner table set with silver and crystal, the pheasants and beautiful ladies in luxuriant clothes, the verandas of palaces, the palms and cypress trees in the beautiful pictures in gilded picture frames, and the spines of very old books in glassed-in mahogany bookcases had nothing at all in common with petite bourgeoisie pretentiousness. Luxury was a habit there, not a goal. The ascetic furnishings of the apartment where I grew up contrasted starkly with the interior decoration of the home where Yura grew up.

After the Revolution Yura's family, with its single provider, was hungry even at the best of times. Great plenitude arose in our family after the introduction of the New Economic Policy (NEP). Nanny gave Sim and me extra food on the sly, and I passed on my wonderful breakfasts to Yura on the sly. Yura could invite me home and introduce me to his parents. I could not even dream of inviting Yura to tea and seating him at the same table with my father and stepmother. The knight who was striving to worship the beautiful damsel was the cultured offspring of loving, cultured parents. The damsel lived on the periphery of her famous father's new family and was a barely tolerated encumbrance. In her development the damsel also lagged far behind her admirer. I would guess that it was not without his parents' influence that at that time, the end of the 1920s, Yura realized the whole horror of what was going on, and he understood what was the truth and what was a lie. My consciousness was scarcely aroused. The main thing that must have disappointed the knight was the complete absence of proud inaccessibility. Inaccessibility is as essential as air. Without it there is no role of the beautiful damsel's knight. I have more than enough pride, but my pride lacks femininity. I was not fit to be a damsel. Tormented by my stepmother; pathetically, humiliatingly, hideously clothed; deprived of any domestic training and education at all, I flaunted my vagabond style of life, my contempt for earthly goods, and my unshakable certainty that independently, without anyone's help, I would make my way to a bright future.

I was about fifteen, and he about seventeen when Milusha Denissova introduced us. Our love affair was platonic, and that was just as much Yura's doing as mine. We decided to get married in five years, when we had completed our education. "We'll fulfill the five-year plan in four years!" Yura would say, laughing. The wedding date was set: July 6, 1934. To the barriers separating us was added yet another one—one erected for the builders of socialism. As the daughter of a university teacher, I had the right to enter the university. Yura, however, wound up in the category of people who were punished for their parents' social inferiority. Yura's father did not teach in an institution of learning and did not belong to the class of farm laborers and workers, and only the offspring of those social categories could enroll in institutions of higher learning. That is how things were until 1936, when Stalin did the royal favor of granting freedom to his subjects by allowing children of white-collar workers to enroll in institutions of higher learning and registering it in his Stalin constitution.

I had left home and was earning money and preparing to enter the university while Yura, working as a topographer, went away, first to the North, and then to Tien Shan with Krylenko's expedition. He became a specialist on the topography of glaciers. He discovered an unknown glacier, and Krylenko named the glacier after his young traveling companion. I was a university freshman when the letter announcing our breakup arrived. "Your views are absolutely alien to me. It's as though we're looking at the moon and you assert that it's not the moon, but an onion." What he meant was my faith in the triumph of communism. He listed my shortcomings—grains of sand wearing out the millstones, as he put it.

Deprived by my stepmother of the slightest opportunity to have contact with representatives of the older generation, I was completely in the power of official propaganda. I had not the slightest idea about the bloody atrocities of total collectivization or about the reprisals against the so-called *lishentsy*, people stripped of the right to participate in elections for the Soviets of Worker Deputies because of their social position. The best aspects of Communist doctrine reigned over my mind. An understanding of the flawed nature of the path down which the country was headed in the pursuit of noble goals was only just beginning to penetrate my consciousness. Yura was the son of his intelligent, educated parents. They knew and understood the whole horror of what was going on, and if they ever had revolutionary ideals, as the overwhelming majority of the intelligentsia in prerevolutionary Russia did, then by the end of the 1920s, when the events being described were being played out, they had renounced them.

There were no arguments between Yura and me. He knew of my attachment to official ideology and without suspecting its sad sources, decisively censured me. He expressed his disagreement with me in a letter. I think that my hobo style, my leaving home at the time when my father was being persecuted for his anti-Darwinian publications which were far from official dogma, played a role in Yura's decision to break with me. We didn't see each other, but from time to time Lyutsia Gustavovna and Yura's sisters invited me to drop by. In most cases, though, those visits were timed to coincide with periods when Yura was on a trip.

During the war with the Finns, he was wounded in the leg. "I lost about a half a boot of blood," he wrote Lyutsia Gustavovna from the hospital. "You should thank God that it happened this way," I said to her in response to her lamentations. "A tremendous war is ahead of us. Millions of mothers will lose their sons, but yours will remain unharmed." Apparently something did not mend properly after the operation, and Yura was forced to abandon his profession. He became a planner for railway lines. He married. Later the news reached me that he had gotten divorced. We did not meet or correspond.

I did not answer the letter in which he wrote about the onion. Neither that letter, nor our separation, nor his marriage was capable of altering my triumphal, joyous, all-powerful feeling toward him. There is a great curative power in hopelessness. The death of hope gives birth to freedom. A person acquires freedom in old age, and in an elderly person's knowledge of the inaccessibility of bliss there is an element of bliss. I acquired the freedom of old age from the hands of my former knight at the age of eighteen. I had nothing to lose. Everything was already lost forever, until the end of time. Everything except my immortal feeling. Proud inaccessibility, the feminine concealment of feelings, who needed them now? What could they make possible or destroy?

It was happiness to know that he was alive, that he breathed. I was his paladin and bore on my chivalric helmet the colors of the one I worshiped. Everyone around me knew. I told them: "When I die, put me in a coffin. Let him only come into the room, and I will rise." I was not giving anyone instructions—I was proclaiming my feelings, giving them exact description. I was not the least bit concerned about the impression I made on people around me. He did not love me, and I loved him even for that. He was not suitable material for constructing my theory of correlation pleiades. His spiritual and physical aspects coalesced in a divine harmony where there was no place for me.

We met twenty-three years after the fatal letter. He was a divorced, childless man. I was a divorced woman, the mother of two children. He was suffering from a nervous complaint and had recently been a patient at my longtime friend Nina Alexandrovna Kryshova's clinic. Yura and I lived three blocks from each other, but he wrote me letters reproaching me and everyone for indifference to his sufferings. He wrote one of his letters on birch bark. It was a miracle that I received it. He had misaddressed it. All my energies, money, and connections were thrown into saving my idol. In paying the nurse who came to give him injections, I had to conceal my expenditures from him. He would never have agreed to my financial help, although he himself lacked the means for paying for the nurse's visits.

"Ask Nina Alexandrovna whether I can get married," he said when his course of treatment was coming to an end. Nina Alexandrovna said that he could. He disappeared. Two years passed. It was already 1955. A quarter-century had passed since that horrible letter.

Yura's sister Nina called to ask me to visit. "Raya," I heard over the telephone, "Mother has died, and we want you to be with us." I made a wreath out of leaves from my home plants, forming a tropical garden, wrapped it in a blanket to protect it from the freezing weather, and set off. Lyutsia Gustavovna lay on her bed, very beautiful, very young. We placed the wreath next to her. "And now let's go," Nina said. I was led to another room.

There on a chair sat a short, stout woman who was far from young, wearing large horn-rimmed glasses and feeding a very young baby from a bottle. The woman seemed so old that not only could she not breastfeed the child, but she could not even see it without glasses. She was also a planner for railway lines. She and Yura had met in China, later gone their separate ways, she to her home town and he to his. Still later he had gone to her town to fetch her, and here was the result.

After my expulsion from Akademgorodok and from the Agrophysics Institute, at that period when not all doors had yet been closed to me, I received an invitation to visit the Valters in their new apartment. At the wife's insistence, they had left the apartment in the building where

Blok's Carmen had lived, and were now living in a two-room apartment in a new area. The ceilings there were low, they had to buy new furniture, and the chandeliers could not be installed there either. But the building was on a stretch of wasteland, and they could enjoy the sunsets. They left their little girl with Nina in the old apartment. The last aunt, the only one left alive, had been put in a nursing home.

I arrived at the Valters' straight from my final lecture at the institute, with a bouquet of scarlet peonies that my students had given me. The peonies had not yet blossomed—they were all tight little clumps and leaves.

Yura was alone. His wife had not yet come back from work. He fed me something very healthful and bland—I had a stomachache. We talked over lots of things. Yura expressed his sympathy and sincere regrets with extreme tact. He did not say a word about politics. We talked about the recent events in my life, without saying a word about my participation in the democratic movement. I was left with the impression that Yura did not know of it. His wife arrived. Yura set the table and served her, which she accepted as her due. All her attention was turned to me. "Raya," she said, "I know everything. I condemn you."

"For what?"

"We should love the government."

"Why should we love it?" I asked, perplexed and already sensing the approach of my victory, of my defeat.

"It feeds us and defends us," Atropa said firmly.

"Well, I think that we're the ones who feed *it*, and not badly, and defend *it*. Yura, for instance, lost half a boot of blood defending the government." He did not ask how I knew how much blood he had lost and why I expressed the measurement in such an unusual form. He did not make a sound. He stood behind my back, ready to serve his wife the next dish across the table. She sat opposite me, firm, erect, and stocky. From that moment on and until I left their apartment, the corporeal substance of Yura Valter and his imperious and loyal wife disappears from my memory. I cannot recall either facial expressions or words of good-bye or handshakes. I was shown out by a fiery sunset over a stretch of wasteland, the red clumps of my peonies, the wood carving on the splendid mirror in the tiny entryway, the only thing brought there from Officer Street. I never saw him again. He showed no signs of life. I forgot the name of the goddess of inevitability, as though I had never known it.

No one escaped the all-powerful law of correlation pleiades: neither

Sim nor my father nor the poor lame wretch—the man who had understood everything since childhood, the husband of Moira—my idol, Yura Valter.

When I wrote these lines in 1979, in the United States, I was convinced that our relations would never be renewed. If Yura's wife condemned my dissidence, and he listened to her reproaches in silence, how enraged she must have been by my emigrating, which from her point of view—that of a loyal builder of communism—was treason. But our relations were renewed thanks to my school girl friends, with whom I had never lost touch. In a letter from one of them I learned that Yura Valter stayed in touch with them. His words "How is our exilee doing?" were cited. And that's how I learned of his request that I write him a letter, and that I should send my letter to him in care of my girl friends, not him.

I couldn't write him a letter. It's beyond me to explain why. A letter hung in the air, it was in me, wordless, soundless even. But it didn't even beg to be written on paper. To my great good fortune, I was able to fulfill Yura's request without having recourse to words. I drew him a picture. It was called "The Swallows Fly Away at Dawn." The pink and green gold-edged patterns that are woven into each other are cut through here and there with doubled lilac edges. The news arrived that the picture had been delivered to its addressee. But later came another piece of news. Yura Valter had died. His comrades in high mountain research buried the urn with his ashes and erected a monument over it in Tien Shan, high in the mountains near the end moraine of the glacier bearing his name.

It is with great difficulty that I tear myself away from his grave, descend from the high mountains, and return to the kingdom of the Moirai, to communal apartment no. 6 in building no. 1 on Maklin Prospect.

The second Moira—she passed through in a flash and disappeared—was a character in a play where the main protagonist and director was Maria Ivanovna, my neighbor in the communal apartment, a lawyer with an advanced degree, the wife of a fireman who was also an informer. I do not remember what provoked Maria Ivanovna's attacks this particular time. Perhaps she thought that the lid of my garbage pail located in a kitchen was not secured tightly, or that there was a bad smell coming from my room, but in any event, I told her to leave off with her injunctions, which I called by a harmless Ukrainian word *Vytrebenky*, meaning persistent but groundless demands. "I'll call you as witnesses!" the lawyer yelled to women neighbors crowding the kitchen: "We'll call her [i.e., me] to account for her foul language." I had retired from the scene without saying a thing, and those final words of Maria Ivanovna's, whose husband used almost nothing but foul language, flew after me. I was terribly amused by the selectivity of the lady lawyer's hearing—she heard only the tail end of the Ukrainian word, only the *"ebenky,"* which struck her as obscene—and I was also amused by the syllables in question, because as far as I can tell on the basis of my limited knowledge, the "*ebenky*" are not part of the arsenal of standard Russian obscenities.

A couple of days later, in the kitchen, Maria Ivanovna introduced me to a short, stocky, old-fashioned woman—a representative of the building management. The witnesses were there, too. "Before the matter of obscene and abusive language is transmitted to the management's comrades' court, try to give some explanations." I proposed that the representative of the building management hear the testimony of the plaintiff and the witnesses in my absence and then come to my room afterwards to talk. I left.

The Moira presented herself at my door. After I gave her a seat on the divan, I noticed that this short-legged woman was wearing worn-down low shoes that had once been black but were now gray, matching the rest of her.

"Precisely what obscenity did I use, according to Maria Ivanovna's testimony?" I asked the representative of law and order. She repeated the Ukrainian word. I asked her what was obscene about it. She asked me what the word meant. "Demands," I said. And then I told her about several of the nasty tricks that Maria Ivanovna and her husband had played. To be more accurate, I tried to tell her. She interrupted me and, with an admonishing sigh, asked, "And what are we going to do now?"

"Keep on living just as we did before. After all, there's nowhere to go."

She lost all interest in the conversation. And by the way that she hurried to leave and refused to give her name when I asked her for it, I realized that she was not a representative of the building management, but a Moira sent by Maria Ivanovna with the aim—actually, I can only guess about the aim of the visit—to frighten me, provoke a scandal, find out something about my plans for the future. The door closed behind the Moira. The curtain fell. This daughter of Themis severed another thread.

The third Moira appeared in 1970, when I was expelled from the

Agrophysics Institute and I was applying for my pension. She was an inspector for the Department of Social Security. Her name was Prokhorova. Her job was to process applications for pensions. Setting up my pension was one of the most vivid episodes of my life, but its participants, my comrades in misfortune—retired people—seem pale shadows to me. Against their unsteady background the employees of the Department of Social Security—vampires covered with the blood of their emaciated victims—stand out sharply.

They extort bribes there with extreme virtuosity. And how could they not extort them! Retirement is always a fiasco. Where and to whom will a shadow that has already outlived its time go to complain? Nothing depends on it any longer anywhere. The slight danger that the shadows will unite to repulse the extortionists is out of the question. The future pension recipient has to produce a pile of documents dealing with his or her complete work history. The receipt of those documents is linked at times with unbelievable difficulties. The establishments that employed the shadow figures who used to be people are scattered all over the enormous country. It depends on the inspector whether a document is accepted or rejected. And time is passing. Nowhere does time play as dramatic a role as in the Department of Social Security.

Judge for yourselves: I put in a claim for a pension of 160 rubles per month. In order to receive that enormous monthly sum—and only corresponding and full members of the academy receive more—one needs to have a doctoral degree, work experience of at least twenty years since receipt of the degree, a salary of not less than 400 rubles per month, a position no lower than professor or head of a laboratory, and a final year of work without a single sick day. Failure to comply with any one of the points results in a pension reduction of tens of rubles.

I was especially amazed by the last point. I hadn't guessed about the existence of that absurdity, just as I had not about many others of which I have already spoken. After all, a person is not retiring from the good life. He or she is old and sick. But no! Please submit a document showing that for the last year not a single day of work has been missed on account of illness. I kept thinking that I was mistaken, delirious, that it was all a dream. No. It turned out that everything really was as it was.

But by great good fortune the draconian law did not touch me. For the final year before applying for my pension I had not missed a single day of work because of illness. According to the inspector's way of thinking, I was fabulously wealthy. An inspector's salary is 90 or 100 rubles a month. The pile of documents that I would have to submit gave the inspector almost unlimited power over me. Holding up the pension would leave me without a kopeck. If the inspector held up the pension for three months, I would lose 480 rubles. Wouldn't it be easier to give the inspector 100 rubles right away?

I remember every detail of my visits to the kingdom of shadows. The Department of Social Security is located in the District Executive Committee building on Sadovaya Street, where I had occasion to find myself more than once. The building is old, grand, built without taste. The first barrier was the coat check. You had to take off your coat. The attendant—the gatekeeper of the underworld—controls entry to the bank of the Styx. He has been raised onto a pedestal by the absurd design of the grand building. There was a permanent line of people waiting just to check their coats. But there was no place to hang coats, no one was leaving, and there were no tokens, which the attendant gives the visitor who has checked his or her coat. I went wearing the coat. "Don't come in wearing your coat. It's stuffy enough in here as it is," the inspector commanded. I returned to the line. The attendant did me a favor. He allowed me to put my coat on the banister, which was already draped with the rags of other shadows. I put my coat there too. But I was given special attention. So that I would not lose my property, the attendant folded it up, put it on his chair, and sat down on it.

I was allowed to pass by Cerberus. I waited for the inspector to see me, and I was surrounded by shadows, shadows of the insulted and the injured, victims of the political and economic system of socialism. I was allowed to enter the inspector's office. The appointment with the previous visitor had not yet ended. The victim lingered there, quarreling. The woman inspector was in no hurry to cut short the stream of complaints and was relishing the chance to demonstrate her adherence to the letter of the law. I stood at the door, and although I could not see the participants' faces, I could hear every word of the hideous verbal skirmish between the world of the beyond and full-blooded Soviet reality. "The stamp was legible, but it has faded. I've been keeping this piece of paper for forty years, and you know very well what the ink was like back then. And the office that issued it no longer exists."

"That is no concern of ours. The years mentioned in that document cannot be counted toward the sum of your pension."

Wailing, crying, the grinding of teeth. Shadows grinding their shadow teeth. And everything was quite right, in full correspondence with the letter of the law. Why did I not stand up for the victim? All I had to say was, "Let me see whether I can make out the stamp," and things could have turned out quite differently. I kept quiet—a shadow in a herd of shadows, submitting to the crack of a quite real whip, constrained by spiritual paralysis, the companion of fear. The scene being played out by the inspector before my eyes did not lack design; its subtext was: "Watch and listen. The greater sympathy you feel for the victim, the wider you'll open your purse in order to thank the inspector ahead of time for making out illegible stamps on *your* documents."

And so I appeared before a Moira named Prokhorova. She was not without color. She was wearing a black frock made from a dress after its sleeves had worn through. The corners of her lips and eyes were bent downward. One of her eyes was surrounded by a bruise. Its crimson-and-lilac streaks were crossed obliquely by a strip of adhesive bandage that followed the eyebrow downward. She was the goddess of inevitability, the inevitability of rejection, although for the time being only of delay. They would call me back when a decision had been reached.

Intoxicated by his royal favor, the coat-check attendant got my coat from under his rear end. But that was not all. He stood up, gallantly unfolded my coat, and handed it to me, and then stood there, as though rooted to the spot, while I extracted my scarf and hat from the sleeve. I gave him twenty kopecks in witness of the fact that we had appreciated each other for our worth. Such money used to be called by the insulting term *na chay*, for tea, a synonym of the English "tip," but has now turned into a sign of proletarian solidarity. The coat-check attendant thanked me. The point was not the money. We had had enough contact with each other for human relations to arise between us, and the twenty kopecks were a symbol of those human feelings.*

The thunderstorm broke at the next visit. Without waiting to be called, I went to see Prokhorova. The bruise had faded. The streaks had turned yellow and green. The adhesive bandage had not yet been removed and still slanted downward, along with the corners of the eyes, intensifying

* In the USSR, when giving cloak-room attendants and taxi drivers a tip, I always asked them to forgive me and said that it was a "sign of proletarian solidarity." I do the same thing here in America. I should point out that in the first years after the Revolution in the USSR many people proudly rejected such money as a humiliating handout. No service personnel in the Soviet Union reject it now, but drivers were glad for my apologies and laughed understandingly when I said that in giving them money, I was expressing my proletarian solidarity with them. In the United States, offering a dollar, I tell the postmen who deliver my personal mail the same thing. The majority refuse the money until the words about proletarian solidarity are pronounced. Then I hear the same understanding laughter.

her doleful look. My application had been rejected—not permanently, of course, but until such time as I presented a certificate stating that my position as the head of a laboratory corresponded to the qualification of a doctoral degree. Those who now governed over my fate, the officials of the department, knew that in institutions of higher learning heads of laboratories had neither advanced degrees nor, in some cases, regular university or institute diplomas and were not obliged to have them. These laboratory heads are sometimes like warehouse dispatchers—they oversee the sending off of laboratory equipment for repair. An endless argument began. "I headed a laboratory at an academy institute." I was talking about Novosibirsk, of course. It was a million miles away. "I have a doctoral diploma. There is a copy of it among my papers."

"All right. Present a document stating that the Academy of Sciences Siberian Division Institute of Cytology and Genetics is not an institution of learning." That was enough to make a cat laugh. A document of denial. But there was no room for jokes here. The matter threatened to get dragged out interminably. They had not sent me any written notification; that way the start of the paper chase would not be documented. They knew that I would come without any notice. The shadows are obliged to channel an endless wave of documents into the poisonous waters of the Styx. The extortionists make do with a system of communication using sound signals. I went to see the chief of the realm of the dead.

Plouton, a representative of the Soviet elite who was sprightly, full of Ukrainian lard, and who spoke with a Ukrainian accent, was not at all a shadow. Like his Greek analogue, he was the god of wealth and fertility in the underworld. I found him in lively conversation with a circle of his fellow employees. He had sunflowers growing at his dacha. Some of them were very tall but seedless, while the small ones were providing a harvest. My case would be transferred to a senior inspector.

I had another appointment. The case was returned to Prokhorova, and I would have to do as she said. I needed to bring a document. I had a better idea. "Liza," I said to my elder daughter, "we're going to go to the Department of Social Security. You'll stand nearby while I talk to Prokhorova. Act as though you don't know me. I'll tell her that I'm going to make a complaint, and I'll ask you, 'Will you agree to be a witness for me?' And you'll agree. But things won't get to that point. It will be enough for you to show interest, and Prokhorova won't dare indulge in her lawlessness. I won't have to procure a certificate proving that I'm not a camel." Liza said that it would be better to pay a bribe, that everyone either paid in person or acted through intermediaries from their place of work or through lawyers. But she went. Just as I had foreseen, Liza had an effect on the Moira.

Moirai! They are an inseparable feature of the Soviet landscape. Together with Maria Ivanovna, Filatych, and many, many others who secretly hold the threads of our fates in the marionette theater, the Moirai not only severed threads, they sliced through my anchor cables; they were the wind that swelled the sails of the ship that took me out to sea. I am obliged to them for the resolve to break out. They save me from nostalgia. I do not want to go where everything is enmeshed in webs spun by them.

Huge numbers of the most diverse and extremely repellent Moirai were needed in order to overcome my great attachment to the beautiful city on the Neva, to the graves of people dear to me—Sim and my father—to the wrought-iron lace of the fences that suggested the designs on my pictures. There proved to be more than enough Moirai to foster the desire to leave. To get away not only from my own trouble, but from the kingdom of the lie, violence, absolute lawlessness, and slavish vegetable existence. But an outside stimulus was needed for the desire to turn into a decision—into a passionate, unrestrainable urge. A drama played out in the family of Savely Yuryevich Dudakov served as the trigger mechanism.

I became acquainted with Savely and his wife Inna immediately after my return to Leningrad from Akademgorodok. Savely's father, a Jew, had been a member of the Bund, the Polish faction of the Bolshevik party. Yury Dudakov was the father's party nickname. When Poland resisted the onslaught of the Red Army and separated from Russia, a split occurred in the Bund. One faction was in favor of the unification of Poland with Russia. Savely's father belonged to that group. He had settled in Petrograd.

We had only just become friends when I learned that the signatures of Savely and his brother were among those on a printed appeal to Jews. The appeal, dictated by political wisdom with extreme tact, urged Jews to emigrate to Israel. It did not contain so much as a drop of anti-Soviet propaganda. The Soviet Union's role in the United Nations' decision to create the State of Israel was noted with gratitude. The appeal was distributed through *samizdat* channels. No matter how diplomatically the appeal was written, it was clear that Savely and Inna were in for a bad time. Jews were harassed everywhere even as it was, but here there was the charge of sedition to boot.

Savely is an engineer, an authority on Russian poetry, the owner of a very rich library (his father had begun collecting it), and a chess player of the highest class. In 1971, he no longer worked as an engineer but as a chess intructor at a Pioneer Palace in one of the districts of the city. Inna is a pediatrician, precisely what I needed for studying the frequency of the appearance of mutations in man. In 1969, Inna tried to enroll as my graduate student at the Agrophysics Institute. She performed brilliantly on the entrance examination, and her paper—one of the tests of a candidate's suitability—had all the earmarks of an article ready for publication. Her name—Inna Ilyinishna Dudakova—would not have aroused any suspicion. The head of the personnel department came to see me at the laboratory in order to fulfill her police function, the only purpose for which a personnel department in a Soviet institution exists. She informed me that Dudakova's candidacy had been rejected. "Have you seen her photograph?" she exclaimed.

"Why do I need her photograph when I know her personally?" I asked, understanding full well the meaning of the question. Inna's name sounds Russian, but the photograph exposed her Jewish origins.

The job of a clinic pediatrician resembles penal labor. The doctor makes endless house calls, going from patient to patient, climbing up to the fourth, fifth, or sixth floor of apartment buildings that lack elevators. Toward the end of a workday, Savely would go from apartment to apartment with Inna and drag her by the arm up the stairs. They lived in a communal apartment that was much worse than the one I lived in. They were subjected to the most hideous outrages and were witnesses to hideous villainies committed by their neighbors upon each other. Things had even gone as far as murder there. Their family of five lived in a single room partitioned by dressers and cupboards. On one side of the partition lived Inna's parents, and Savely, Inna, and their little daughter lived on the other side. Savely very much wanted to go to Israel. Inna, unhappy, harassed, rejected by everyone, resisted with all her might.

Ornithologists know that in preparing for takeoff, some birds of the same species rehearse a group flight and then fly off. The other part migrates, attracted by the spectacle of the flock having taken wing. I turned out to be a bird of the second category. But to a certain extent, everything was just the reverse. It was not Savely's passionate desire to break out but Inna's no less passionate desire to remain that made up my mind. She came to see me with a request that I dissuade Savely from what she thought had turned into an obsession. And suddenly I felt an unrestrainable, passionate desire to follow Savely. I told Inna that I envied Savely's decisiveness, and that it was better for them to leave. She continued to resist until the KGB threatened Savely with arrest if they didn't leave.

They departed in 1972. The circumstances of their emigration bordered on a pogrom. At the meeting where Inna was branded a traitor, the head of the clinic turned to her and said, "Divorce your husband and stay here. You don't want to? You're a prostitute—you need a man." Savely received permission to take out his library, but it perished on the way. The boxes were drenched in some sort of liquid.

I knew all of this, but nothing could restrain me any longer.

The preparation for leaving took two years. I copied my expedition journals in order to have them microfilmed. At times it seemed that the door to the kingdom of freedom which had been just slightly opened was leading me not to freedom abroad, but into a Soviet trap, a trap in the literal sense of the word—prison. I committed one crime after another, forced by the laws either to break those laws or give up the idea of leaving. Making a microfilm copy of the journals was a crime. There was an even higher crime—transporting the microfilm out of the country. It is also a crime to take pictures out without receiving permission from the appropriate authority. And it's senseless to appeal to the appropriate authority; you won't be granted permission anyway. It is permitted to take out pictures done in the style of socialist realism. An artist has to pay the appropriate authority fabulous sums of money for each picture. Pictures that depart even the slightest bit from socialist realism in the direction of impressionism, expressionism, or any other "ism" within the realm of representational art are not allowed to be taken out. It is forbidden. And there's no point even in mentioning abstract paintings. The authorities have obviously taken upon themselves the task of protecting the West from the corrupting influence of Soviet nonconformism. If I had appealed to the proper authorities, the fate of my appeal would have been as lamentable as one could imagine. The road to the West for my abstractions was completely blocked.

You have to get an infinite number of documents after having overcome an infinite number of barriers, and bribes are involved—another crime. And moreover, there's no money for taking out one's own books, for the visa, the ticket, baggage. By committing crimes, risking getting caught, giving bribes, I managed to get all the necessary documents and send my pictures and microfilm to the West. I got part of the money legally and part through criminal paths. I should start with the story about those criminal means of raising money in order to dispel any suspicion that I raised money through robbery.

Through the will of fate and the play of circumstance, by living in Leningrad and thinking about fleeing the Land of Mature Socialism, I wound up as a member of a Jewish family, in the capacity of a foster grandmother. My "family" lived and lives to this day in Miami, Florida. Although it's a very amusing story, I won't tell it. Please try to understand me. My new relatives sent me packages and money. They called me on the phone and when, after eleven o'clock at night, the phone would ring in the hall of my communal apartment and a voice from the other side of the ocean would say, "We are going to appeal on your behalf, dear Raissa, to our senators, to congressmen, to President Ford," I would freeze, terrified that at any moment the fireman–KGB informant would unplug the phone on the legitimate grounds that after eleven o'clock at night it is forbidden to use the telephone in the hall of a communal apartment. The money that my "family" sent me was transmitted to me by the bank in the form of ruble certificates that gave me the right to make purchases in special stores intended only for foreigners and which stocked all sorts of items in short supply. At speculators' prices I would exchange those ruble certificates on the black market for regular rubles and save them in order to pay for a visa. That was the criminal means of raising money.

There were also legal means. An enemy of the people, a traitor to his native country, if he is a retired person, is given a pension for six months, stripped of his citizenship, and after that the state is free of any obligations toward him (or her). The money is given out after the receipt of the visa, but in order to receive it you have to pay money in. My friends collected that money for me. That was the legal source of funds.

In my desperate struggle I suffered only one defeat. The three of us were to leave together—Masha, her daughter Marina, and I. Liza had gone to Siberia to join her first husband in exile. One of the humiliations to which people emigrating from the USSR are subjected is the necessity of receiving permission from any relatives remaining in the USSR. Masha's former husband, Sasha Shenderov, refused to give permission for his daughter to leave. Shenderov's refusal, which was inspired by the KGB, was supposed to transfer me automatically to the category of *otkaznik*, those permanently denied exit. We received indubitable proof of Shenderov's collaboration with the KGB during our second visit to OVIR inspector Borovikova. That heavy woman was in charge of our "case." We were still only at the door of her office when she turned to Masha and said: "Your husband writes us su-u-u-uch letters, telling us about your letters"—she turned sharply to me—"about your conversations."

"I address my conversations to the Central Committee and sign my name to them, so he's wasting his time," I said, "but allow me to doubt his epistolary zeal. You can't get a line out of him."

"I'll show you," the dumpling said and left. She was gone for a long time and came back with a packet of papers. She went up to Masha and showed her the first sheet in the packet. I couldn't see anything. "Is that his handwriting?" she asked.

"Yes, it is."

Borovikova quickly departed.

The tender parent who had not given a single kopeck toward the child's support until the court garnished his salary for alimony now declared his readiness to take in his daughter. If Masha would give him the child, he would be willing to give her permission to leave.

We were not just denied permission to leave. We were misinformed, sent from office to office. The absurdity of the demands we were making and the officials' goal of turning us into permanent *otkazniki* finally became obvious. We decided that I would leave on my own and would later try to get Masha and Marina out. It was nearly impossible to make that decision, but we saw no other way out of the situation that had been created.

When I told Borovikova that I was submitting my documents to leave without my daughter and granddaughter, her mouth literally fell open. The bird was getting away. Borovikova's responsibilities included giving me a list of documents that I was to attach to my application. There was no printed list. She started dictating the addresses of the Telephone Administration, the Housing Department, the Office of Rental of Such-and-Such Goods, and the address of other offices that rented other things. I was supposed to bring documents certifying that I was not a client of theirs. Many, many documents. She gave me the business hours for each of these offices. The list of establishments was the truth. Their addresses and their business hours were complete misinformation. She concluded her dictation by saying that if all the necessary things were not done by such-and-such a date, a visa would be denied.

She was certain that if I followed her instructions, I would never manage to complete the task. In my simplicity of heart I told her that I would manage, that I had instructions, that she had wasted time with her dictation. I had known everything that she was supposed to tell me, and which she had purposely distorted, even without her. All the people in Leningrad who had sought the right to emigrate had formed a sort of fraternity. Typed instructions were transmitted from hand to hand. They contained a list of the required documents, a description of where and how to get them, the addresses and business hours of the establishments, the names of the people in charge of giving out documents. Changes were entered into the instructions and they were retyped. Masha did the retyping. I managed to get all the documents in time.

Before giving me the visa, Borovikova put a document before me on the table and suggested that I sign it. I was signing my recognition of the fact that at the moment I crossed the border I would lose my citizenship and along with it the right to return to the Soviet Union or to visit it. I said that she had no right to give people such documents to sign. The Soviet Union was a member of the United Nations and was obliged to abide by the principles proclaimed by that organization in the Declaration of Human Rights.

She objected: "Yes, we observe all the declaration of the United Nations, but we make exceptions for Jews and Spaniards. We offer them our native land, but they betray it."

I had received an enormous dose of antihomesickness medicine. Oh, you police bitch, I thought. A country belongs to the people who love it, wherever they may be, who beautify it with themselves, not to people like you, who trample its merits in the dirt. I signed the document without any further objections.

When I tell people in the West—Americans, Germans, emigrants from the Soviet Union, whether Jews or Russians—about Borovikova, I am always asked the same question: Why did she name Spaniards in addition to Jews? She had in mind the children of Republican parents who died during the Spanish Civil War. Orphans were brought from Spain to the Soviet Union in 1936. Thirty years later, these elderly people began wanting to return to their native land. Franco was still alive. The Soviet Union allowed them to return. Now I had learned from Borovikova that they were stripped of their citizenship and stigmatized as traitors to the homeland. Borovikova's admission that the Soviet Union makes exceptions in observing the laws of the United

Nations is also quite important. A crime, after all, is simply an exception in the obeying of the law.

On December 13, 1974, the plane bearing me to Vienna took off.

Masha, whom I had abandoned to the mercy of the KGB, went through terrible times. By harassing her they punished me.

Masha had parted from her mother in order not to be parted from her child. The criminal desire to leave was, however, there and had to be punished. The child was to be taken away and given to the parent who had not even thought of abandoning the socialist paradise. Shenderov took Masha to court. The case was tried in Leningrad. Shenderov called the residents of the communal apartment as witnesses to testify that Masha was not worthy of being a mother to her child. Maria Ivanovna; her husband, the KGB informer; the rat king; and other stone women were all called to appear. The charming Sasha was already married by then and living in a one-room apartment with his wife and two children. And a one-room apartment in the Soviet Union is far from what a one-bedroom apartment in the United States is. A one-bedroom apartment in the United States consists of two rooms, but a one-bedroom apartment in the USSR has precisely one. Shenderov worked as a lab assistant. The court may take a child from its mother and give it to the father only if the father has sufficient space to house the child and if his salary is sufficient for supporting it. Shenderov presented false documents certifying that he had all that. Unfortunately for him, he had the job of lab assistant in the laboratory headed by Zoya Sofronyevna Nikoro. She learned of the false documents, flew to Leningrad specially from Novosibirsk, and testified at the trial that they were false. Nor was Shenderov able to prove that Masha had a depraved style of life. Not a single living soul turned up at the trial to testify against Masha. At apartment no. 6 in building no. 1 on Maklin Prospect they all were in some way or other wrongdoers, they sensed trouble and preferred to stay home.

It was six years before Shenderov gave his permission, but my friends finally forced him to grant it. In January 1981, Masha, Marina, and Masha's second husband, Zhenya, whom she married soon after my departure, arrived in the United States.

Liza did not have the difficulties that Masha did. She has been living in Paris since 1976, teaches genetics at the university, and is relishing the joys of family life and raising two sons with her second husband, Jacques Reperant.

I so want to tell about the great help given to me by my friends, by those people who risked landing in prison in order to microfilm my expedition journals, to send my pictures to the West, to transmit my microfilm across the border, who forced Shenderov to give his permission, who continued troubling themselves for my sake when I was already here, out of danger. But I will be silent about the people dear to my heart as long as they are still alive. They are over there—in the web. My lips have been sewn shut by the Moirai's thread.

# *Appendix*

Transcript of the Closed Meeting of the Science Council of the Institute of Cytology and Genetics of the Siberian Branch of the USSR Academy of Sciences, April 4, 1968

*Participants:*

- D. K. Belayev, chair of the Science Council, director of the institute, not a party member
- L. A. Antipova, administrative secretary of the institute, member of the CPSU
- O. A. Monastyrsky, junior staff researcher at the Laboratory of Ecological Genetics, secretary of the Party Bureau
- R. I. Salganik, deputy director for Science Affairs, head of a laboratory, member of the CPSU
- G. F. Privalov, same as above

Members of the Science Council, heads of laboratories, or group leaders:

*Members of the CPSU:*

- Yu. Ya. Kerkis
- V. B. Yenken
- I. I. Kiknadze
- V. V. Shumny
- O. I. Moystrenko
- V. N. Tikhonov
- G. A. Stakan

*Nonmembers:*

N. B. Khristolyubova
V. V. Khvostova
A. N. Mosolov
L. I. Korochkin
N. N. Vorontsov
A. N. Lutkov
Z. S. Nikoro
Yu. O. Raushenbakh
R. L. Berg

*Specially Invited:*

V. Ternovskaya, chair of the Institute Trade-Union Committee
N. Dymshyts, secretary of the institute's Komsomol bureau
N. V. Tryasko, senior researcher for the Laboratory of Evolution Genetics (head of the laboratory—D. K. Belayev), district deputy

BELAYEV: On the agenda for today's closed meeting of the Science Council we have a single item of business: a letter written by several employees of institutes of the Siberian Division, Raissa Lvovna Berg among them, in regard to the trial of four people convicted of anti-Soviet activity and foreign currency speculation. The letter found its way to the West and was broadcast over the Voice of America. Both the latter fact and the letter itself have been condemned by party organizations. Oleg Alexandrovich Monastyrsky will give us more detailed information.

MONASTYRSKY: From January 8 to 12, 1968, a trial of four people was held in Moscow. They were Galanskov, Ginzburg, Dobrovolsky, and Lashkova. These people were convicted of foreign currency speculation and of links with subversive anti-Soviet organizations in the West, in particular, with the terrorist organization NTS.* On January 16, an article appeared in *Izvestiya* about who these people are and how their link with Brooks-Sokolov, an agent from NTS, was uncovered. On January 18, *Komsomolskaya Pravda* carried a description of these people's political activity and unmasked their political profile. On January 29, the same newspaper printed a collection of letters from readers. On March 27, the *Literary Gazette* printed an evaluation of these people's activity and remarks about the trial by people who had attended it: professors, professionals from the world of science and technology, and workers. On March 26, American newspapers, including the *New York Times,* published the contents of a petition

* NTS—Narodniy Trudovoy Soyuz (People's Labor Union), is a Russian émigré association located in West Germany, known for its anti-Soviet activity.

signed by forty-six employees of the Siberian Division of the Academy of Sciences who live in Akademgorodok, demanding the court's decision be rescinded. On March 27, the petition was broadcast over the Voice of America. (*He reads the text, but instead of "unproven accusations" reads "illegal accusations."*)

BERG: Your copy of the text has a misprint. The letter says "unproven accusations," not "illegal accusations." That alters the matter.

MONASTYRSKY: (*Nods*) Ginzburg and Galanskov had been tried for felonies even earlier. This time they were convicted of a link with subversive Western organizations, to which they transmitted anti-Soviet material distorting our Soviet reality. The reactionary journals *Posev* and *Grani* published those materials. Reactionary radio stations such as the Voice of America and others broadcast them. They rendered serious assistance to anti-Soviet propaganda. A few days ago *Komsomol Pravda* printed Mikhalkov's report, according to which these people are not writers. Leonid Ilich Brezhnev mentioned them in a speech to party activists and roundly condemned them. *Pravda* has published an article by Mstislav Vsevolodovich Keldysh that roundly condemns the petition. Those in attendance at the Presidium of the Siberian Division of the Academy of Sciences and at plenary meetings of District Party Committees have condemned the letter and the people who signed it. Workers and members of the engineering and technical personnel at Sibakademstroy, Postal Branch 100, and many establishments in Novosibirsk have held crowded meetings at which the people who signed the letter were unanimously condemned.

BELAYEV: How many scientists signed the letter?

MONASTYRSKY: Four with doctoral degrees and nine with candidate degrees.

BELAYEV: Thirteen scientists.* Who are the others?

MONASTYRSKY: Graduate students, laboratory technicians, engineers. Many signatures were illegible.

BERG: You don't have any illegible signatures. The signatures on all the letters, except for the original, which was sent to the procurator, were typed.

MOYSTERENKO: Raissa Lvovna, was your letter sent to the District Committee of the party?

BERG: No, it wasn't.

BELAYEV: Was it received by the government?

MONASTYRSKY: No. The letter was registered at the Office of the Procurator General, the original. Other offices don't have it.

BERG: I received postal notifications of receipt by all seven offices to which the letter was sent.

* It is amusing that Belayev considers to be scientists only those who have a governmental certificate bestowing their right to be scientists.

TIKHONOV: Is the letter by any chance a forgery?

MONASTYRSKY: The District Committee thinks that it's genuine.

KERKIS: Raissa Lvovna, how many copies did you sign?

BERG: One.

KERKIS: I have heard that during the process of preparing the letter many comrades argued against writing it. Is that so?

MONASTYRSKY: Yes, there was such a conversation. Respectable people argued against it. They said it would wind up abroad.

KERKIS: The enemies knew where to look!

BERG: What do you mean?

KERKIS: I'll explain in my own good time.

TERNOVSKAYA: What was the extent of Raissa Lvovna's participation in this business?

BELAYEV: Give us your explanations. Our discussion is of a comradely nature.

BERG: Can alarm be provoked by the conviction of four people accused of links with anti-Soviet Western organizations, foreign currency speculation, and subversive activity? The circumstances of the case produce doubts. The people were under investigation for a year. In Moscow there was a demonstration protesting their arrest. Information about the demonstration got through to the West. Young people—the participants in the demonstration—were arrested. They were kept under investigation for more than half a year. They were tried and convicted. Information about their fate got through to the West. And then, after all this, the witness Brooks-Sokolov turns up, as a liaison man between subversive organizations in the West and the young people who had already been in prison for a year, and appears as the main witness for the prosecution. That provokes doubts. The very description of the trial does not give a persuasive, legally sound picture of the events or of the corpus delicti, although it is known that Ginzburg and Galanskov did not admit their guilt, nor does it give a description of the motives for their actions or the speeches made by the defense.

The one-sided nature of the treatment provokes doubts. The rumors that reached me about Ginzburg did not paint him as a mercenary person, and the accusation of foreign currency speculation does not fit with the other information that I have. He seems a selfless person, not at all a criminal. The letter that I sent to high offices expresses fear that lawlessness may transpire behind closed doors. Was the trial open? It is possible to make a trial open and prevent the entry of all undesirable elements. Simply taking an interest in the trial can cast a shadow on the people who are interested in it. Make every person who goes to a trial on his own initiative black and all those invited white, and the courtroom will turn into a gathering of angels.

I am well acquainted with the trial of Brodsky and its treatment in the press.

You may object that in a legal matter one should not judge by analogy. Brodsky could have been guilty of nothing, and Ginzburg, Galanskov, and the others guilty. But where the possible violation of rights is concerned, analogies are absolutely justified. If it was possible to take a frail young man absolutely innocent of any wrongdoing, the only son of two workers, a poet and translator recognized in literary circles, and accuse him and sentence him to five years of exile and hard labor in accordance with a decree about parasitism, then lawlessness is possible, and while loving one's country, worrying about it, jealously concerned about its prestige, one may ask for attention to a case that in fact was tried behind closed doors.

If one person is tried today and everyone keeps silent, tomorrow yet another one is tried, and everyone is silent again, many people will wind up in prison. The next generation will condemn those who were silent.

May one petition the Supreme Court and the Central Committee to reexamine a case that has provoked doubt? Such is the right of every citizen, and no one contests that right, and it was not the petition itself that provided the reason for convening this closed session of the Science Council.

Are collective actions on the part of citizens allowable? That depends. Collective criminal actions are punishable. But any organizations of people pursuing legal goals are provided for by law and are a condition of democracy. A collective request or a joint petition, if they are just, cannot be condemned.

Can you lay the blame on Soviet citizens for the fact that the foreign press or radio stations came into possession of a petition or a letter?

An information leak is always possible. We are not talking about divulging state secrets or information related to defense. Here one needs to distinguish two possibilities: 1) The person himself transmitted the information, and 2) it wound up there without his knowledge. The approach in each case should be different. But even the first is not in and of itself a criminal matter. And of course, the same goes for the second, and even more so. Let us now imagine a person who fears raising his voice against lawlessness because he fears that the information may be leaked abroad. But then there's no chance at all of fighting evil. A violation of the law within the country is one hundred times more dangerous for us than all our enemies' propaganda. Our principles are unshakable and do not depend on what our foes say.

We are accused of having transmitted a letter to the enemy camp, a letter in which collective distrust of Soviet justice was expressed.

I can speak about myself. I see no particular harm in the fact that the letter was broadcast over the airwaves. The West does not consist exclusively of enemies. All the anticapitalist forces are for us, and we need to take care not to alienate them. Surely the trials of Brodsky, of Sinyavsky and Daniel, of Ginzburg, Galanskov, Dobrovolsky, of Bukovsky, Khaustov, Delauney, and

Kushev, and the condemnation of Pasternak at one time lower our country's prestige. And a petition to the Central Committee and Supreme Court by a group of scientists merely requesting a public trial? It seems to me that as things stand now, friends of the Soviet Union have become convinced that democratic principles in our country are in force. Citizens enjoy all the blessings of freedom. With absolute impunity, they may join together and address their protest to the highest possible offices. They are not threatened by arrest or exile or loss of work. No one can incite their comrades to hound them. Their civic feeling will not be affected by any possible legal consequences. Their request will be satisfied, and I am waiting impatiently for the response to our letter.

I fully recognize the validity of the interest with which comrades have come here in order to learn what led me to place my signature on the letter. It was concern for democratic principles. I have seen how they can be trampled in our time. But I have also seen how they can be reinstated. Brodsky is living with his parents in Leningrad, and his poetry and translations are being published. I hope that our request will be satisfied and that the letter will play that role for the sake of which it was sent.

SALGANIK: The analogy with Brodsky is, after all, only an analogy. As a scientist, you do not have the right to use it.

BELAYEV: Was the trial of Brodsky open or closed?

BERG: It was a de facto closed trial conducted under the guise of a public trial.

BELAYEV: And how did you get in?

BERG: I went with Brodsky's parents.

BELAYEV: No one invited you. That means the trial was open and there's nothing more to talk about. We aren't interested in Brodsky.

RAUSHENBAKH: And everyone present at Brodsky's trial came of his own desire. Raissa Lvovna wants to persuade us that she was presented with tickets of admission and warrants. Raissa Lvovna, do you admit that they conducted foreign currency operations and transmitted libelous information to the West?

BERG: The trial was fictitious. Galanskov and Ginzburg had gathered the materials of the trial of Sinyavsky and Daniel, and those materials wound up in the West. The foreign currency operations amounted to changing a fifty-dollar bill that Dobrovolsky had received from religious circles.

KHVOSTOVA: What religious circles?

RAUSHENBAKH: How do you know that?

BERG: From acquaintances.

RAUSHENBAKH: So you recognized that they were receiving money for the information they were sending to enemies, that they were engaging in foreign currency speculation, and that they were conducting subversive activities. Nevertheless, you demanded the sentence be rescinded on the grounds that the accusations were illegal. So do you think that no one should be tried? No, don't try to get off the hook. In your version, that's how it looks.

BERG: If this perversion of every word and the attribution to me of what I have not said bears the name of a comradely discussion, I refuse to answer. Yuly Oskarovich is provoking me. I ask that he be removed from the proceedings. Otherwise, I'll leave.

RAUSHENBAKH: (*Falls silent, but with a look of satisfaction, remains.*)

ANTIPOVA: When I was working in Germany after the war, I had the chance to check what and how newspapers and radio stations of various countries reported the events that were transpiring before my eyes and about which I had information on the basis of classified documentation. I became convinced that the most intelligent, honest, accurate, and fullest documentation is Soviet. I know that the employees of NTS are anti-Soviet people. They were interested in us, too. They hunted us out and tried to recruit us. The link with NTS alone sufficiently characterizes the comrades who were tried.

KHVOSTOVA: What sort of comrades are they to us?

ANTIPOVA: These citizens, these people who have been convicted, and of course, they are not our comrades.

BELAYEV: Raissa Lvovna, do you regret what you did?

BERG: I am sorry that the letter wound up abroad. But I would sign it again if I hoped that I could help those young people. The newspaper the *New York Times* did not have the right to publish a letter that was not addressed to it. I am ready to enter a protest.*

*Credulousness is the sister of truthfulness. It is much easier to deceive a truthful person than a liar. I believed that the *New York Times* had published our letter. That was why I said that I was ready to enter a protest in the international arena against its publication. Doubt crept into my soul when a person deserving of trust told me that the *New York Times* publishes documents only if it has the original. There was only one original copy of our letter, and it had been sent to Rudenko, the procurator general of the USSR. That meant that if the letter wound up at the editorial office of the *New York Times*, it was sent there from Rudenko's office. That version seemed plausible. The publication abroad provided grounds for legal action and could have been fabricated. I am aware of an instance in which the Voice of America broadcast the text of a letter that had been sent to the Supreme Court, the Supreme Soviet, and the Central Committee. The author of the letter had taken measures to prevent the letter from winding up abroad, and he had not told even his own mother, let alone foreign correspondents, about the content or even the fact that he had written the letter. He told me the whole story when the letter was broadcast over the airwaves. He had been guided by patriotism—the readiness to conceal his country's crimes from the world's eyes. That person's letter and ours were as similar as two peas in a pod, and like us, he had protested the absence of public political trials, not the fact of their existence. Other than as a governmental provocation, there was no way the letter could have landed abroad. The Voice of America is one thing, however, and the *New York Times*, another.

RAUSHENBAKH: Ha-ha-ha!

KHVOSTOVA: That's funny!

BERG: Dmitry Konstantinovich, ask Yuly Oskarovich to leave.

RAUSHENBAKH: (*Falls silent and remains.*)

BELAYEV: The whole issue has an ugly coloration. We educate our comrades and should educate them in the spirit of Communist morality. Each of us bears responsibility for the views of the other. One for all and all for one. That's how! And if someone writes, there's nothing bad about that, but it's important in what regard he writes. And did you protest the arrest of the national Greek hero Manolis Glezos? And did you protest with regard to Spock's trial in America?

Who wishes to make a statement about the letter that was signed? Zoya Sofronyevna, perhaps you'd like to?

NIKORO: No, I'll wait.

KHVOSTOVA: Hee-hee.

KIKNADZE: Tell me, please, from which Communist newspapers did you, as you put it, extract your information?

BERG: The *Morning Star* and *l'Humanité*. On January 28, 1968, the *Morning Star* reported that not a single one of the correspondents from Communist newspapers was admitted to the trial and that the trial was a de facto closed one.

BELAYEV: It's a good thing that the correspondents weren't admitted. Who else wishes to state his opinion?

KERKIS: You realize, comrades, that it is not easy to talk about this issue. I have always had and still have nothing but good feelings toward Raissa Lvovna. And when I found out that such a thing had happened, I was upset. For me it is beyond question that one may address any office about any question and that there is nothing reprehensible about that, but in the given instance, I have no reason to doubt the correctness of the characterization of the actions of these people that the organs of our press have made. Many people whom I know well have not expressed doubts about the correctness of the publication. Raissa Lvovna has information from "AWTM" ["a woman told me"] and, moreover, in the circumstances of a large public market. We need to take measures to assure that no one knows what has been sent to the government and that it is

---

After finding myself in the United States, I decided to take a look at that publication. Unlike the Saltykov-Shchedrin Public Library in Leningrad, which I had such difficulty gaining access to for the sake of my history of the merchant Krivozhikhin and his misadventures while sailing the Sea of Aral, the Madison (Wisconsin) Public Library offered me a microfilm of the newspaper without the slightest difficulty. The newspaper does not contain our letter. Monastyrsky's report turned out to be an educational canard.

not passed from hand to hand. I believe that it was not your fault that the letter wound up in the wrong place, but it wasn't for nothing that I said that American counterintelligence knew where to look for compromising information. It looked for it around you. Your tendency toward patronage of the arts got you into trouble. Raissa Lvovna concentrates around herself people who consider themselves unrecognized writers, people with a chip on their shoulders. I heard your Delauney at the banquet when he said that Galich had given back to poetry its quality of bread. I wanted to smack him in the face.

BELAYEV: In another place.

BERG: Dmitry Konstantinovich in his home repertory.

KERKIS: The worst thing is that the letter wound up abroad. If gross caricatures wind up there, then everyone ought to try to prevent that. We have very many failings, and, in particular, in the way of public access to information. But these questions need to be raised within the country, not abroad. I regard your actions in the most negative way. You've read the Criminal Code, but I couldn't care less about the Criminal Code, and you'd do better to occupy yourself with science.

YENKEN: Yury Yakovlevich is quite right. Two worlds now exist. They are doing everything possible over there to cause us trouble in the way of ideology. We need to conduct ourselves in such a way that our behavior strengthens the idea that we are doing everything in the right way. These rogues and scoundrels were not punished so very severely. They should have been given longer terms. All our sources of information are based on opinions, and opinions are based on political views. You should have been certain that there would be no leakage of information abroad. The sending of that letter undermines our authority, that of our institute, and that of all of Akademgorodok. A city was created for scientists, and what do they do? Write a letter that produced nothing but harm. Why did you, an important scientist, who ought to know what's what, intercede on behalf of obvious scoundrels, encourage foreign currency speculation and the publication of caricatures of the Soviet regime? You are looking after oppressed people—

BELAYEV: What oppressed people? People who consider themselves oppressed.

KHVOSTOVA: Vadim Borisovich meant oppressed in quotation marks.

YENKEN: You should have sent the letter via the Regional Party Committee in order to exclude the possibility of its winding up in America.

BELAYEV: Who would like to speak?

BERG: You're performing a play that you've learned by rote memory. Vera Veniaminovna will speak now. She has already written down an outline of her remarks.

KHVOSTOVA: No, it isn't an outline. Here's what I'm doing. (*She shows a piece of paper with a drawing of mytosis in the anaphase on it.*)

BERG: If I'm guilty, then I should be tried according to the Criminal Code, but these sorts of proceedings are illegal and unprofessional in and of themselves, and nonprofessionals have no right to them. You've organized a courtroom here!

SHUMNY: We haven't gathered here to try anyone. Raissa Lvovna's declaration is an offense to the people present. Many people here will stand up for truth.

KHVOSTOVA: And have!

SHUMNY: One should cry out if an injustice is done, but the case here is not the sort for which one needs to stir up public opinion. If they were honest people, America would not have taken up their defense. They linked themselves to NTS, and after all, former Gestapo members work there. The court had every reason to try them. What difference does it make that the courtroom was small?

KHVOSTOVA: In Raissa Lvovna's opinion they're worthy of the Palace of Congresses.

SHUMNY: Through Raissa Lvovna's fault a bad thing has resulted. The Americans have driven a wedge between our workers and members of the intelligentsia. Raissa Lvovna gave in to her desire to be revolutionary.

KHVOSTOVA: What do you mean—"revolutionary"? We've always stood for truth.

NIKORO: Allow me to speak. I do not know at all what these people are like, but the press has not been providing any information either. Issues of an ideological nature should be treated fully, not the way this trial has been treated. The impression has been created that the foreign currency speculation and the link with anti-Soviet organizations have been tacked on in order to conceal the true reasons for the trial. When that is done abroad, we are told about it, but we don't know anything about our own affairs. These people had lawyers, but there's nothing in the newspapers about the defense. One cannot consider the information that was in our newspapers exhaustive. If these people were punished by statute 70,* then what does foreign currency speculation have to do with anything?

KHVOSTOVA: But what's statute 70? Explain!

BELAYEV: Let Zoya Sofronyevna speak!

NIKORO: We are the masters of our life, and we have the right to information from judicial organs, not from journalists. If they had come to me with that letter, I would have signed it. I think that the fact that Soviet citizens petition their government does honor to Soviet citizens.

* In 1958, statute 70 replaced statute 58 of the Criminal Code dating from Stalin's time. The new statute provided punishment—seven years of labor camps plus five years of internal exile—for all forms of action, organization, or discussion hostile to the government.

BELAYEV: The workers have a different opinion.

NIKORO: Don't speak for the workers, Dmitry Konstantinovich. Speak for yourself. Under Nikita Sergeyevich Khrushchev I was "worked over." Let's remember that incident now. The people who organized it ought to be ashamed. What if the record of this meeting wound up abroad? Our enemies would be delighted. So don't do what you'll have to blush about later.

BELAYEV: Without having information, how could you sign a letter that categorically demands the rescission of a verdict reached on the basis of "illegal accusations"?

BERG: You're quoting a misprint, which alters the matter.

NIKORO: Why would I sign? I often see an illegal action and don't do anything about it. When a collective letter is being written, it's impossible to arrive at a formulation that would satisfy everyone. After others have already signed, it's too late to change anything. As long as I was in agreement in principle, I would sign and reconcile myself to the wording of the letter.

BERG: Dmitry Konstantinovich, why did you distort the meaning of the letter?

BELAYEV: You're the one writing down a summary, Raissa Lvovna. The others aren't doing that.

KHRISTOLYUBOVA: I would not have signed that letter. I don't find the people who wrote *The White Book* appealing. One should do everything within the country. In the newspapers you often read about injustices, and those injustices are corrected. But these people looked for readers abroad. The propaganda that the Voice of America carries could end in a war. It was unpleasant for me to hear Raissa Lvovna's name broadcast over the Voice of America. We ought to enter a protest against the publication of the letter abroad.

KERKIS: (*maliciously*) Via our press, in open form via party offices!

DYMSHYTS: Raissa Lvovna will think that I have been programmed. But what I'm going to say is the voice of young people. I have consulted the Komsomol committee. The young people believe that the reason for the letter was concern for democracy. But the people who made 1937 can play on that. They are alive, and they're ready to act.

TIKHONOV: The letter and all the acts connected with it—the transmittal abroad and everything—deserve the most comradely condemnation. Any miscalculation by Soviet people, by people of all democratic countries, attracts the malicious attention of our enemies. That letter provides grist for our enemies' mills. It confuses young people, especially youth. The fact that the letter left the country is no accident.

KIKNADZE: I am obliged to say that I am breaking off my acquaintance with Raissa Lvovna and Zoya Sofronyevna for a certain time. They have insulted all of us. I don't want to be regarded as a dull sheep. I don't consider myself a dull sheep. The trials of writers disturb me, and something needs to be changed.

There is a complex conflict there linked to the mutual relations between politics and art. But the letter does not bring it to light. It does not contain facts. Either the illegality of the actions of the court and press should have been based on facts or they should have requested information. One cannot demand information.

I too want to participate in the struggle for truth, but I would not have signed that letter. It's obvious that the letter is being used for evil ends. Raissa Lvovna is a respected person, but I condemn this thoughtless action of hers, and I am angered that she refuses to listen to the voice of her comrades.

LUTKOV: I'll say a few words. In 1956, I wrote a letter in connection with the unmasking of Lysenko's activity. But I knew very well what I was writing. Raissa Lvovna acted thoughtlessly. Signing that letter was a mistake. I wouldn't sign a letter like that containing a demand, not a request.

VORONTSOV: Alexander Nikolayevich mentioned the letter in regard to Lysenko. There were five such letters. Hundreds of people, Raissa Lvovna among them, signed them. They were all addressed to the Central Committee. The matters in question were much more important than this trial—they concerned the restoration of an entire science, the fates of hundreds of thousands of people, the teaching of vitally important scientific facts that had been perverted. Kurchatov himself delivered the letters to the Central Committee. Information about them did not leak out to the West and across the ocean. The letters produced action. Oparin was removed and Engelgardt was appointed. A department of genetics was organized at Leningrad University. First under the direction of M. S. Navashin, and then Lobashev, and only when Zircle's article describing the fate of genetics in our country appeared in the United States did the situation worsen again. The enemies of science abroad and here at home took advantage of that article to strike at genetics and people of science. We have enemies abroad, and we also have friends. The enemies are always happy to sow discord among us. For example, Yuly Oskarovich [Raushenbakh] and Raissa Lvovna have adopted irreconcilable positions with regard to each other. Iya Ivanovna [Kiknadze] wants nothing to do with Raissa Lvovna and Zoya Sofronyevna. But the strife is spreading beyond the walls of the institute and beyond the city limits of Akademgorodok. Novosibirsk has not gotten used to Akademgorodok. We're not Moscow and Leningrad, with their centuries-old cultural traditions. Novosibirsk hates us, and now they've taken advantage of this letter to express their discontent over our privileged position.

No one is ensured against mistakes, but I'm not certain that the letter wound up abroad through the fault of one of the signers. It's possible that it was transmitted there from the editorial office of *Komsomolskaya Pravda*.

I think that a protest against the publication of the letter in the foreign press without the authors' permission should be sent.

RAUSHENBAKH: Raissa Lvovna misunderstood the smiles of our comrades and me with regard to her proposal to enter a protest against the publication of the letter in the bourgeois press. If the enemy has a weapon placed in his hands, he uses it. It is incomprehensible to me how knowing about the foreign currency speculation and about the transmittal abroad of information defaming our system, Raissa Lvovna could write that letter. There is an intense battle going on, and bombs are not the weapons in this struggle. The main thing is the ideological struggle. It's very bad that the letter fell into enemy hands, although Raissa Lvovna considers that nothing bad has happened. How naïve and silly it is even to think that it won't be used. You yourself told us about the criminal activity of their people, and knowing all of that, you considered it possible to demand their release.

SALGANIK: Raissa Lvovna thinks that we're programmed, but in actuality she's the one who's programmed and, moreover, with an inaccurate program designed by herself. You think that we're reactionaries, and that you alone are capable of mounting the bonfire. Without information, reckoning ahead of time that these people were not guilty, you signed a letter in their defense. We don't doubt that you did that without any evil design.

BELAYEV: The irresponsibility is outrageous.

SALGANIK: You express distrust of the court and the government. You should have heard with what wicked glee that statement was broadcast over the Voice of America. Write a letter to the newspaper protesting the publication of your letter.

BELAYEV: Who would like to add anything? No one? Apparently not. If not, then I'll say a few words. The issue is rather clear. The excuse for the letter was the trial. For some reason or other the conviction of a group of anti-Soviet agitators produced doubts about democracy. I thought Raissa Lvovna had some information, but it turns out she didn't. She thinks that she alone has civic feelings. And are we not citizens?

KERKIS: From her point of view we're rats of the lowest rank.

BERG: No, the highest.

BELAYEV: At a certain time a group of comrades and I protested against N. G. Portnova's arrest, trial, and sentence of five years. We all knew her. She was a livestock specialist. Her case was reexamined, and she was given fifteen years instead of five and tried this time on a political matter. The times were not what they are now. Comrade Stalin was alive. I visited her in prison. We continued to intercede, and three years later she was released. She now works with sable.

The subtext of the letter that Raissa Lvovna signed is clear. Doubt is cast on the judicial system itself with the goal of defaming it. I do not have the slightest reason to doubt the information that was in the newspapers. Chakovsky's information is exhaustive. Can one write in a categorical form? They don't request—they demand. And they don't do that because they are interested in the fate of Galanskov and Ginzburg or in information. They need to cast a shadow on our legal system. That is their goal. Precisely that, if you want to know.

Raissa Lvovna lacks any goodness—she was ruled by irresponsibility alone. If the letter had not wound up abroad, we would regard the matter differently. We know how the letter wound up abroad. Two people from Moscow came here and acquired dominance. That's where the program was. Someone took the letter to Moscow. It's clear that the primary address that the organizers had in mind was precisely the one where the letter wound up.

There is information that Paustovsky, while in the hospital, signed a similar letter, but after leaving the hospital, removed his signature. I understand Iya Ivanovna, but nonetheless I'm continuing to speak to Raissa Lvovna in a comradely way. But we will be talking in quite a different way if Raissa Lvovna doesn't change her opinion.

KHVOSTOVA: How did Raissa Lvovna dare to insult us? (*yelling*) You have gotten the institute in trouble with your irresponsibility and stupidity.

KERKIS: As long as you think that the actions here are programmed, you won't understand anything.

BERG: It's well-known that the play you're performing was rehearsed.

BELAYEV: And the forty-six of you didn't consult with each other? If you don't change your position, this will end badly for you. We are condemning your action as irresponsible, we're asking you to reexamine your position, but we'll speak differently if you continue to be stubborn. We're going to vote now. Who is for the resolution to condemn the irresponsible actions expressed in the signing of the letter?

VORONTSOV: We didn't gather here to vote.

KHVOSTOVA: Let's vote.

NIKORO: I am not offering up my opinion for a vote, but if there's to be a vote, I declare my special opinion and will ask that it be entered into the record.

MOSOLOV: You said yourself that this is a comradely discussion, so what place does a vote have?

BELAYEV: We're all for condemnation—unanimously except for Zoya Sofronyevna. This is a meeting of the Science Council and you, Alexander Nikolayevich, were invited through a call to the meeting. Did you receive the call? (*to Antipova*) Was the call sent?

MOSOLOV: Yes, I received it.

ANTIPOVA: Yes, it was sent.

KERKIS: Zoya Sofronyevna has a special opinion. She agrees with Raissa Lvovna. How can we not vote? We should enter into the resolution a point about an open letter to our newspapers condemning the bourgeois newspaper that published the letter. Have Raissa Lvovna write it.

BELAYEV: We have established that on Raissa Lvovna's part irresponsible actions were taken, the support of those people who fabricated the fraud for America, concealing it under the guise of an appeal to the Central Committee.

BERG: You do not have the right to try people for felonies, and there was no crime in the participation of which you suspect me. You can express your suspicion and condemn me if a court confirms that your suspicions are justified.

NIKORO: Do we have the right to judge, although Raissa Lvovna was not guided by evil aims? We can express our difference of opinion.

RAUSHENBAKH: Dmitry Konstantinovich says that the letter was dictated by the desire to undermine the Soviet system. How can we not condemn her?

BERG: Bear in mind that Raushenbakh has had some experience with these sorts of matters. In 1937, didn't he give testimony about the enemies of the people who had done irreparable harm to horse breeding? He testified that the enemies of the people spread an infection under the guise of vaccinations.

BELAYEV: So you're also going to throw those sorts of things at us!

BERG: And don't you think that's exactly how that mass psychosis began, and that its peak was precisely the terror of 1937?

KHVOSTOVA: But you're the one who's guilty. Who else as far as you're concerned could be guilty? You're an old child.

BELAYEV: I propose a resolution to condemn the political irresponsibility that was expressed in the signing of the letter. Who votes "aye"? Everyone except Zoya Sofronyevna. Who is opposed? Only Zoya Sofronyevna.

There is, of course, a lack of information in our country, but it will be overcome by our leadership. There is already a change for the better. In the institute we have a good, healthy collective, good young people. We're all for the Soviet system. The real point of the letter is clear to all of us. It's the undermining of faith in the Soviet system. Today they doubt the legality of Soviet courts, tomorrow, the one-party system of rule.

KERKIS: We need to conduct explanatory work among the young. The lack of information could give cause for discontent.

BELAYEV: Trofimuk and two other comrades wrote a closed letter to Brezhnev about a certain matter, and that was an act of great valor, not like your miserable little appeal, and about an important matter, not because of some anti-Soviet activists.*

* The "closed letter" sent by scientists of Akademgorodok to Brezhnev was an appeal to prevent the pollution of Lake Baikal and the destruction of cedar forests in its surroundings. Brezhnev refused to stop the desolation. For a person not affected by a severe mental illness, it is impossible to grasp why and from whom the contents of the letter had to be hidden.

BERG: What was the occasion?

BELAYEV: I said that it was a closed letter. And right away you rush in to find out everything. Why do you need to know? (*yelling*) For what purpose? You just go ahead and keep writing, and if this information winds up abroad, we'll know who passed it on.

MONASTYRSKY: The party organs of Akademgorodok have petitioned the District Committee to petition before the Central Committee for increased information.

BERG: Did you protest against the lack of freedom of the press?

MONASTYRSKY: (*Ignores the question.*)

KERKIS: (*throwing up his hands*) Oh, but you're a dangerous child, Raissa Lvovna!

BELAYEV: You've already said that I accommodate myself to the Soviet system.

BERG: No, I haven't.

BELAYEV: Yes, you have, and I'm not the only one you've said it about. (*paternally*) The meeting was a closed one. Bear that in mind, Raissa Lvovna, when disseminating information about it.

# Index